Developmental Cognitive Science Goes to School

This book addresses core issues related to school learning and the use of developmental/cognitive science models to improve school-based instruction. The contributors comprise a veritable "who's who" of leading researchers and scientists who are broadly trained in developmental psychology, cognitive science, economics, sociology, statistics, and physical science, and who are using basic learning theories from their respective disciplines to create better learning environments in school settings.

Developmental Cognitive Science Goes to School:

- Presents evidence-based studies that describe models of complex learning within specific subject-area disciplines
- Focuses on domain knowledge and how this knowledge is structured in different domains across the curriculum
- Gives critical attention to the topic of the ability to overcome errors and misconceptions
- Compares instructional models across disciplinary boundaries
- Discusses tools that can be used to rapidly code complex real world information within a discipline and within a specific instructional framework
- Addresses models that should be used to begin instruction for populations of children who normally fail at schooling

This is a must-read volume for all researchers, students, and professionals interested in evidence-based educational practices and issues related to domain-specific teaching and learning.

Nancy L. Stein is Professor, Department of Psychology and National Opinion Research Center, University of Chicago.

Stephen W. Raudenbush is Chair, Committee on Education, and Lewis-Sebring Professor, Department of Sociology, University of Chicago.

Developmental Cognitive Science Goes to School

Edited by

Nancy L. Stein
Department of Psychology
University of Chicago
National Opinion Research Center

and

Stephen W. Raudenbush
Department of Sociology
University of Chicago

First published 2011
by Routledge
270 Madison Avenue, New York, NY 10016

Simultaneously published in the UK
by Routledge
2 Park Square, Milton Park, Abingdon, Oxon OX14 4RN

Routledge is an imprint of the Taylor & Francis Group, an informa business

Typeset in Minion Pro by Prepress Projects Ltd, Perth, UK.

Library of Congress Cataloging in Publication Data
Developmental cognitive science goes to school/edited by Nancy L. Stein & Stephen W. Raudenbush.
p. cm.
Includes index.
"Based on a conference held in the Fall of 2007, in Chicago."
1. Cognitive learning—Congresses. 2. Science—Study and teaching—Congresses. 3. Mathematics—Study and teaching—Congresses. 4. Content area reading—Congresses. I. Stein, Nancy L. II. Raudenbush, Stephen W.
LB1062.D497 2010
370.15′23—dc22
2010021699

ISBN 13: 978-0-415-98883-4 (hbk)
ISBN 13: 978-0-415-98884-1 (pbk)
ISBN 13: 978-0-203-83753-5 (ebk)

Dedication

This book is dedicated to Tom Trabasso, whose lifelong interest in learning and understanding served as a guideline for addressing issues related to developmental psychology, learning, and schooling. Tom's interest in learning was apparent in his earliest work in the 1960s, with Gordon Bower and Rochel Gelman on discrimination and concept learning, in his work with Peter Bryant on transitive inferences, learning and development, in his work with Peter Ornstein on organizing, learning, and remembering, and in all of his work on models of understanding, thinking, and development. His later work on narrative and causal understanding was fueled by an attempt to account for language, memory, and thinking that went beyond simple word and sentence understanding. That is, he wanted to study complexity, and the ways in which complex systems impacted people on a daily basis.

The feature that characterized Tom the most was his passionate quest for answers and discoveries that would lead to a better understanding of how children think, reason, and remember. He was a consummate scientist and always sought evidence, no matter what the issues were, and no matter whose theories were being tested, to explain and account for data. The pursuit of answers and evidence often got him into trouble, especially with those whose theories were being tested. Tom persevered, however, until he got the answers he was seeking, often at variance with current belief systems related to how children learn.

Tom was as good at teaching as he was at doing research. His tenacity, goal directedness, and razor-sharp intellect enabled him to impart a sense of discovery and delight to students and colleagues who were engaged in studying learning and memory. Tom was faster than just about anyone in discerning confounds and problems with an approach, devising ways of explicating and testing an issue, advancing a theory that was far more robust than the one with which he started, and pointing to broad implications that theories of learning had in regard to developmental issues.

Had Tom survived, he would have been an integral part of the efforts to become more intimately involved in science and school learning. Although Tom was mathematically gifted, he was never trained in the physical sciences, and so his new venture required that he take time out to actually learn the content of physics, chemistry, and earth sciences.

Several of the chapters in the present volume do what he would have done, if he could: make developmental and cognitive science relevant and understandable to those who teach young children on a daily basis.

Dedication

Contents

Preface

In the last 10 years, the goals for educating our young children have become more complex, more urgent, and less forgiving. What will it take to keep the United States abreast of all of the adaptive changes that we will undergo? As a nation, we can no longer make decisions and act in isolation. The speed with which we can communicate with people in all parts of the world has broken down walls and borders in all parts of the world. We even make contact on the Internet, despite many governments' desire to block or edit what is being said. If there is a way to forge communication and contact, people will eventually find it.

The rise of China and India as very well-educated super-powers, each having three times the population of America, is a profound obstacle as well as a Herculean challenge to the United States. Both countries select and educate children who are considered gifted much earlier than we do, and both countries are unrelenting in terms of the importance of education to their survival and advancement. Both have programs for the development of excellence in an academic domain far earlier than we do. And both countries are already ahead of us in terms of identifying and developing talent in their young children.

The publication of the data from the first TIMSS study[1] resulted in a wake-up call that few took seriously. The fact that the TIMSS exam is given every four years to children across the globe, however, continually pointed to the fact that the United States is the leader in neither math nor science, nor in any other discipline. What appears to be persistent evidence, however, is that three Asian countries, Singapore, Taiwan, and Hong Kong, stand above the rest of the world, in terms of producing students who score well in science and math.[2,3] The only exception to the Asian block is Finland, which scored very high on the PISA assessment, another international exam across various academic disciplines.[4]

The outstanding performance of the Asian countries is further underscored by the fact that the world agenda is shifting mightily in terms of the importance of math and science skills for the survival of every country on our planet. The major problem that the world faces is twofold. First, we are using up natural resources at an alarming rate. In doing so, we are also causing a serious climate change process—one of warming. The use of carbon and the production of greenhouse gases need to stop or be lowered significantly, and alternative sources of energy need to be pursued.

Children and adults need to understand the problems we face, not only in terms of the social and moral consequences of global warming, but also in terms of the scientific basis of halting or stabilizing climate change. Massive amounts of time and talent need to be directed to sustainability issues and to problem-solving strategies. The National Institutes of Health predict that over half of all new jobs will somehow be related to the energy sector, to the pursuit of clean energy sources.

In order to accomplish this feat, both children and adults need to be trained in the physical sciences, and children clearly need to master mathematics and measurement concepts far earlier and in more depth than they do now. More important, the change in agenda to a more science, technology, engineering, and mathematics (STEM)-oriented education, even in the early elementary years, forces us to reconsider the role that reading will play in allowing children to acquire scientific content at far earlier ages.

We are used to considering reading as a general ability that affects everything that we do. The ability to attend to new words and to understand the meaning of these words is considered a general ability that predicts children's performance across many different domains. The general ability and vocabulary skills evidenced by some readers and learners, however, may not be a powerful predictor of academic success over the long run. Science learning is domain specific, with many of the sciences being linked to and dependent upon the goodness of understanding of core concepts in math and measurement.

Thus, one of the questions motivating the 2007 conference, Developmental Learning Sciences Go to School: Implications for Education and Public Policy Research, focused on the role of vocabulary and reading in predicting developing interest and competence in science. Another concerned the role of visualization in combination with verbal representations in gaining more accurate representations. A third issue was the usefulness of oral representations in acquiring different types of content.

But the growth of mathematical and scientific thinking remained a theme throughout most of the presentations. The various approaches to scientific and mathematical thinking offered a glimpse of some of the problems we face in constructing a new agenda for schooling. Despite the fact that we have access to many math and science curricula, much of how we teach in both of these disciplines needs to be rebuilt. We do not have the proper curricula in place. Further, in almost anything we do, teachers will need to undergo retraining. The content and conceptual organization of both science and math have not been considered or acquired by most elementary school teachers.

The suggestions made by many of the scientists attending this conference, however, were thought provoking, and many broke new conceptual ground. The diversity of opinions also pointed to some of the more contentious issues associated with education and learning. While some of the researchers focused on the conceptual organization of domain knowledge in successful teaching, others focused on the methods of teaching materials as well as the subjective awareness of learning that children develop as they become less of a novice in a domain and more of a participating learner.

All of these issues were important, and all were discussed. The most important contribution made by all of the participants is that each one entered school and the classroom as a scientist, using the classroom as a laboratory. At the same time, all of the participants were aware that studying development and learning in real world classrooms challenged many of their own theories of understanding and learning. Thus, reading all of the chapters offers a look at some of the more serious issues faced by developmental scientists as they proceed with courage and inventiveness in classroom and school settings.

Nancy Stein
April 2010

Notes

1 Martin, M. O., Mullis, I. V. S., Beaton, A. E., Gonzalez, E. J., Smith, T. A., & Kelly, D. L. (1997). Science Achievement in the Primary School Years: IEA's TIMSS. Chestnut Hill, MA: Boston College.

2 Gonzales, P., Guzmán, J. C., Partelow, L., Pahlke, E., Jocelyn, L., Kastberg, D., & Williams, T. (2004). Highlights from the Trends in International Mathematics and Science Study (TIMSS) 2003 (*NCES 2005–005*). U.S. Department of Education, National Center for Education Statistics. Washington, DC: U.S. Government Printing Office.

3 Martin, M. O., Mullis, I. V. S., & Foy, P. (2008). TIMSS 2007 International Science Report: Findings from IEA's Trends in International Mathematics and Science Study at the Fourth and Eighth Grades. Chestnut Hill, MA: TIMSS & PIRLS International Study Center, Boston College.

4 Baldi, S., Jin, Y., Skemer, M., Green, P. J., and Herget, D. (2007). Highlights From PISA 2006: Performance of U.S. 15-Year-Old Students in Science and Mathematics Literacy in an International Context (*NCES 2008–016*). Washington, DC: National Center for Education Statistics, Institute of Education Sciences, U.S. Department of Education.

Acknowledgments

The Spencer Foundation funded the conference upon which this volume is based. The conference was held from 10 to 14 October 2007, in Chicago, Illinois. We are indebted to the Foundation for the opportunity to consider and talk about issues that underlie research efforts with respect to schooling and learning. The amount of collaborative activity that took place among the participants was heartening. Many of the conversations turned into collaborations for the future, which was one of the main goals of the conference.

1 Developmental and Learning Sciences Go to School

An Overview

Nancy L. Stein

Theories of Learning for School- and Domain-based Achievement

The chapters in this volume resulted from a conference held in the fall of 2007, in Chicago, supported by the Spencer Foundation. The purpose was to bring together scientists from different disciplines in order to address issues of learning across three domains: reading, science, and mathematics. Although government committees, books, and publications have attempted to address some of the issues related to learning in school, many of the issues, especially those related to science and mathematics, have not been addressed fully, especially in terms of obstacles that block successful school learning.

The same holds for advances to be made in reading (National Reading Panel Final Report, 2000; Trabass & Bouchard, 2000). Although we have made good progress in understanding the different components of reading, we have made less progress in understanding that success in reading, when decoding skills have been acquired, is a function of domain-specific knowledge. What works in one domain does not necessarily transfer to another domain. Learning is situated, and we need domain-specific models of the important concepts and ideas in each domain (Pitt, 1976).

In order to address learning issues across domains, we invited researchers from several different disciplines: psychology, sociology, computer science, cognitive science, education, and science. Although all contributors study learning, the ways in which learning was defined and studied differed dramatically as a function of the discipline. The communication across disciplinary boundaries was almost non-existent. Many of the reading researchers knew each other but did not know the contributors from math and science. The same held for researchers in math and science. Integrating approaches across disciplinary boundaries was essential.

The lack of interdisciplinary work is a major stumbling block for research on successful learning. We can teach reading by teaching science and math. We can teach core principles in science at much earlier ages if we understand and can teach specific mathematical concepts earlier than previously taught. If we are successful at using interdisciplinary approaches to school learning, we will be able to reconsider the way we teach the disciplines of reading, science, and mathematics. What is taught at each grade level, and how it is taught, need reconsideration.

The goals of the chapters in this book, then, were to address issues in learning and development that impact school-based instruction and public policy research. We focused on theories that advocate and produce evidence about learning and schooling in science, mathematics, and reading. As a result, six issues were approached and discussed throughout the volume:

1. ways in which children can engage in science and math learning earlier than currently taught;
2. successful strategies used in mathematics and science that also increase skill in reading and writing comprehension;
3. ways in which reading skills can be increased above the existing level of skill;
4. obstacles that prevent developmental and learning scientists from successfully carrying out innovative research in school settings;
5. myths that abort or stop learning attempts, especially in the scientific and mathematical domains;
6. ways in which ethnicity and culture impact children and their ability to carry out school learning.

Making Learning and Understanding Accessible to Young Children

A major theme throughout the book is centered on ways in which relatively complex concepts that underlie learning, especially in science and math, can be made accessible to young children and to the novice. The transitions from home to school in kindergarten, from second to third grade in elementary school, and from elementary to middle school are particularly important because new task demands and expectations are introduced at each of these stages, and children must master these task demands if they are to succeed in school. We focus on children from preschool through 14 years of age, some of whom experience failure at critical transition points, especially when their earlier development has been devoid of any real preparatory training. We ask questions about the types of learning that would better prepare children to succeed and advance in different disciplines, and when these types of learning should start.

The thinking and reasoning skills children bring to school learning are especially important. In general, successful children ask more why questions, seek explanations, express dissatisfaction when their questions are not answered, become more autonomous in their learning, monitor ways in which they learn, and accomplish learning outside of the classroom, with tutorial help from parents, grandparents, siblings, friends, and professionals. All of these factors play a critical role in determining how much and how fast children advance intellectually in school.

A content and conceptual coherence issue is germane, however, in addition to focusing on the acquisition of successful strategic skills. The concepts that children are taught, the order and depth with which they are taught, and the ways in which these concepts are embedded in specific content domains account for a significant part of the variance, almost 50 percent, in predicting successful learning (Stein, Anggoro, & Hernandez, 2010a). These content issues are especially important for science, mathematics, history, and social studies. We can teach children all of the sophisticated instructional strategies that we deem appropriate. If the relevant content is not present, and if the content is not organized in an "accessible" fashion, however, critical concepts will not be learned, despite all of our efforts to encourage and support learning. In other words, we can use all of the mapping activities possible, all of the questioning strategies, and all of the memory and retrieval strategies that are known to result in a deeper understanding of the materials (Newcombe et al., 2009; chapters 4 and 6). However, if critical concepts are left out, or if critical causal information and content are deleted, children will not learn the relevant concepts, no matter how well the instructional sessions are organized and no matter how bright and smart the children are (chapter 6).

Schmidt (Schmidt, Wang, & McKnight, 2005) has begun to show the importance of content in mathematical learning, but the truth is that content and its organization are as important in every other domain that exists. The content and organization overpower and control the success of teaching strategies. Trabasso and Bouchard (2000; 2001) showed that many different reading comprehension strategies work, as long as the instructor keeps an eye on the mastery of content. As Trabasso and Bouchard pointed out, however, the content of a lesson is rarely discussed, so we rarely get a glance at what is being learned by students and whether or not they can use their new knowledge in novel situations to understand complex ideas that require building on previous concepts. Brown (1990) also points out the relevance of specific knowledge and contexts in determining the power of a strategy. The omission of content and domain-specific knowledge, however, continues to be a problem for psychologists (e.g., Newcombe et al., 2009) who focus primarily on the power of generalized instructional strategies rather than on an awareness of the interaction between specific content and successful strategies.

Studying the role of conceptual content in a domain, such as math or science, is seriously constrained by the fact that public school science and math learning programs focus on content covered in standard-based state achievement tests. For example, each state decides what students should learn and be exposed to at each grade level. A yearly exam is then carried out for each specific discipline, where the items in the test supposedly correspond to the standards set by the state. The problem for researchers is that states disagree on what concepts and content should be taught at each grade level (Cutting & Scarborough, 2006), and not all of the state and the national exams are theory based (Klahr & Nigam, 2004). The most frequent strategy used to determine state standards is the use of an integrative meta-analysis of existing studies in each disciplinary area. The average age at which each concept is learned then becomes the criterion for assigning mastery of the content to a particular age.

Although empirical generalizations are always helpful, using a meta-analysis strategy can be disastrous for setting policy and improving skills in a domain. The derivation of age norms is always contingent on the types of tasks and data that currently exist. What the meta-analysis can never tell us is whether the chosen concepts are the appropriate ones or whether children can learn concepts earlier than thought, if other disciplinary knowledge is introduced. This issue is especially germane for science and mathematics (chapter 7), but it is also true for literacy, reading, and writing (Morphy & Graham, 2010). Critical learning tasks and content are often deleted or skipped over when school district personnel deem children to be slow learners, especially if they have a disability, such as dyslexia, or when they come from impoverished backgrounds (chapter 20).

A problem that compounds the introduction of new and more complex school content is that teachers are rarely trained in the new content or discipline. Many of the researchers studying science and math are also not trained in the disciplines, except at the very beginning levels on par with content introduced in elementary grades 1 through 3. Thus, a strong constraint on introducing innovative curricula in different disciplines is that both teachers and researchers need more training (chapter 13). Neither the instructors nor the researchers collecting data have the requisite content knowledge to either propose or judge innovative research.

The ways in which we deal with this lack of knowledge will prove critical to the success of future research in this area. Although disciplinary experts can act as collaborators, very few understand or have used theories of learning to motivate their contributions to curriculum development, especially at the elementary school level. And yet, the creation and innovation of curricular content is one of the most necessary outcomes in many different disciplines. Learning about certain topics and core concepts determines whether or not children will be able to take advantage of opportunities that require more difficult and complex learning. If children have not been exposed to the topics and themes that scientists consider essential, they cannot benefit from more advanced training.

Further, the issue of when concepts are introduced becomes another concern. Although adults learn new materials with success, the reality is that often they do not learn as well as children who have been introduced to a topic very early in development. The fields of mathematics and music, certain types of science, different types of language learning, both oral and sign, and the development of expertise in sports and dance are particularly prone to early development effects (Bloom, 1985; Feldman, 1993). Older children and adults can learn new domains, but the majority do not reach the levels of expertise that early learners reach. This phenomenon has been documented broadly (e.g. Bloom, 1980; Feldman, 2003), but has been ignored by the educational community, especially in math and science.

School learning is further complicated by the fact that we live in a multicultural society, and a significant and growing proportion of public school students possess widely varying levels of English proficiency. Anyone who has had sustained contact with education, teaching, and learning, especially in large urban centers, must consider the impact of education on children from different cultural and language environments. The presence of children from different cultural backgrounds impacts school policy, the choice of topics, and the school instruction that get carried out on a daily basis. In a growing number of public schools, government policy requires that children be given initial instruction in their native language. The focus, however, is on learning to read and speak English as a second language, so that successful transitions can be made to school environments where only English is spoken.

Learning to use a second language, however, is just one of the many hurdles that children must overcome when they enter school. The demands placed on children today, even those at the preschool and elementary school levels, are quite daunting. The new goal of teaching science in the elementary school years involves a new domain for teachers to master, even those teaching in private schools or in gifted programs. The necessity for ensuring that children receive a better science education has resulted from several factors: shifts in the job market toward technology and energy-related occupations, cognizance of the climate changes occurring on earth, and the awareness that the United States may not have a leading edge in the production and creation of scientists and scientific agendas to address problems of energy production and climate change. Introducing children to science and math as early as possible and finding ways to facilitate learning with technology must be considered. All of these concerns serve as the motivational force for this book.

Our focus on school learning is not a new endeavor. The educational centers created by the United States Department of Education in the mid-1970s (e.g., the Center for the Study of Reading at the University of Illinois, the Learning Research and Development Center at the University of Pittsburgh, Wisconsin Research & Development Center for Cognitive Learning) are good examples of the successes that educational scientists can accomplish when they turn their attention to instructional issues and the necessity of creating better learning environments. Each of the five Research and Development (R&D) centers established by the U.S. Office of Education had a major impact on schooling and learning and, in turn, each exerted a significant influence on the ways in which researchers carried out instructional studies. The impact of these R&D centers on the broader educational community can still be seen in conferences and volumes supported and published by the National Academy of Science.

In 1999, a book edited by John Bransford, Ann Brown, and Rodney Cocking, entitled *How People Learn: Brain, Mind, Experience, and School*, tried to summarize and address many issues relevant to this volume. One such issue focused on making different types of subject matter accessible to young children. Bransford et al. (1999) stated:

> Learning research suggests that there are new ways to introduce students to traditional subjects, such as mathematics, science, history and literature, and that these new approaches make it possible for the majority of individuals to develop a deep understanding of important subject matter. This committee is especially interested in theories and data that are relevant to the development of new ways to introduce students to such traditional subjects as mathematics, science, history, and literature. (p. 6)

The agenda for this volume focused on this goal and attempted to make even deeper connections between domain-specific knowledge, basic learning, developmental and cognitive science, and classroom learning. Although many issues discussed in the last 30 years remain the same, new concerns and new technology have forced a reconsideration of how we go about designing learning environments in the early years. The changes and advances made with young children will affect all other decisions that get made, and they will have a profound impact on whether programs designed for older children succeed at the level that we desire. Sequences of instruction are critical, as is the time at which children are exposed to different content matter.

The availability of technology and computers in teaching environments has impacted just about everything that gets carried out in the classroom. The computer can do many things that written texts and teachers cannot do, and the impact of this advance needs consideration, especially in the early teaching of science and mathematics. The computer's ability to convey and transmit information in oral language, visual graphics, and different auditory modes makes interactive tutoring very appealing, especially for children from impoverished or restricted environments. Further, discipline-specific learning modules can aid a teacher who is initially limited in knowledge of a specific domain. If computer-based learning modules are used in the correct way, teachers can become learners of the material before they are forced into the role of instructor.

The computer can also track acquisition of specific skills in an automatic fashion in ways that teachers cannot. Individualized instruction for each child is a distinct possibility (see chapters 1, 7 and 11) with the computer, because of its ability to automatically code errors as well as correct responses. Were more classrooms able to take advantage of this technology for a range of subjects, improvement might occur rapidly because of the ability of the computer to point out and integrate the type and number of errors that occur when learning a new concept. If the errors can be classified and understood properly, the knowledge that is missing and needs to be taught can be surgically targeted and remediated.

The organization and content of each discipline, how we sequence ideas, and how we build connections across different disciplines, such as links between history and reading or links between science and reading, all need to be addressed. At the moment, 90 percent of the research on reading, especially in young children, is carried out in the domain of literature and narrative understanding. Rarely are the domains of science and mathematics considered relevant to the topics of advancing reading skills. Yet, the data that we have on effective science learning (chapter 9) show that when scientific concepts are mastered with learning strategies that increase understanding, reading skills increase within and possibly across domains.

To improve school learning across domains requires that we think at the comparative and systems level in terms of learning and instruction (chapter 3). A systems-level approach is one that focuses on sequences of learning and the links within and among the components of a sequence. Analyzing and determining the content and sequence of ideas in one domain would be an example of a systems approach. An important question that needs answering is whether or not specific sequences of instruction within a domain are more powerful than other sequences of instruction. Thus, we are focused on causal sequences of ideas and the necessity or lack of necessity in ordering ideas within a sequence. A systems analysis also allows us to determine whether or not concepts from one domain are requisite for learning in another domain.

Although much work has been carried out on children's early learning in domains such as language, the acquisition of number skills, and early understanding of physical causality, little work has addressed the ways in which models of early development can be used to advance later classroom learning. It may well be that models of early learning of language and number concepts are not the models to be used in the learning of language, mathematics, and scientific concepts in elementary school classroom settings. We know from work on young children (Siegler & Mu, 2008; chapters 8 and 12) that measurement concepts which underlie quantitative learning have not been taught well at the younger grades. Further, when these measurement concepts are taught, the strategies for using tools such as a ruler are rarely taught adequately. We would never think of learning how to play golf or tennis without learning how to hold a golf club or tennis racquet. Learning how to measure things also requires learning how to use the tools that provide the results. However, both Huttenlocher (chapter 12) and Stein (chapter 7) show that children may not receive adequate instruction in tool use.

Rulers are not the only tools with which children experience difficulty. Fourth-grade children often do not know how to use a protractor or a calculator. Many schools do teach children basic procedural routines associated with the use of a tool. What they do not teach children is how to use the tool when abnormal conditions exist. For example, fourth-grade children know how to measure the length of a candy bar, if they are given a ruler that starts at the 0″ point and ends at 12″, where the candy bar is 4″ long. Far fewer children can arrive at the correct solution when asked to measure the candy bar offset from the edge of the ruler (chapter 12), or when they are given a broken ruler that begins at 6″ rather than 0″. Children experience difficulty with these problems even when they are in selective enrollment schools, signifying that they are in the upper 15 percent of the student population.

Huttenlocher, Levine, and Ratliff (chapter 12) and Siegler (Siegler & Mu, 2008) show that when instruction is explicit, young children can understand certain components of measurement. The question remains, however, as to whether the measurement knowledge will transfer to situations requiring quantification in a domain that is novel and as yet unlearned. Further, as Lucariello and Tine (chapter 17) point out, we need to focus on difficulties experienced when more complex forms of mathematical concepts are introduced. The introduction of more complex systems is a critical area of study.

Each discipline must consider the progression of instruction and the ways in which a sequence of instruction affects the ability to learn. The need to compare models of instruction across content domains is critical and as yet an untapped frontier. The content domain in each area may dictate different models and different instructional strategies. Conversely, the types of learning strategies and problems that arise in each discipline may have remarkable similarities, such that studying learning environments in one discipline may truly benefit models in other disciplines.

We need evidence-based studies that describe models of complex learning within and across disciplines such as science, mathematics, and reading. We also need to recognize that the different sciences may require different models of learning, depending upon the specific types of problems introduced by a specific domain. It has been recognized by many researchers (see Brown, 1990; Brown, Collins, & Duguid, 1989; Cobb & Bowers, 1999; Pitt, 1976; Stein, Trabasso, & Liwag, 1993) that learning strategies are often specific to the domain being mastered and that expertise in one domain does not transfer to another domain. Thus, any model of learning that succeeds in describing core concepts in that domain needs to address the generality or lack of generality of the specific content learned. Thus, scores on general vocabulary or comprehension tests may not predict comprehension success for every domain.

As an example, our initial studies of physical science learning in fourth-grade children (chapter 7) show a moderate to significant correlation between vocabulary performance and rate of learning. However, the correlation disappears as children acquire more vocabulary and knowledge associated with the domain. Further, children who have high scores on general vocabulary measures may not be the ones who have high scores on mathematical achievement or science achievement. Doing well in the physical sciences depends very much on mathematical understanding, certain forms of spatial ability, the ability to remember specific types of visual arrays, and the ability to reproduce and talk about physical events. Thus, even though vocabulary is important in acquiring an initial understanding of a domain, it does not remain as the critical predictor of successful learning in science, especially when all learners are at a moderate level of being able to use and express new scientific concepts verbally.

The lack of a correlation between vocabulary and success in a specific domain poses a problem for those researchers who believe that understanding academic language (chapter 5), which constitutes a specific vocabulary, is the distinguishing characteristic that separates successful children from those who experience failure. In the early grades, a high vocabulary score may predict reading and writing achievement, simply because the reading materials at the lower grades are focused on gaining an understanding of people, the various activities they carry out, their differences, and the different types of motivations that propel people into action. Much of this information is conveyed through narratives and narrative understanding.

Acquiring an initial expertise in specific domains such as the physical or the biological sciences, however, is contingent upon learning the core concepts that define the specific domain. Much of the vocabulary in science is specific to what is being studied. Without exposure to the core concepts in a domain, no amount of vocabulary ability will result in understanding the domain. Further, the number of concepts and words central to each of the scientific domains is limited and highly constrained. The vocabulary can be taught quite easily. The difficult part of learning sciences is often centered on inadequate mathematical or quantitative skills and a failure to define and use core concepts repeatedly in all of the different contexts that define the concept. Thus, it is not the vocabulary of the learner that is limited, per se. It is the number and type of opportunities given to a learner to use specific words and concepts in order to acquire and understand what the words mean.

The ways in which language is used become very important when the meaning of concepts is unpacked. Brown and Ryoo (2008) have shown that explanatory language in teaching physics and science is extremely important in increasing understanding. Alibali, Nathan, and Fujimori (chapter 15) show that gestures are also important in teaching, especially if we are to track how a teacher gets children to attend and orient to critical information. Ornstein, Haden, and Coffman (chapter 6) show that teachers who make children aware of remembering during learning gain better performance from those children. The one thing missing is the "what" of learning and studies to determine whether or not these procedures are situated and constrained to a limited set of problems.

In a recent article, Newcombe and her colleagues (Newcombe et al., 2009) argued for a more systematic use of learning strategies that have been shown to be successful in math and science studies across different domains. Examples of the different strategies would (a) be space learning over time; (b) include worked examples with problem-solving exercises; (c) combine graphics with verbal descriptions; (d) connect and integrate abstract and concrete concepts; (e) ask deep explanatory questions; and (f) use quizzing to promote learning.

Our assessment of what goes on in physical science teaching is that teachers *already* use many of these strategies, with slim but moderate success. We have also shown that dynamic visual graphics increase comprehension, over no graphics (chapter 7). However, graphics are not always successful, as we have shown, especially when they interfere with or provide evidence that disproves a strongly held belief of a learner. We have further shown that the need for spatial representations in understanding physical events and phenomena is imperative, but quite different from a study of individual differences in processing spatial information. In order for children and adults to understand specific molecular processes to which they have never been exposed, all learners need to see the process modeled with visual graphics displaying the molecular processes. This requirement is independent of a learner's skill at remembering or manipulating visual forms on tests of visual rotation and discrimination ability. When dynamic graphics unfold and explain a causal sequence of molecular movement and speed across different physical states, performance on spatial rotation tests does not predict who will master the scientific module. Once children see and have some concrete representation of a visual sequence, especially one that has causal significance, mental rotation ability becomes a non-factor.

The real issues of science learning focus on the identification of the critical concepts to be learned, ways in which these concepts can be learned, and the combination of skills and abilities that are needed to acquire the requisite knowledge. Performance in the physical sciences requires not only a minimum amount of spatial ability, but also an understanding of mathematical concepts as well as an understanding of the ways in which concepts are causally connected to one another. In organizing a sequence of instruction, some concepts are causally related and must be taught in a specific temporal sequence. Other concepts in the domain may not be causally related, and therefore can be introduced at different times during learning. The specific concept and content domain regulate the order and structure of an instructional sequence.

The ability to carry out instruction that overcomes errors and misconceptions is also critical. It is probably one of the most important teacher behaviors and instructional components that we can examine. Although many developmental scientists have described the types of misconception that children have when they attempt to learn new materials (Vosniadou, 2008), very few have addressed the ways in which misconceptions can be corrected. One reason for the lack of studies that focus on correction of misconceptions is that researchers may believe that misconceptions are too deeply rooted and very resistant to change (chapter 18).

On the other hand, some misconceptions can result because children have not been introduced to the critical concepts that would change their representation of the problem and then change the errors that they make. Wilkening (chapter 10) reviews some of the past work in which children have been shown to have strong incorrect biases about area and volume, because of a failure to take into account more than one dimension (Wilkening, 1979; 1980). Thus, for area, two dimensions, height and width, are necessary. For volume, three dimensions are necessary: height, width, and depth. Children make integration errors even in the middle and late elementary grades.

In exploring measurement and conceptual errors associated with area and volume, one reason for these errors is that children have never been taught that the square is the unit of measurement for area, and the cube the unit of measurement for volume (Stein, Anggoro, & Hernandez, 2010b). When fourth-grade children were asked to give the correct unit of measurement for area and volume, the clear majority of them chose the inch as the unit of measurement. Further, most did not know what a square or cubic inch was. All of them knew the algorithm for area, but they did not know that the algorithm meant that height was multiplied by width. They also did not understand why you could measure the sides of a square and compute the area of the square. The same held for volume. That is, these children, all of whom were magnet school children in the upper 15 percent of the population in the Chicago Public Schools, did not know that area could be measured with square inches. They also did not know how to compute the length and width of the edge of a square or rectangle by computing how many squares were adjacent to the sides of the square for the height and width.

Thus, in the case of measurement errors and failure to integrate across two or three dimensions, children do not necessarily hold deep misconceptions about either area or volume. Rather, they have not been taught the proper unit of measurement and the procedures used for determining area and volume. Using and understanding three dimensions associated with volume can be taught thoroughly only by illustrating how volume is calculated, by filling a three-dimensional space with cubes. Children can then be taught the association between the length of the side of a cube and how counting the number of sides of cubes is the actual length of each of the sides of a cuboid.

Some misconceptions that children (and adults) have acquired may be more difficult to eradicate. For example, we know that adults make the error of thinking that a ball will continue on a circular path once the ball has exited piping that is circular in nature. We also know that both children and adults make the mistake of believing that hot water gets cooler in the freezer because cold air has "gotten into" the warmer water, when in fact, just the reverse is true (Stein, Hernandez, & Anggora, 2010). No one has carried out learning studies to determine whether or not misconceptions vary in their malleability, in terms of a learner being able to supplant the incorrect misconception with the more accurate information. Much work has been carried out on different types of strategies for change, including direct confrontational strategies (Vosniadou, 2008). However, no one has actually varied the types of corrective information given to the learner. It may be that almost all misconceptions are changeable, provided that the information to form an alternative representation is provided.

The work by Richland (chapter 14) as well as that by Wilkening (chapter 10) show the necessity of providing children with additional information about mathematical reasoning in order to understand and solve new problems. Richland elegantly shows the significant role that analogical thinking plays in allowing children to solve new problems. This strategy could also be used to get children to explore their misconceptions and supplant old information with new. The difficult part is to get children to use the information they derive from an analogy instead of the information they already have because of their misconception.

The work by Massey, Kellman, Roth, and Burke (chapter 16) approaches mathematics learning by focusing on the perceptual processes involved in learning different mathematical strategies. This work is important in that a theory of learning is espoused, and the connection to the science studies (chapters 7, 9, and 10) is important. Although many investigators are focused on understanding and improving math and science, rarely do they use a theory of learning to test whether or not they have been effective in getting children to attain and maintain an increased knowledge and perceptual awareness of specific ideas. The use of the word "perceptual" in Massey's chapter (chapter 16) may be somewhat confusing because other researchers employ similar techniques to Massey's but focus on the properties that have been learned to discriminate one concept from another. Thus, the more traditional way of referring to the procedures used in Massey's approach is to focus on the concept being learned and what a theory of concept learning specifies in terms of increased and accurate performance. Massey argues that the most effective way of teaching a concept (e.g.

fractions in this case) is to expose children to as many instances and features of the concept as possible.

Were we to use Klausmeier's (Klausmeier & Hooper, 1974) theory of concept acquisition, he would argue that all dimensions of the concept must be made explicit. Furthermore, the concept needs to be distinguished from other concepts that are similar but do not have all relevant dimensions of the concept under consideration. The one thing that Klausmeier (Klausmeier & Hooper, 1974) does not specify is how frequent or how many times a dimension needs to be taught in order for a learner to retain a correct representation. Massey and her colleagues (chapter 16) do not distinguish between a one-time presentation and multiple presentations of all critical features. It may be that the perceptual training is effective because all variants of the concept are made explicit. How much repetition of each feature needs to be presented is a second issue. The more important point, however, is that breadth, repetition, and discrimination of exemplars and types of problems are necessary for fluid achievement and learning of these math concepts.

The use of the computer to aid children in the rapid learning of phonological rules, mathematics, and science has been shown to have strong robust effects. Massey et al. (chapter 16) illustrate how the computer can be used to rapidly focus children on acquiring the fractional patterns necessary to make the correct discriminations. Morrison and Connor (chapter 2) showed similar types of success for learning different phonological patterns during reading. Stein et al. (chapter 7) showed the success of using computer-generated dynamic graphics in increasing children's understanding of molecular speed and movement, while Graesser et al. (chapter 11) show how interactive computer programs enable students' success with learning Newtonian physics principles.

Computers can further aid in the teaching of a relevant content area by increasing the frequency of assessment as children progress through a learning sequence, and by making the assessment outcomes and necessary remedial strategies available to teachers in a very rapid fashion. Computer programs can also aid in the rapid analysis of complex language used by children when attempting to learn a new concept or solve a new problem. Although the number of answers and ways to solve problems may appear to be infinite, the set of solutions that learners produce is actually quite constrained.

The time-consuming part of creating a computer program to analyze answers to different types of questions is in the initial coding of the data. Dictionaries that abstract out the invariant properties of all responses can be carried out, such that the second time through the problem or lesson, a computer program can be used to code the responses. The advent and sharing of these computer programs would provide teachers with a more in-depth analysis of each child's performance, plus an analysis of what each child still needs to learn in regard to each concept.

The types of findings that the majority of contributors made need to be considered in a broader framework than that used by most developmental and cognitive scientists. The ways in which studies on reading, science, and math impact the broader educational community is a serious issue, and one that influences the type of research that gets carried out. How can we ensure that all children have equal access to educational opportunities? How do we entice qualified teachers to get involved with teaching, even at the preschool level? How do we begin to introduce topics related to math and science at the early grades, when our teachers and (sometimes) researchers are not trained in the specific domain?

Finally, if we do envision a different type of curriculum and school setting evolving, how do we best describe it? What are the strategies we can use to create and test new programs? How do we begin and continue the task of schooling children differently because of the new and changing demands put on them. Cook et al. (chapter 20), Pellegrino (chapter 19), and Raudenbush (chapter 21) all wrestle with these issues. They have yet to provide us with the clarified answers that we will eventually seek, but they certainly have raised the appropriate issues so that we can address and begin to solve the problems of learning in a better fashion than currently exists.

References

Bloom, B. (1980). *All our children learning: A primer for parents, teachers, and other educators.* New York, McGraw-Hill.

Bloom, B. (Ed.) (1985). *Developing talent in young people.* New York: Ballantine Books.

Bransford, J. D., Brown, A. L., & Cocking, R. R. (1999). *How people learn: Brain, mind, experience, and school.* Washington: National Academy Press.

Brown, A. L. (1990). Domain-specific principles affect learning and transfer in children. *Cognitive Science: A Multidisciplinary Journal*, 14(1), 107–133.

Brown, B. A., & Ryoo, K. (2008). Teaching science as a language: A "content-first" approach to science teaching. *Journal of Research in Science Teaching*, 45, 529–553.

Brown, J. S., Collins, A., & Duguid, P. (1989). Situated cognition and the culture of learning. *Educational Researcher*, 18(1), 32–42.

Cobb, P., & Bowers, J. Cognitive and situated learning perspectives in theory and practice. *Educational Researcher*, 28(2), 4–15.

Cutting, L. E., & Scarborough, H. S. (2006). Prediction of reading comprehension: Relative contributions of word recognition, language proficiency, and other cognitive skills can depend on how comprehension is measured. *Scientific Studies of Reading*, 10(3), 277–299.

Feldman, D. H. (1993). Child prodigies: A distinctive form of giftedness. *Gifted Children Quarterly*, 37(4), 188–193.

Feldman, D.H. (2003). *Cognitive development in early childhood.* In R. Lerner, A. Easterbrooks, & J. Mistry (Eds.), Developmental Psychology. New York: John Wiley. pp. 195–210.

Klahr, D., & Nigam, M. (2004). The equivalence of learning paths in early science instruction: Effects of direct instruction and discovery learning. *Psychological Science*, 15(10), 661–667.

Klausmeier, H. J., & Hooper, F. K. (1974). Conceptual development and instruction. *Review of Research in Education*, 2, 3–54.

Morphy, P., & Graham S. (2010). Word processing programs and weaker writers/readers: A meta-analysis of research findings. *Journal of Educational Psychology*, submitted.

National Reading Panel Final Report (2000). Teaching children to read: An evidenced assessment of the scientific research literature on reading and its implications for instruction (NIH Publication No. 00-4769). Washington, DC: U.S. Government Printing Office.

Newcombe, N. S., Ambady, N., Eccles, J., Gomez, L., Klahr, D., Linn, M., Miller, K., & Mix, K. (2009). Psychology's role in mathematics and science education. *American Psychologist*, 64(6), 538–550.

Pitt, R. B. (1976). Toward a comprehensive model of problem-solving: Applications to solutions of chemistry problems by high school and college students. San Diego: University of California, unpublished doctoral dissertation.

Schmidt, W. H., Wang, H. C., & McKnight, C. C. (2005). Curriculum coherence: An examination of US mathematics and science content standards from an international perspective. *Journal of Curriculum Studies*, 37, 525–559.

Siegler, R. S., & Mu, Y. (2008). Chinese children excel on novel mathematics problems even before elementary school. *Psychological Science*, 19(8), 759–768.

Stein, N. L., Anggoro, F. K., & Hernandez, M. W. (2010a). *A developmental study of physical science learning: The importance of starting early.* University of Chicago, IL, unpublished manuscript.

Stein, N. L., Anggoro, F. K., & Hernandez, M. W. (2010b). *Children learning about physical science: The importance of visual modeling and describing physical processes.* University of Chicago, IL, unpublished manuscript.

Stein, N. L., Hernandez, M. W., & Anggoro, F. K. (2010). Understanding complex concepts: Unpacking the causal mechanisms for physical events and state changes. Unpublished manuscript, University of Chicago.

Stein, N. L., Trabasso, T., & Liwag, M. (1993). The representation and organization of emotional experience: Unfolding the emotion episode. In M. Lewis & J. M. Haviland (Eds.), *Handbook of Emotions*. New York: Guilford Press. pp. 279–300.

Trabasso, T., & Bouchard, E. (2000). 'Teaching children how to comprehend what they read: A review of experimental research on direct instruction of reading comprehension'. In Report of the National Reading Panel. Teaching Children to Read: An Evidence-Based Assessment of the Scientific Research Literature on Reading and Its Implications for Reading Instruction (NIH Publication No. 00-4769). Washington, DC: U.S. Government Printing Office. pp. 39–118.

Trabasso, T., & Bouchard, E. (2001). Teaching readers how to comprehend text strategically. In C. C. Block & M. Pressley (Eds.), *Comprehension instruction: Research-based best practices.* New York: Guilford Press. pp. 176–200.

Vosniadou, S. (2008). *International handbook of research on conceptual change.* New York: Routledge.

Wilkening, F. (1979). Combining of stimulus dimensions in children's and adults' judgments of area: An information integration analysis. *Developmental Psychology*, 15, 25–33.

Wilkening, F. (1980). Judgment experiments with area, volume, and velocity. In F. Wilkening, J. Becker, & T. Trabasso (Eds.), *Information integration by children*. Hillsdale, NJ: LEA.

Part I

Reading, Learning, and Teaching

2 Instructional Influences on Growth of Early Reading

Individualizing Student Learning

Frederick J. Morrison and Carol M. Connor

In the ongoing effort to understand and improve the literacy skills of American children, several important insights have begun to focus and shape theoretical and empirical work. First, it is becoming increasingly evident that meaningful individual differences in important language, cognitive, literacy, and social skills emerge before children begin formal schooling in kindergarten or first grade (Morrison, Bachman, & Connor, 2005; Shonkoff & Phillips, 2000). Second, this early variability is influenced by a number of factors in the child, family, preschool, and larger socio-cultural context [National Institute of Child Health and Human Development Early Child Care Research Network (NICHD-ECCRN), 2004]. Third, these contributing influences do not operate in isolation, but interact with each other in complex ways to shape children's trajectories (Connor, Son, Hindman, & Morrison, 2005). Fourth, recent work has discovered that the early schooling experiences of American children are highly variable, in some cases exacerbating the degree of difference found among children prior to school entry (NICHD-ECCRN, 2002; Pianta, Paro, Payne, Cox, & Bradley, 2002). Finally, longitudinal work has revealed the lasting effect of early experiences on reading acquisition, grades earned, and dropout rates (Entwisle, Alexander, & Olson, 2005; Juel & Minden-Cupp, 2000). These trends have directed attention to the importance of individual child variability in programming effective reading instruction during the early elementary school years.

Beyond the Reading Wars

There has been a long-standing controversy regarding the best way to teach children how to read (Ravitch, 2001). The debate has centered on the efficacy of phonics or code-based instruction versus whole language or meaning-based instruction (Rayner, Foorman, Perfetti, Pesetsky, & Seidenberg, 2001). Code-based instruction concentrates on helping children learn to crack the code of reading—that letters have sounds that combine to make words in fairly predictable ways. Meaning-based instruction views learning to read as a more natural process (Goodman, 1970) that requires abundant experiences with books and other print within a literature-rich environment. However, in contrast to the polarized views, mounting evidence shows that most children develop stronger reading skills when they are provided with explicit decoding instruction in combination with meaningful reading activities—a more balanced approach (Cunningham & Hall, 1998; Pressley, 1998; Rayner et al., 2001).

But even a balanced approach to teaching reading leaves in place an implicit assumption that a curriculum or even a specific instructional practice will be equally effective for all children (National Reading Panel, 2000; Ross, Smith, Slavin, & Madden, 1997). Accumulating research indicates that the efficacy of a particular instructional practice will depend on the skill level of the student. Instructional strategies that help one student (e.g., with weak vocabulary skills) may be ineffective for another student (e.g., with stronger vocabulary skills). This phenomenon has been called aptitude by treatment interactions (Sternberg, 1996) or, more recently, child by instruction interactions (Connor, Jakobsons, Crowe, & Meadows, 2009; Connor, Morrison, & Katch, 2004; Connor, Morrison, & Petrella, 2004). However, as this research has relied on the naturally occurring variability in children and classroom instruction, the causal implications of the interactions, while compelling, remain unclear.

The present chapter first describes our efforts in recent years to understand the nature of effective instruction for children in the early stages of reading acquisition. Findings from that work demonstrated quite clearly that the most effective instruction differs depending on the initial skill level of the child. These child by instruction interactions led us to hypothesize that individualizing instruction would yield strong outcomes for all children. In the second part of the chapter we describe a program of intervention to individualize student instruction and present results from the first randomized controlled trial (Connor, Morrison, Fishman, Schatschneider, & Underwood, 2007). The findings point to the promise of individualizing student instruction in the hope of helping all children learn.

Conceptualizing and Measuring Instruction

One challenge in research investigating individualized instruction is conceptualizing and measuring the complexities of students' classroom experiences in ways that are meaningful for teachers while also permitting statistical analysis of their relations to student outcomes. Much of reading research has compared the relative effectiveness of two broad curriculum types, meaning-focused and skill- or code-based instruction (Rayner et al., 2001). However, during any given school day, teachers may provide a variety of activities that incorporate both of these major curriculum types. The teacher may use code- or skill-based instruction for teaching phonological skills, and a meaning-focused approach as he or she sets up the library corner and encourages students to read independently.

Comparing one curriculum with another may overlook the variety and complexity of learning activities students actually experience, including who—teacher or child—manages the activity and how instruction shifts over the course of the year. Additionally, accumulating evidence indicates that literacy is multidimensional (Hoover & Gough, 1990; Sénéchal & LeFevre, 2001) rather than a more global construct. Literacy is distinct from (though connected to) oral language and metalinguistic awareness (Mason & Stewart, 1990; Whitehurst & Lonigan, 1998).

Different components of the home literacy environment—formal versus informal emergent literacy experiences—differentially predict the distinct components of literacy (Sénéchal & LeFevre, 2002; Storch & Whitehurst, 2002). For example, reading to children promotes primarily vocabulary and oral language skills, whereas direct instruction in letter/sound knowledge and word decoding promotes those skills without much effect on vocabulary. Using a parallel argument, if literacy has multiple dimensions, then examining sources of classroom influence on literacy multidimensionally should prove more informative than examining their impact globally at the curriculum or classroom level. In our work, we have conceptualized the content of classroom literacy activities across five dimensions: (1) code- versus meaning-focused; (2) teacher- versus teacher–child- versus peer- versus child-managed; (3) explicit versus implicit; (4) classroom- versus student-level; and (5) change in instruction over the school year.

Code- versus Meaning-Focused Activities

The code- versus meaning-focused dimension captures the content focus of language and literacy activities. Code-focused activities (Foorman, Francis, Fletcher, Schatschneider, & Mehta, 1998) include teaching children how to name and write letters, rhyme words (Torgesen et al., 1999), relate letters to the sounds they make, and sound out words (phonological decoding). In contrast, activities designed to help students understand words and passages, comprehend what is read to them and what they read, and enhance receptive and expressive language skills including listening comprehension (Scarborough, 1990) are considered meaning focused (Beck, McKeown, & Kucan, 2005; Foorman, Francis et al., 1998; National Reading Panel, 2000; Whitehurst et al., 1994).

Teacher-, Teacher–child-, Peer- versus Child-Managed

Recent findings reveal that one important element of classroom learning reflects who is focusing the child's attention—the teacher, the peers, or the student (Connor, Morrison, & Katch, 2004; Morrison et al., 2005)—or whether the attention is jointly focused (teacher and students interacting). These studies, conducted with preschool and first-grade children, suggest that those with higher levels of vocabulary benefit most from more child-managed learning, while those with developing skills profit from teacher- and teacher–child-managed experiences. In our framework, child-centered or child-initiated learning may be teacher- or child-managed or managed jointly in a teacher–child-managed situation; the critical issue is who is directing the child's attention: the teacher and/or the child. For example, a teacher reading a book to students without discussion, even if the children selected the book, would be considered teacher-managed because the teacher is focusing the children's attention and is doing most of the talking. Sharing, scaffolding, or interactive read-alouds are considered to be teacher–child-managed within this framework because the teacher is actively involved with and responsive to the child. In contrast, activities in which the child is working with peers without immediate teacher guidance, such as playing a phonics game or reading with a peer, is considered peer-managed. In one recent study (Connor, Morrison, & Slominski, 2006), we found that organized peer play activity significantly enhanced the vocabulary growth of preschool children who started the school year with relatively low vocabulary skills. Finally, independent, individual activities, such as working alone to complete a journal or workbook page, are considered child-managed since the student is directing his or her own attention without the support of others.

Explicit versus Implicit

The explicit versus implicit dimension incorporates the idea that activities can be centrally or incidentally focused on promoting a specific outcome. This dimension is defined relative to the outcome being explored. For example, if the target outcome is reading comprehension, activities that directly focus on the extraction and construction of meaning from text (Snow, 2002), such as teaching comprehension strategies, would be considered explicit. During a book reading, a teacher might read text aloud to children; this would be considered an implicit comprehension activity. This kind of activity might be expected to build comprehension skills, but implicitly rather than systematically and explicitly. On the other hand, weaving in instruction about testing one's understanding while reading or looking up unfamiliar words in the dictionary would represent explicit comprehension instruction.

Classroom- versus Student-Level

The classroom- versus student-level dimension considers the extent to which instruction is the same or different for each child in the classroom. Literacy activities may be at the classroom level, such as when the teacher is reading aloud to the entire class. All of the children are doing the same thing at the same time. Even if children are working in small groups or individually (e.g., centers), if they are all doing substantially the same thing (e.g., completing a phonics activity), then that is classroom-level instruction. In contrast, for student-level instruction, children are engaged in substantially different activities at the same time. Teachers may provide student-level instruction in small groups or they may work with children individually (e.g., centers with different activities, tutoring one child while the rest do other activities). A video-coding system developed in our labs permits coding of each individual student's participation in specific literacy activities so that we can examine this dimension of instruction.

Note that specific types of activities (e.g., book reading) can occur on both the classroom and student levels. For example, in one classroom, a teacher might read aloud to the entire class (teacher-managed, meaning-focused, implicit, classroom-level), while in another, a teacher might read aloud to a small group of children (teacher-managed, meaning-focused, implicit, child-level) while the rest of the children engage in substantially different activities such as writing in their journals (child-managed, meaning-focused, explicit, child-level).

Change in Instruction over the School Year

The final dimension cuts across the other four and highlights changes in instructional activities over the school year. One provocative finding in the Juel and Minden-Cupp (2000) study revealed that some teachers changed their instructional emphasis over the course of the first-grade school year. For example, one teacher began the year with a strong focus on explicit, teacher-managed decoding instruction that tapered off as the year progressed and as children mastered basic skills. In this class, children with weaker fall reading skills (i.e., children in the low reading group) achieved stronger spring decoding scores than did children in the low reading group in other classrooms with less initial teacher-managed decoding focus.

Putting the Dimensions Together

A key feature of these dimensions is that they operate simultaneously. Thus activities might be fundamentally designated as teacher-, teacher–child, peer-, or child-managed, but then assigned three other modifiers to reflect the focus of the content (code versus meaning), the nature of the content (explicit versus implicit), and the level (classroom versus student) at which it operates; together, these dimensions can be combined to produce a total of 16 possible designations.

Child Factors

Notwithstanding the importance of the instructional environment and other teaching dimensions, children's characteristics play independent and interactive roles in shaping literacy trajectories during the early elementary years. Accumulating evidence reveals that a number of literacy skills in the early grades consistently predict children's later reading and academic success, including alphabet knowledge, phonological awareness, letter–word recognition, and phonological decoding (Rayner et al., 2001; Scarborough, 1998; Schatschneider, 2004; Snow, Burns, & Griffin, 1998). Language skills, particularly vocabulary and metalinguistic awareness, are also consistent predictors of later reading success, especially as comprehending what is read becomes important (Scarborough, 1990, 2001; Snow et al., 1998). Additionally, self-regulation plays a central role in early literacy development and academic success (McClelland, Morrison, & Holmes, 2000; Morrison et al., 2005). Students with weak self-regulation present classroom management challenges and demonstrate less growth in reading skills than do children with stronger self-regulation (Cameron, Connor, Morrison, & Jewkes, 2008; McClelland et al., 2007).

Unearthing Child by Instruction Interactions

In recent work, we have examined child by instruction interactions in first and second grade (Connor, Morrison, & Underwood, 2007) and third grade (Connor, Morrison, & Petrella, 2004) and more recently in preschool (Connor et al., 2006). Other work by Connor and colleagues using a Reading First sample and the same dimensions-of-instruction framework has examined child by instruction interactions on decoding, comprehension, and fluency measures over the first three years of elementary school (Connor et al., 2009). Two examples illustrate the pattern of findings emerging from this work that form the foundation for the proposed studies. In one study in first grade (Connor, Morrison, & Katch, 2004), we found that, for children who started the year with relatively low word decoding skills, more versus less time spent in teacher-managed code-focused instruction yielded greater reading gains by spring. In contrast, for students with higher initial skills, more time in teacher-managed code-focused instruction made no difference. But for students with stronger vocabularies, more time in child-managed meaning-focused activities yielded greater improvement over the year, although it had the opposite effect for children with initially lower scores. A separate investigation followed these children into third grade, focusing on growth of reading comprehension skills (Connor, Morrison, & Petrella, 2004). Here we found that, for children starting with lower comprehension scores, more time in teacher-managed explicit instruction in comprehension strategies (e.g., summarizing and inferring) was associated with greater gains. For students with higher initial skills, more teacher-managed explicit instruction had no effect whereas greater amounts of child-managed explicit activities significantly improved comprehension among these children. Overall, across a broad spectrum of grades (from preschool to third grade) and skills (letter and word reading, fluency, and comprehension), child by instruction interactions emerge as a pervasive feature of early literacy development.

An Intervention to Individualize Instruction in First Grade

These and related findings (Foorman, Francis et al., 1998; Juel & Minden-Cupp, 2000) imply that the most effective pattern of instruction (amount, type, and change over time) from preschool to third grade differs depending on the initial skill levels of the child. Thus, educational efforts to individualize (or personalize or differentiate) instruction for each child could prove highly effective. Toward that end we are currently engaged in an intervention study aimed at assessing the impact (using a randomized controlled trial design) of individualizing instruction on first graders' early reading growth.

The intervention includes the interaction algorithms, the Assessment-to-Instruction (A2i) software, the multiple dimensions framework, and practice-based professional development. The components work together to enable teachers to better individualize instruction in the classroom. This design allowed us to test the hypothesis that child by instruction interactions were causally related to students' reading growth.

Computer Support for Teachers' Instruction and Student Learning

Given the complex dimensions of instruction outlined above, as well as the goal of tailoring instruction for each child, teachers face a substantial challenge as they implement individualized instruction. In our project, computer technology is used for two primary functions: (1) as a way to facilitate individualized learning through child assessment and monitoring as well as activity planning and delivery; and (2) as a tool for teacher training.

First, recent research has identified effective ways to use computer software to facilitate the monitoring of child progress through assessment (Mengeling, 2000). An example of this is the Dynamic Indicators of Early Literacy System (DIBELs) (http://dibels.uoregon.edu, www.fcrr.org), which has been used successfully in a variety of school districts. The Center for Academic and Reading Skills (http://cars.uth.tmc.edu/) has also used technology to support assessment and reading instruction (Foorman, Santi, & Berger, 2007; Foorman, Fletcher et al., 1998). The University of Michigan Center for Highly Interactive Computing in Education (hi-ce; http://www.hi-ce.org) is continuing to develop effective learning systems for urban schools which use hand-held computers.

In addition to gathering data on children's learning, software has the potential to help teachers translate this information into classroom-ready activities to foster further growth. A2i software uses the information provided by child assessments to compute recommended daily amounts and types of instruction tailored to each child using the interaction algorithms based on our research. This information is delivered to teachers in a lesson plan format (e.g., for each child, how much teacher–child-managed meaning-focused or child-managed meaning-focused instruction to provide that day and when; how children with similar learning goals might be grouped, etc.). A2i also displays predicted and observed assessment results to aid teachers in monitoring student progress.

Yet Mengeling (2000) observes that, while "these products hold great promise in increasing the quality and efficiency of classroom assessments, . . . they have the potential to cause harm because each product assumes an assessment-literate user . . . the promise of these assessment tools can only be realized in a school culture proficient in assessment" (p. 277). As a result, technology is used in conjunction with professional development (PD).

In an earlier study (Connor, Morrison, & Katch, 2004), hierarchical linear models (HLM; Raudenbush, Bryk, Cheong, Congdon, & du Toit, 2004) were developed to predict students' letter–word reading outcomes with a surprisingly high degree of precision (Morrison et al., 2005). These models form the foundation of the A2i algorithms.

Two of the most difficult tasks facing teachers as they endeavor to individualize their students' instruction are using assessment to guide instruction and deciding how much time to spend in specific types of activities (Wharton-McDonald, Pressley, & Hampston, 1998). Taking into account child by instruction interactions makes planning for each student in the classroom highly complex, because each student has a unique pattern of instruction that will be optimal for him or her, based on assessed reading and vocabulary scores. Moreover, these amounts will change from month to month. A2i software was designed to make this process more transparent to teachers. Fortunately, complex calculations are something computers do quite well and so, for each student in the classroom, A2i algorithms compute recommended specific daily amounts (in minutes) of different types of instruction as a function of students' fall letter–word reading and vocabulary scores and target child outcome. For all children, the target outcome is at least grade-level performance and a minimum of one school year's growth in reading achievement by the end of first grade. In this way, instructional goals for all children—both high- and lower-performing students—are addressed. For the A2i algorithms to work, the target spring reading outcome (i.e., grade level and at least nine months of progress) is set, based on students' initial reading score. For students reading above grade level, the target outcome equals their grade equivalent (GE) plus .9, which would be the minimum amount of expected reading skill growth in one school year. (e.g., initial GE $1.5 + .9 = 2.4$). In this way, adequate yearly growth is anticipated for all students, with more than a year's growth expected for students beginning school with lower reading scores. The minimum target reading outcome for this study was set at a grade equivalent of 2.1, based on district means; in other words, regardless of children's skills at the beginning of first grade, they should finish first-grade reading just above second-grade level. If they start first grade above grade expectations, the amounts of teacher-managed code-focused and child-managed meaning-focused instruction required to achieve each child's target outcome are computed using the most recent letter–word reading and vocabulary scores. The month of the year is included in the algorithms so that recommended amounts automatically change monthly, which captures the dimension of change over time. Additionally, when updated scores are entered, the recommended amounts of instruction and small group assignments change.

The A2i software provides easy access to students' scores, including graphs of expected and observed reading and vocabulary scores, and a catalog of activities indexed to the schools' core reading curriculum and other literacy activities using the multidimensional framework (teacher- versus child-managed, etc.). By using the dimensions of instruction, in principle, any evidence-based language arts core curriculum can be indexed and used with A2i along with supplemental and teacher-created activities. Thus, A2i represents not a new reading curriculum but rather a new way of implementing current reading programs, supplemented with other literacy activities. A2i also incorporates lesson-planning software so that teachers can schedule, plan, and print daily lesson plans. In this way, A2i software encourages lesson planning, which is associated with improved student achievement (Borko & Niles, 1987; Fuchs, Fuch, & Phillips, 1994) while providing easy access to individual student assessment results and their interpretation, which is also associated with more effective instruction (Taylor, Pressley, & Pearson, 2000; Wharton-McDonald et al., 1998).

First-Grade Study

In the first-grade study (Connor, Morrison, Fishman et al., 2007), teachers in the treatment schools ($n = 22$) received PD (Garet, Porter, Desimone, Birman, & Yoon, 2001) on how to individualize student instruction, including how to use the A2i software. PD topics were (1) using assessment to guide instruction; (2) planning for effective instruction using A2i; (3) classroom management and organizing classrooms using small groups based on learning goals; (4) implementing effective reading instruction; and (5) using research to inform instruction. During this first year, PD efforts focused primarily on topics 1, 2, 3, and 4. Teachers participated in two workshops, in spring 2005 and fall 2005, and received ongoing support using a mentor model (Vaughn & Coleman, 2004) while actively implementing the intervention. Researchers ($n = 6$) were assigned to schools and met with individual teachers at the schools about every other week; during these meetings, researchers served as participant observers in the classroom and taught the teachers how to use A2i during planning time. Additionally, teachers met monthly after school in collaborative professional development groups.

The teachers in the control schools ($n = 25$) participated in one introductory meeting at their school, which presented the purpose of the study and A2i software and later received the results of student assessments (fall, winter, and spring). All teachers had access to the school district training, which included Reading First reforms (Connor et al., 2009; US DOE, 2004). Of note, all schools in the district were required to provide a 90- to 120-minute block of uninterrupted language arts instruction, of which 45 minutes was supposed to include small group instruction.

Results of the First-Grade Individualizing Student Instruction Study

The results of the random field trial, using hierarchical linear modeling, revealed that children in the treatment group ($n = 289$) made significantly greater gains overall on the Woodcock-Johnson III (WJ-III) passage comprehension test than children in the control classrooms ($n = 327$), controlling for fall status (Connor, Morrison, Fishman et al., 2007). The fitted mean difference in achievement between the students in the treatment and control classrooms represented about a two-month difference in grade equivalents (translated from W scale scores used in the analyses). Additionally, there was a fidelity or dosage effect. We observed substantial variability in teachers' implementation of the intervention; however, the more teachers individualized their students' instruction and the more time they spent using A2i (i.e., total number of minutes from September to May, mean = 180 minutes, range 15–374), the greater was their students' passage comprehension score growth. While, overall, students made gains in intervention classrooms, as hoped, the effect of treatment was greater for children who began the year with weaker vocabulary scores. Children who began first grade with lower vocabulary scores made substantial gains in classrooms where teachers used A2i for moderate to high amounts of time, so these students, not just students with strong vocabulary scores, were achieving, on average, a grade equivalent of 2.0. Using A2i, supported by PD, appears to increase teachers' ability to individualize student instruction (fidelity of implementation and A2i use correlation $r = .39$), thereby providing more optimally effective instruction for each child and greater student reading skill growth. Highly similar results were obtained for students' letter–word reading skills (WJ-III letter–word identification subtest).

Findings from classroom observations also produced a profile of effective individualized instruction in practice. First, we found that teachers who fully individualized instruction used multiple student grouping configurations, including homogeneous skill groups, in order to address the unique needs of the individual students in their classrooms. Amounts and types of instruction were aligned with A2i individual student recommendations. Students worked independently at literacy-focused centers, using activities designed to meet their learning objectives, while the teachers worked with small groups of students. Virtually the entire language arts block was spent in meaningful literacy activities. More generally, teachers used an observable organization system (e.g., center chart, daily schedule) to facilitate transitions and efficiently paced instruction. Further, they used A2i software for lesson planning and for easy access to individual student assessment results and interpretation, both of which are associated with student achievement (Borko & Niles, 1987; Fuchs et al., 1994; Taylor et al., 2000; Wharton-McDonald et al., 1998).

Finally, no single characteristic emerged as the defining teacher, classroom, or school variable that predicted whether or not teachers would successfully individualize instruction in their classrooms. Neither years of teaching experience, years of education, nor the school's average student socio-economic status systematically predicted successful implementation of the intervention.

Summary and Conclusions

Our results revealed that algorithm-guided individualized instruction, using A2i, promoted stronger student reading growth compared with the control group in first grade. One obvious question is the degree to which these effects will generalize to other grades. In our ongoing project we are replicating the efficacy of the first-grade intervention and extending it to the second and third grades, including assessment of the cumulative impact across grades. Moreover, the first-grade intervention combined three separate components: professional development; individualizing instruction through small groups; and A2i support. We are also planning to experimentally examine the value added by each of the components to children's reading growth. Finally, the significant individual differences across teachers in fidelity of implementation in the first-grade study require closer examination. An important future goal will be to evaluate the extent, nature, and sources of variation in teacher response to the professional development to individualize instruction.

In conclusion, our work on early literacy growth has revealed the presence of intricate child by instruction interactions across normally varying classroom environments from preschool to third grade. The inference that individualizing instruction could be a powerful educational tool led us to develop and test one strategy for providing each child with an instructional package designed for their specific literacy needs. Our initial work yielded positive results and, if successful, scaled-up efforts could prove powerful in our efforts to improve literacy for all American children.

References

Beck, I. L., McKeown, M. G., & Kucan, L. (2005). Choosing words to teach. In E. H. Hiebert & M. L. Kamil (Eds.), *Teaching and learning vocabulary: Bringing research to practice*. Mahwah, NJ: LEA. pp. 209–222.

Borko, H., & Niles, J. (1987). Descriptions of teacher planning: Ideas for teachers and research. In V. Richardson-Koehler (Ed.), *Educators' handbook: A research perspective*. New York: Longman. pp. 167–187.

Cameron, C. E., Connor, C. M., Morrison, F. J., & Jewkes, A. M. (2008). Effects of classroom organization on letter-word reading in first grade. *Journal of School Psychology*, 6, 173–192.

Connor, C. M., Jakobsons, L. J., Crowe, E., & Meadows, J. (2009). Instruction, differentiation, and student engagement in Reading First classrooms. *Elementary School Journal*, 109, 221–250.

Connor, C. M., Morrison, F. J., Fishman, B. J., Schatschneider, C., & Underwood, P. (2007). Algorithm-guided reading instruction. *Science*, 315, 464–465.

Connor, C. M., Morrison, F. J., & Katch, E. L. (2004). Beyond the reading wars: The effect of classroom instruction by child interactions on early reading. *Scientific Studies of Reading*, 8(4), 305–336.

Connor, C. M., Morrison, F. J., & Petrella, J. N. (2004). Effective reading comprehension instruction: Examining child by instruction interactions. *Journal of Educational Psychology*, 96, 682–698.

Connor, C. M., Morrison, F. J., & Slominski, L. (2006). Preschool instruction and children's literacy skill growth. *Journal of Educational Psychology*, 98(4), 665–689.

Connor, C. M., Morrison, F. J., & Underwood, P. (2007). A second chance in second grade? The independent and cumulative impact of first and second grade reading instruction and students' letter-word reading skill growth. *Scientific Studies of Reading*, 11(3), 243–268.

Connor, C. M., Son, S., Hindman, A., & Morrison, F. J. (2005). Teacher qualifications, classroom practices, family characteristics and preschool experience: Complex effects on first graders' vocabulary and early reading outcomes. *Journal of School Psychology*, 43, 343–375.

Cunningham, P., & Hall, D. (1998). The four blocks: A balanced framework for literacy in primary classrooms. In K. R. Harris, S. Graham, & D. Deshler (Eds.), *Teaching every child every day: Learning in diverse schools and classrooms*. Cambridge: Brookline Books. pp. 32–76.

Entwisle, D., Alexander, K. & Olson, L. (2005). First grade and educational attainment by age 22: A new story. *American Journal of Sociology*, 110, 1458–1502.

Foorman, B. R., Fletcher, J. M., Francis, D. J., Carlson, C. D., Chen, D., Mouzaki, A., et al. (1998). *Texas primary reading inventory (1998 edition).* Technical report prepared for the Texas Educational Agency. Houston: Center for Academic and Reading Skills, University of Texas-Houston Health Science Center, and University of Houston.

Foorman, B. R., Francis, D. J., Fletcher, J. M., Schatschneider, C., & Mehta, P. (1998). The role of instruction in learning to read: Preventing reading failure in at risk children. *Journal of Educational Psychology*, 90, 37–55.

Foorman, B. R., Santi, K., & Berger, L. (2007). Scaling assessment-driven instruction using the internet and handheld computers. In B. Schneider (Ed.), *Scale Up in Practice*. Lanham, MD: Rowan and Littlefield Publishers, Inc. pp. 69–79.

Fuchs, L. S., Fuch, D., & Phillips, N. (1994). The relation between teachers' beliefs about the importance of good student work habits, teacher planning, and student achievement. *Elementary School Journal*, 94(3), 331–345.

Garet, M., Porter, A., Desimone, L., Birman, B., & Yoon, K. S. (2001). What makes professional development effective? Results from a national sample of teachers. *American Educational Research Journal, 38*(4), 915–945.

Goodman, K. (1970). Reading: A psycholinguistic guessing game. In H. Singer & R. B. Ruddell (Eds.), *Theoretical models and processes of reading*. Newark: International Reading Association. pp. 259–272.

Hoover, W. A., & Gough, P. B. (1990). The simple view of reading. *Reading and Writing*, 2(2), 127–160.

Juel, C., & Minden-Cupp, C. (2000). Learning to read words: Linguistic units and instructional strategies. *Reading Research Quarterly*, 35(4), 458–492.

McClelland, M. M., Cameron, C. E., Connor, C. M., Farris, C. L., Jewkes, A. M., & Morrison, F. J. (2007). Links between behavioral regulation and preschoolers' vocabulary, literacy and math skills. *Developmental Psychology*, 43, 947–959.

McClelland, M. M., Morrison, F. J., & Holmes, D. L. (2000). Children at risk for early academic problems: The role of learning-related social skills. *Early Childhood Research Quarterly*, 15, 307–329.

Mason, J. M., & Stewart, J. P. (1990). Emergent literacy assessment for instructional use in kindergarten. In L. M. Morrow & J. K. Smith (Eds.), *Assessment for instruction in early literacy*. Englewood Cliffs, NJ: Prentice-Hall. pp. 155–175.

Mengeling, M. A. (2000). Computer software products for classroom assessment purposes. In W. G. Wraga, P. S. Hlebowitsh, D. Tanner, & National Association of Secondary School Principals (U.S.) (Eds.), *Research review for school leaders* (Vol. III). Hillsdale, NJ: LEA. pp. 277–300.

Morrison, F. J., Bachman, H. J., & Connor, C. M. (2005). *Improving literacy in America: Guidelines from research*. New Haven, CT: Yale University Press.

National Reading Panel (2000). Teaching children to read: An evidence-based assessment of the scientific research literature on reading and its implications for reading instruction (NIH 00–4769). Washington, DC: U.S. Department of Health and Human Services, Public Health Service, National Institutes of Health, National Institute of Child Health and Human Development.

NICHD-ECCRN (National Institute of Child Health and Human Development-Early Child Care Research Network). (2002). The relation of global first grade classroom environment to structural classroom features, teacher, and student behaviors. *Elementary School Journal*, 102(5), 367–387.

NICHD-ECCRN (2004). Multiple pathways to early academic achievement. *Harvard Educational Reviews*, 74(1), 1–29.

Pianta, R. C., Paro, L., Payne, K., Cox, C., & Bradley, R. H. (2002). The relation of kindergarten classroom environment to teacher, family and school characteristics and child outcomes. *Elementary School Journal*, 102(3), 225–238.

Pressley, M. (1998). *Reading instruction that works: The case for balanced teaching*. New York: Guilford.

Raudenbush, S. W., Bryk, A., Cheong, Y. F., Congdon, R., & du Toit, M. (2004). *HLM6: Hierarchical linear and nonlinear modeling*. Lincolnwood: Scientific Software International.

Ravitch, D. (2001). It is time to stop the war. In T. Loveless (Ed.), *The great curriculum debate: How should we teach reading and math?* Washington, DC: Brookings Institutional Press. pp. 210–228.

Rayner, K., Foorman, B. R., Perfetti, C. A., Pesetsky, D., & Seidenberg, M. S. (2001). How psychological science informs the teaching of reading. *Psychological Science in the Public Interest*, 2(2), 31–74.

Ross, S. M., Smith, L. J., Slavin, R. E., & Madden, N. A. (1997). Improving the academic success of disadvantaged children: An examination of success for all. *Psychology in the Schools*, 34(2), 171–180.

Scarborough, H. S. (1990). Very early language deficits in dyslexic children. *Child Development*, 61, 1728–1743.

Scarborough, H. S. (1998). Early identification of children at risk for reading disabilities: Phonological awareness and some other promising predictors. In B. K. Shapiro, P. J. Accardo, & A. J. Capute (Eds.), *Specific reading disability: A view of the spectrum*. Timonium: York Press, Inc. pp. 75–119.

Scarborough, H. S. (2001). Connecting early language and literacy to later reading (dis)abilities: Evidence, theory, and practice. In S. B. Neuman & D. K. Dickinson (Eds.), *Handbook of early literacy research*. New York: Guilford Press. pp. 97–110.

Schatschneider, C. (2004). A multivariate study of individual differences in performance on the reading portion of the Florida Comprehensive Assessment Test: A brief report. Online. Available at: http://www.fcrr.org/TechnicalReports/Multi_variate_study_december2004.pdf (accessed August 4, 2006).

Sénéchal, M., & LeFevre, J.-A. (2001). On refining theoretical models of emergent literacy: The role of empirical evidence. *Journal of School Psychology*, 39(5), 439–460.

Sénéchal, M., & LeFevre, J.-A. (2002). Parental involvement in the development of children's reading skill: A five-year longitudinal study. *Child Development*, 73(2), 445–460.

Shonkoff, J. P., & Phillips, D. A. (Eds.) (2000). *From neurons to neighborhoods: The science of early childhood development*. Washington, DC: National Academies Press.

Snow, C. E. (2002). *Reading for understanding: Toward an R&D program in reading comprehension*. Arlington, VA: RAND.

Snow, C. E., Burns, M. S., & Griffin, P. (Eds.) (1998). *Preventing reading difficulties in young children*. Washington, DC: National Academies Press.

Sternberg, R. J. (1996). Matching abilities, instruction, and assessment: Reawakening the sleeping giant of ATI. In I. Dennis (Ed.), *Human abilities: Their nature and measurement*. Hillsdale, NJ: LEA. pp. 167–181.

Storch, S. A., & Whitehurst, G. J. (2002). Oral language and code-related precursors to reading: Evidence from a longitudinal structural model. *Developmental Psychology*, 38(6), 934–947.

Taylor, B. M., Pressley, M. P., & Pearson, P. D. (2000). *Research-supported characteristics of teachers and schools that promote reading achievement. Reading Matters Research Report*. Washington, DC: National Education Association.

Torgesen, J. K., Wagner, R. K., Rashotte, C. A., Rose, E., Lindamood, P., Conway, T., et al. (1999). Preventing reading failure in young children with phonological processing disabilities: Group and individual responses to instruction. *Journal of Educational Psychology*, 91, 579–593.

US DOE (2004). *No child left behind: A toolkit for teachers.* Washington, DC: US Department of Education, Office of the Deputy Secretary.

Vaughn, S., & Coleman, M. (2004). The role of mentoring in promoting use of research-based practices in reading. *Remedial and Special Education,* 25(1), 25–38.

Wharton-McDonald, R., Pressley, M., & Hampston, J. M. (1998). Literacy instruction in nine first-grade classrooms: Teacher characteristics and student achievement. *Elementary School Journal,* 99(2), 101–128.

Whitehurst, G. J., Epstein, J. N., Angell, A. L., Payne, A. C., Crone, D. A., & Fischel, J. E. (1994). Outcomes of emergent literacy intervention in Head Start. *Journal of Educational Psychology,* 86, 542–555.

Whitehurst, G. J., & Lonigan, C. J. (1998). Child development and emergent literacy. *Child Development,* 69, 335–357.

3 Literacies for Learning

A Multiple Source Comprehension Illustration

Susan R. Goldman, Yasuhiro Ozuru, Jason L. G. Braasch, Flori H. Manning, Kimberly A. Lawless, Kimberley W. Gomez, and Michael J. Slanovits

Jean Chall distinguished between acquiring reading skills (learning-to-read) and using them to acquire information (reading-to-learn) (Chall, 1983). Learning-to-read referred to mastering the basic tools of reading, including symbol–sound relationships (phoneme–grapheme correspondence) and print-based competencies such as word recognition and basic sentence understanding. Reading-to-learn meant using the basic tools to acquire information about the physical, social, and personal world from print-based sources of information. In the United States, the transition from learning-to-read to reading-to-learn was pegged at third or fourth grade (eight or nine years of age).

Chall (1983; Chall & Jacobs, 2003) did not expect that reading-to-learn was a process that students would simply develop once they were presented with the new demands of content area materials. Rather, they would *learn* to read-to-learn (Snow & Biancarosa, 2003). Instruction for reading-to-learn has focused on generic strategies, including summarizing, finding main ideas, learning vocabulary in context, and making inferences (Alvermann, 2001; Beck & McKeown, 1985; Guthrie, Anderson, Aloa, & Rinehart, 1999; Meltzer, Smith, & Clark, 2002; Palincsar & Brown, 1984; Pressley, 2002). Generic strategies address some of what students need if they are going to be able to acquire new concepts, relations among concepts, and ways of thinking critically about the content they are learning. However, the successful application of generic strategies often depends on students already knowing the content they are supposed to learn. That is, in the absence of content knowledge, how can a learner determine the reasonableness of an inference, the appropriateness of meaning inferred from context, or whether a summary captures the important ideas from a text? Content area texts also use a wide array of disciplinary genres and conventions for communicating with which students may not be familiar. Thus, they cannot bootstrap their understanding of the content by relying on genre knowledge in the way they may be able to in narratives about everyday life (cf. Goldman & Rakestraw, Jr., 2000).

The conclusion that generic reading strategy interventions do not go far enough is supported by national and state reading achievement data. Persistent trends show that many students who are "on track" up through grade three experience a decline in their reading achievement scores in fourth grade (Chall, Jacobs, & Baldwin, 1990), with the slump escalating into the "eighth grade cliff." The eighth-grade cliff reflects the inability to engage in complex literacy tasks (e.g., inference, interpretive comprehension), especially in content areas such as science and history, and contributes to increased school dropout rates (Chall & Jacobs, 2003; de León & Carnegie Corporation of New York, 2002).

These achievement trends have stimulated research examining the cognitive and social demands of reading-to-learn in specific content areas (e.g., Lee & Spratley, 2006; Moje & O'Brien, 2001; Shanahan & Shanahan, 2008). The research points to the importance of recognizing that each discipline has a consensually agreed upon set of literacy practices that govern how members of the discipline establish knowledge claims and communicate these claims within their disciplinary communities (Goldman & Bisanz, 2002; Moje, 2008; Shanahan & Shanahan, 2008). In most disciplines, these literacy practices fundamentally involve dealing with multiple sources of information, although the how and why of source comprehension and reasoning differ across disciplines (Bazerman, 1998; Goldman, 2004; Wineburg, 1991). Students need opportunities to learn disciplinary literacies in conjunction with learning established disciplinary concepts and principles if they are to be prepared to subsequently and successfully engage with complex and technical disciplinary information (Moje, 2008).

Compounding efforts to address the fourth-grade slump and avoid the eighth-grade cliff are literacy demands engendered by 21st-century technological advances in information accessibility. Available at the click of a mouse are not only vast amounts of traditional print-based information but also multimodal forms such as complex visuals and animations (Lawless & Schrader, 2008). "Reading" these multimodal forms of information is anything but transparent, and learners are often not very successful at it (cf. Lowe & Schnotz, 2008). The knowledge and skills required to read-to-learn in the 21st century need to encompass both multiple sources of information and multiple media forms (Kress, 2003; Lemke, 1998; New London Group, 1996).

In this chapter we first elaborate on the need for expanded conceptions of literacies for learning. In particular, we stress the importance of multiple source comprehension, by which we mean selection, coordination, and synthesis of information that comes from more than one source. Second, we illustrate one approach to research on expanded conceptions of literacy by reporting selected aspects of our work on multiple source comprehension. In conclusion, we consider the instructional implications of developmental research on multiple source comprehension in the context of schooling.

Literacy Demands of the 21st-Century Knowledge Society

We live in a knowledge society in which increases in the availability of, and access to, different kinds of information mean that emphasis must be placed on a broader array of literacy skills and competencies, specifically those associated with synthesizing, integrating, and evaluating the quality of information (Hartman, 1993; Orr, 1986). These literacy skills enable critical analysis of information and constitute the cognitive activities of critical reading. They make clear that reading comprehension is an explicitly intertextual practice in that understanding a *set* of texts or sources means understanding the relations across texts at basic meaning and advanced interpretive levels (Goldman, 2004; Orr, 1986).

This expanded set of literacy skills in no way obviates the importance of already recognized reading-to-learn skills. Research on reading-to-learn from single texts shows that successful comprehenders rely on multiple types of knowledge (e.g., of words, concepts, text structures, genres) as they try to interpret print. Successful readers achieve deep comprehension by actively engaging with the text to construct coherent representations; connect ideas within a text with each other and with relevant prior knowledge; and explain the ideas and connections (Chi, de Leeuw, Chiu, & LaVancher, 1994; Coté & Goldman, 1999; Magliano & Millis, 2003; van den Broek, Risden, & Husebye-Hartmann, 1995). Poorer comprehenders tend to paraphrase or restate verbatim the information presented in the text rather than explain it, and when they do make connections they tend to be at a surface level (Coté, Goldman, & Saul, 1998; McNamara, 2004; Magliano & Millis, 2003).

Skills that are sufficient for single text comprehension are necessary but not sufficient for successful learning in multiple source learning situations. Resources available on the web vary in terms of trustworthiness and task relevance, making critical thinking skills such as the ability to evaluate the credibility, validity, and usefulness of information more important than they ever were before (Coiro, Knobel, Lankshear, & Leu, 2008; Goldman, 2004; Gomez & Gomez, 2007). At the same time, learners need to develop skills for evaluating the sufficiency of information to address their questions and how to connect information from different sources. Thus, the knowledge society spotlights the role of inquiry and problem-solving processes for learning and productivity (Goldman, 1997, 2004; International Adult Literacy Survey, 1997; Scardamalia & Bereiter, 1996).

Contemporary Research on Reading Comprehension and Learning in Content Areas

Research indicates that more robust learning results when students are actively engaged in learning as contrasted with more passive, transmission models of learning (cf. Bransford, Brown, Cocking, Donovan, & Pellegrino, 2000; Greeno, Collins, & Resnick, 1996; Sawyer, 2006). As discussed above, more successful comprehenders engage in more active processing of text, often by generating self-explanations during reading (Chi et al., 1994; Coté et al., 1998). In classrooms, inquiry questions are frequently used to engage students in active learning, although there is wide variability in what counts as inquiry and the degree to which students need to engage in critical reasoning about the subject matter (cf. Krajcik et al., 1998; Magnusson & Palincsar, 2005; Trumbull, Bonney, & Grudens-Schuck, 2005; VanSledright, 2002a,b). That is, merely providing students with a question to answer, some sources to consult, or some activities does not ensure understanding or critical thinking. Tasks must be carefully designed to require synthesis, integration, and evaluation of information (e.g., Magnusson & Palincsar, 2005).

Well-designed, multiple source inquiry instruction has the potential to provide students with opportunities to learn an expanded set of literacies. As well, it exposes students to processes akin to those in which disciplinary experts engage in the process of "doing" their work—in the practices of the disciplinary community (Gee, 1990; Lave & Wenger, 1991). In both history and science, experts routinely engage in selection, analysis, and synthesis within and across multiple sources of evidence (Chinn & Malhotra, 2002; Wineburg, 1991), although the specific manifestation of these activities differs depending on the discipline. Research also indicates that these processes differ depending on disciplinary expertise (Bazerman, 1998; Berkencotter & Huckin, 1995; Janick-Buckner, 1997; Rouet, Britt, Mason, & Perfetti, 1996; Shanahan & Shanahan, 2008; Wineburg, 1991, 2001; Yarden, Brill, & Falk, 2001). Thus, learning the discipline is just as much about learning ways of knowing and forms of communication that govern that discipline as it is about learning disciplinary concepts and vocabulary, procedures, and principles. The two are intertwined, with students needing instruction in both (Moje, 2008).

Studies that engage students in developmentally appropriate forms of disciplinary practices suggest that there are benefits to such instruction. For example, when adolescent students are given tasks requiring the construction of historical narratives from information found in multiple documents, they learn to think more critically about what they read (Hartman, 1993; Hynd-Shanahan, Holschuh, & Hubbard, 2005; Lee & Ashby, 2000; VanSledright, 2002a,b) and engage in deeper processing of sources (Wolfe & Goldman, 2005). Similarly, in science there are improvements in elementary students' skills at using data as evidence and making sense of multiple representations (Hapgood, Magnusson, & Palincsar, 2004). Findings such as these help establish the importance and potential value of engaging students in developmentally appropriate forms of disciplinary literacies involving multiple sources of information.

We turn now to the work we are doing on literacies for reading-to-learn in content areas. Our work is situated in the context of development of an assessment system that will inform classroom instruction in multiple source comprehension in science and history inquiry contexts. To make the research and assessment development tractable, we have circumscribed the sources to text with some supporting diagrams and pictures and are focusing on upper elementary/middle school students, approximately 10–14 years of age. In focusing on this age range, we build on previous research on multiple source inquiry, the bulk of which has been with students 10 years of age or older (Goldman, 2004; Goldman & Bloome, 2005; VanSledright, 2002a,b). Furthermore, this age range reflects those most at risk for the eighth-grade slump. While it would be quite useful to include fourth- and third-grade students, younger students may need different forms of instructional activities that include both text-based sources and "hands-on" inquiry activities (e.g., Hapgood et al., 2004). Learning content from single text sources that involve both texts and dynamic visuals is also feasible with younger students when those texts are conceptually clear and complete (chapter 7). However, learners cannot count on always having such texts. Thus, we are developing a prototype assessment system for those in middle school, with hopes of later expanding downward and upward.

An Illustrative Research Example: Assessing Multiple Source Comprehension

Opportunities to Learn Multiple Source Comprehension

We have conducted microethnographic observations in urban schools and classrooms that reflect a range of reading achievement levels, as assessed by standardized tests, and in which teachers have intended to provide inquiry opportunities for their students (Goldman, 2004; Goldman & Bloome, 2005; Goldman et al., 2010; Manning et al., 2008). Across studies, we have generally found that students did have access to a wide range of information sources but the tasks they were asked to do frequently required little critical thinking about the sources or integration across sources. Types of available sources sometimes included websites, videos, various types of visuals, and small group discussions, but teachers and students tended to rely most heavily on traditional sources, such as textbooks and teacher-led discussions. In one study, these two types of sources were observed four times more frequently than any other source type (Manning et al., 2008). Background knowledge was also a common source, often elicited prior to engaging a text about a new topic. However, there was limited facilitation of students' integration of prior knowledge with new information provided by another source (Goldman, 2004).

In addition, students were directed to use multiple sources serially rather than in conjunction with one another. There were two classrooms that were notable exceptions to this general tendency. In these two cases, the teachers specifically took on the challenge of creating opportunities for intertextuality (Goldman & Bloome, 2005). In both classrooms, we observed teachers and students actively using what they knew to make sense of new information in the context of interpreting a new source. They selected, evaluated, and synthesized information from several different sources. As well, they volunteered connections to sources they had read or experienced outside of the classroom.

In more typical classrooms the focus on one source at a time and lack of connection across sources was further compounded by instructional activities that were largely teacher-directed searches for "known answers." This was the case even when the task instructions suggested otherwise. For example, in one fifth-grade social studies class, students were to use their textbook to get information to construct a timeline for a particular historical period. However, the textbook itself contained exactly the timeline the teacher had instructed the students to construct.

This phase of our work suggested that students were not, for the most part, being provided with opportunities to engage with multiple sources in contexts that required critical thinking. Accordingly, our assessment development work took on a second goal: to convey to teachers the knowledge, skills, and types of tasks that might constitute content area inquiry. We also realized that student performance on the assessments we are developing will reflect baseline data, in the sense that the assessment may well be the first opportunity students have to select and integrate multiple sources.

Providing Opportunities to Engage with Multiple Sources

We focus our discussion in this chapter on two studies pertinent to selecting and integrating information across multiple sources. These are two components of a student model of multiple source comprehension, the development of which is integral to our assessment design approach—evidence-centered design (ECD; Mislevy, Steinberg, & Almond, 2003). In ECD, the student model specifies the knowledge and skills that define or constitute the construct being assessed.

Multiple Source Inquiry in the Classroom: Study 1

Having found in the first phase of our work that students had relatively impoverished experiences with multiple source comprehension in typical urban classrooms, we designed a modest inquiry task about the history of Chicago, the city in which we are conducting this work. We specifically chose this topic so that students would have some knowledge of it and would be interested in the topic. The task was designed in collaboration with a classroom teacher and extended over several days. The inquiry task design enabled us to examine multiple components of the inquiry process, specifically selection of sources, analysis of individual sources, and synthesis across multiple sources. The study was conducted in one fifth-grade intact, urban classroom in a school in which 61 percent of the students met or exceeded grade-level proficiency for reading as assessed on the district-administered achievement test. We had complete data on 20 students (11 girls, nine boys), who constituted a representative sample of the demographics of the school as a whole.

Students were given a packet of sources and told to use them to answer a question about Chicago's emergence as a big city. The question was introduced on the first day of the activity by an anchor text and an accompanying video that described present-day Chicago as the third largest city in the United States, even though in the early 1800s no one wanted to live there. In whole class and small group discussions, students brainstormed possible answers. On day two, they were given a packet of resources consisting of texts, charts, graphs, and tables and were told to use the information in the packet to address the inquiry question. They read each source individually, and in small groups discussed the two sources they each thought would best answer the inquiry question. On day three, students individually wrote an essay addressing the inquiry question. They had the whole packet of sources available to them and were told that they could use any of the sources.

The source packet was intentionally constructed to present sources that corroborated and elaborated one another and, in one case, presented opposing views. There were several primary sources and two secondary sources (Table 3.1). All of the sources contained information relevant to the inquiry question, but they differed in terms of how much and what type of relevant information they contained. We categorized the sentences containing big ideas about Chicago's growth with respect to the different disciplines that constitute the social studies (Table 3.1). For a complete explanation of Chicago's growth, information from multiple sources was needed. Repetition of ideas provided an opportunity for students to engage in a developmentally appropriate form of the expert practice called corroboration (Wineburg, 1991, 2001). Importantly, none of the sources contained a statement such as "The reason Chicago became a big city is . . . ," although sources 3 and 4 explicitly mentioned growth in Chicago. In addition, the "main idea" of each passage, as it might be traditionally defined, was not necessarily the information in the source most useful to addressing the inquiry question (for discussion of this issue see Goldman, 1997). Thus, students needed to read the sources with the particular inquiry goal in mind and "find" relevant information whether it was the main idea or details in the source.

The discussion here focuses on the likelihood that students actually did use multiple sources to construct their essays. We found that the big ideas about which there was the most information across the sources were the ones included most frequently in the essays (Goldman et al., 2010). This repetition effect is consistent with research with college and high school students showing that the frequency of occurrence of particular claims and evidence is one variable that governs its selection and use (Britt, Perfetti, Sandak, & Rouet, 1999; Rouet et al., 1996). We cannot be sure in our case

Table 3.1 Sources of Information in the Resource Packet for the Classroom Inquiry Study and Types of Content in Each

	Big Ideas Relevant to Inquiry Question				
Source (Number of Sentences)	*Economics*	*Politics*	*Location*	*Social/ Financial Support*	*Growth*
1. A personal letter from a woman living in Georgia to a church in Chicago (17) (primary source)	5[a]	5		6	
2. An editorial and newspaper advertisement dated 1915 that appeared in the *Chicago Defender*, a newspaper published by the Chicago African American community (20) (primary source)	2	3		8	
3. A textbook segment on the rise of Chicago as a railroad hub along with a map showing the main rail lines between the North and the South (12) (secondary source)	1		6		2
4. Segment from a website describing immigration to Chicago, why immigrants came, where they were from; included chart showing immigrant groups each decade from 1830 to 1970 (13) (secondary source)	9	2			6
5. An op/ed page attributed to the *Chicago Tribune*, April 1925, that contained two opposing editorials about the stockyards	10	11		4	
5a. one, authored by Michael Armour of the Armour meat packing family, touted the stockyards as a good place to work (11) (primary source)	6	1		4	
5b. one, authored by Jane Addams, described as a person who helped immigrants, deplored working conditions in the stockyards (15) (primary source)	4	10			

[a] Numbers refer to number of sentences in the source that conveyed information of this type.

whether the inclusion of topics mentioned across the sources is simply an effect of the repetition or reflects some deeper form of comprehension. We also had no assessment of comprehension of each individual source, a situation we addressed in the second study, discussed below.

We further analyzed source integration by first tracing the ideas in the essays back to the texts from which they came and then examining the essays for evidence of conceptual synthesis of the information. Information that could not be traced to any of the sources was tagged as prior knowledge and information that occurred in multiple sources was attributed accordingly. Of course, it is possible that students knew some of the information in the sources which we had provided. However, we were not able to obtain estimates of this. We then calculated the number of sources from which students had "taken" information to use in their essays. For the 20 students, ideas in the essays were drawn from an average of 3.7 sources, with a range of 1–6. However, merely including information from all of the sources does not mean that students actually conceptually integrated the ideas, so it is informative to look at the students' responses from this perspective.

The contrast between two students, Annie and Fiona, each of whom included information from all six sources, reflects the kind of variation we observed. Annie's response included the source sequence 3/(2,4,1)/5b/1/2/3/(4,5a)/2/3/5b/4. (Note that the slashes reflect different ideas in her response, and grammatical and spelling errors are those of the student.)

ANNIE'S RESPONSE

> /I have learned that they build railroad tracks to get places./They had lots of jobs there/but it was vary bad place to work./Many came her and wrote letters that the wrote to the church./If they came they would provid them with jobs, help find a place to live, help getting around also they would give you guidance./Because of the trians the popullation of chicago tripled in the 6 years of the railroad./In Chicago many people were poor in the 1800's. The people that immigrated were Europe; African-Americans and Latino's. The jobs that they got promised a paycheck all year around every week. They had equal pay whethere white or black./ The defenders want the people to leave to the north because the south is bad./In the stockyards many people are sick some have even died./The Chicago trains became the largest train station./In the stock yards the kids had no play ground the used garbage dumps./Many people went from farm to farm to get good crops./

However, there was a lack of topical and conceptual coherence as she moved from information in one source to information in another. In contrast, Fiona's response reflected a more topical organization with information from several sources and prior knowledge (in italics).

FIONA'S RESPONSE

> Chicago became a large city because if all of the years./On the 1830 many immigrants came to Chicago and othere places. Many kinds of people came like German, Irish, Polish, and othere kind./The railroad came that was a big change

> in Chicago that made Chicago even more big. Chicago triple./The jobs that were Chicago people worked in all different kind./Like that stockyards people liked it/and some didn't/*everybody had their reasons.*/Some people write letters that nobody liked the south/so people came to the north and Illinoi, Chicago was part of the north so Africans whent there. Not only then Mexicans whanted a better life there./Some peole write letters to get to Chicago/*and because Chicago is well knowed all over the world maybe not but some people know. Chicago was not great but not bad it was just right. Chicago became a large city because years past*/and people wanted a better life./*And because is interesting and has a log of history of all time. All that is how Chicago become a large city.*/

Fiona's sequence, Anchor/4/3/2/5a/5b/*pk*/1/4/1/*pk*/4/*pk*, reflected a more topical organization with information from several sources. She began with information about the different immigrant groups, the role of the railroad and the different jobs available, with stockyard jobs as an example. She then discussed movement from the south to the north and the idea that people wanted a better life.

The responses of students who included fewer sources tended to be more coherent but dealt with fewer big ideas. Thus there was less integrative work to achieve coherence. Rob's essay is illustrative.

ROB'S RESPONSE

> One reason Chicago became a big city was that the stories went through that you would be rich if you came and it would be good to live here./*Also because the world's fair in Chicago showed people that it was a good palce to live.*/Another reason people come to Chicago was they had <u>a lot</u> of jobs./In the South people were paid unfairly,/the only job was farming./Mexican migrants cane here because jobs lasted a bit. [4/*pk*/(2,4,5a)/(1,4)/4]

Rob's essay was credited with including three different sources. Rob's response is dominated by information from source 4 and developed the job availability factor, also discussed in sources 2 and 5a.

Given that students were working across six different sources, it is noteworthy that many were able to piece together the information to capture the push and pull factors contributing to Chicago's growth. The range of responses and approaches to using the source information to answer the inquiry question was not surprising in that students had had no prior instruction in this sort of activity. Their approach to the task is consistent with the tendencies observed in other work with students of this age (e.g., Goldman, 2004; VanSledright, 2002a,b).

Multiple Source Inquiry in the Classroom: Study 2

Our observations of the students working on the task during Study 1 suggested that one source of task difficulty was the amount of material in the packet and differences in comprehension of the individual sources. Some of the students seemed to read one or two sources and then flip through the remainder both on initial reading and when writing their responses to the inquiry question. Others appeared to pick one source to write from, sometimes after carefully reading all of them and sometimes after only a cursory perusal. We decided to look more carefully at comprehension and analysis of individual sources prior to looking at students' integration of information across sources. In Study 2 we reduced the number of sources to two and asked the students to write a response for each source: "Why does the author think that Chicago became a big city and what are the reasons for thinking that?" They then were asked to write their own response: "Why do you think Chicago became a big city?" The questions of interest concerned whether the students could accurately analyze each of the sources—one at a time—in terms of the author's claims and the evidence for those claims, and how they might use one or both of the sources in constructing their own response.

The Study 2 sample comprised 69 middle school students ranging in age from 11 to 13 years, drawn from intact classrooms located in three different schools in Chicago. Across the three schools, approximately 60 percent of the students are African American, 20 percent are Hispanic, and 66 percent of the students in the school met or exceeded proficiency levels for their grade.

We created two text sets for this study, keeping the first text constant across sets and varying the second to create a differential overlap between the two sources. Source 4 from Study 1 was the first source in each set. It was paired with source 1 (set A), source 3 (set B), or source 5a (set C).

We randomly distributed the three sets to the 69 students so that we achieved approximately equal numbers of each text set within each classroom. We introduced the inquiry question "Why did Chicago become a big city?" as we had in Study 1. Students were told they would read each of two passages and after each one would answer the question about what each author thought. The source texts were available to the students throughout the study. When they had completed the second response, they were asked to write what *they* thought—and why—about why Chicago became a big city. In contrast to the first study, this task was done completely individually as a paper-and-pencil whole class task and there was no discussion of the sources themselves.

Students' performance on the questions that assessed single source comprehension generally showed accurate comprehension of each author's main claim regarding why Chicago grew, although accuracy was higher for the texts that had more surface text overlap with growth concepts and causes (e.g., jobs, railroads). Given reasonably accurate comprehension of the individual sources, we looked at responses to the third question for evidence of integration across sources. Table 3.2 shows the four types of relationships that we coded.

"No relation" indicated that the response consisted largely of information from prior knowledge, sometimes accurate, sometimes not. "Related to first source response" or "Related to second source response" indicated that the content of their own response was highly similar to the content of their response for the first source or second source but not both. In both of these cases, responses might also have included prior knowledge but not distortions or inaccuracies. The fourth category reflects integration of information from the two sources, with or without prior knowledge present. The frequency distribution was similar for sets A and C, each of which contained primary sources as the second text: Approximately 45 percent of the students' own explanations for "Why did Chicago become a big city?" drew on information from both sources they had read. These distributions differed from that for set B, the set that contained a secondary source as the second text: only one student drew on both sources when the second source was about railroads (Table 3.2). In summary, the dominant student explanation used information from the first source in the set, in combination either with prior knowledge or with information from the second text. The likelihood of integration of first and second source information was higher when students had read one of the primary sources as the second text than when they had read the secondary historical source about railroads.

Summary

The results of the two studies indicate that most students did not have particularly clear or systematic strategies for using the sources to compose a well-structured essay response. However, the data do indicate that these early approaches to integrating

Table 3.2 Responses to "Why Do You Think Chicago Became a Big City?" Frequency Distribution of Four Response Categories as Function of Condition (Second Text Read)

Relation Between Responses to Own and Authors	*Set A (personal letter,* n = *22)*	*Set B (railroads,* n = *24)*	*Set C (opinion/editorial prostockyards, n = 23)*
No relation; prior knowledge (PK) used	4 (18%)	7 (29%)	2 (9%)
	Example of prior knowledge use: I think Chicago became a big city because when they hosted the world fair many people came. I also think that Chicago became a big city because when the great Chicago fire broke out they had to build more houses and many people came to live in Chicago.		
Related to first source response (with or without PK)	5 (23%)	11 (46%)	6 (26%)
Related to second source response (with or without PK)	0	2 (8%)	0
Integrates first and second source responses (with or without PK)	10 (45%)	1 (4%)	11 (48%)
	Example: Integration of first source (4) and second source (1): I myself think similarly with both Dr. Jones and Mrs. Adams to how Chicago became a big city because I think that better opportunities were open for you if you lived in Chicago rather than places like the South where there were things like segregation where Blacks were not treated as equally as the Whites were. Immigrants were having losts of trouble find jobs and earning enough money to feed their families, but life was easier when living in Chicago. People started moving into Chicago and now there are millions of people living in Chicago. *Example: Integration of first source (4) and second source (3):* Chicago became a big city because people immagrated into america because the needed a well paid job and because rich folks invested money to have railroad and trains built.		
No response	3 (14%)	3 (12%)	4 (17%)

included ideas that appeared in several sources or that were central to one source. These ideas were often simply listed. Organizing such "lists" according to more conceptually based, topical, or thematic considerations constitutes a further refinement of multiple source integration. The basis for the organization will likely depend on the task, with explanation or argumentation (why did it happen?) requiring claim–evidence organization, whereas description tasks (what happened?) require only temporal organization. Understanding these aspects of organizing information to address the task constitute a critical thinking task, precisely the type of activity emphasized in Chall and Jacobs' (2003) discussion of what it means to read-to-learn.

Conclusions

Our work to design assessments of 21st-century reading-to-learn skills is one step toward creating opportunities to learn them. An important part of this work is creating criteria and benchmarks for multiple source comprehension that are developmentally appropriate. Without these it will be difficult for teachers to make effective use of assessment information, although the assessments themselves may give teachers a better sense of multiple source comprehension. Our assessments have potential value for professional development because they define both the construct and expected levels of performance.

Much more research and development is needed. In this vein, we encourage more research around efforts to develop units that encourage inquiry into unsolved questions or disputed phenomena (e.g., Krajcik et al., 1998; Lee & Ashby, 2000; Stevens, Wineburg, Herrenkohl, & Bell, 2005; VanSledright, 2000a,b). Students in elementary school can and do productively engage with multiple sources. However, they need more opportunities to do so. Ultimately, literacies for learning need to take into account a much wider range of information sources than we are presently examining, as well as multiple methods of meaning making and expression of that meaning. Such efforts are needed for our educational system to effectively prepare youths to become productive members of 21st-century society.

Acknowledgments

This chapter is based on a presentation given at the Developmental Science Goes to School conference, October 2007, sponsored by the Spencer Foundation. The assessment project described in the chapter is funded, in part, by the Institute for Education Sciences, U.S. Department of Education (Grant R305G050091). Any opinions, findings, and conclusions or recommendations expressed in this material are those of the author(s) and do not necessarily reflect the views of either sponsoring organization. We gratefully acknowledge the contributions to this work of Shaunna MacLeod and Michael Manderino. For further information about the work contact sgoldman@uic.edu.

References

Alvermann D. E. (2001). Reading adolescents' reading identities: Looking back to see ahead. *Journal of Adolescent & Adult Literacy*, 44, 676–690.

Bazerman, C. (1998). *Shaping written knowledge: The genre and activity of the experimental article in science.* Madison: University of Wisconsin Press.

Beck, I. L., & McKeown, M. (1985). Teaching vocabulary: Making the instruction fit the goal. *Educational Perspectives,* 13, 11–15.

Berkencotter, C., & Huckin, T. N. (1995). *Genre knowledge in disciplinary communication: Cognition/culture/power.* Hillsdale, NJ: LEA.

Bransford, J. D., Brown, A. L., Cocking, R. R., Donovan, S., & Pellegrino, J. W. (Eds.) (2000). *How people learn: Brain, mind, experience, and school* (Expanded edition). Washington, DC: National Academies Press.

Britt, M. A., Perfetti, C. A., Sandak, R., & Rouet, J.-F. (1999). Content integration and source separation in learning from multiple texts. In S. R. Goldman, A. C. Graesser, & P. van den Broek (Eds.), *Narrative comprehension, causality, and coherence: Essays in honor of Tom Trabasso.* Mahwah, NJ: LEA, Inc. pp. 209–233.

Chall, J. S. (1983). *Stages of reading development.* New York: McGraw-Hill.

Chall, J. S., & Jacobs, V. A. (2003). Poor children's fourth-grade slump. American Educator. Online. Available at: http://www.aft.org/pubsreports/american_educator/spring2003/chall.html (accessed January 17, 2009).

Chall, J. S., Jacobs, V. A., & Baldwin, L. E. (1990). *The reading crisis: Why poor children fall behind.* Cambridge: Harvard University Press.

Chi, M. T. H., de Leeuw, N., Chiu, M., & LaVancher, C. (1994). Eliciting self-explanations improves understanding. *Cognitive Science,* 18, 439–477.

Chinn, C. A., & Malhotra, B. A. (2002). Epistemologically authentic reasoning in schools: A theoretical framework for evaluating inquiry tasks. *Science Education,* 86, 175–218.

Coiro, J., Knobel, M., Lankshear, C., & Leu, D. J. (Eds.) (2008). *Handbook of new literacies.* Hillsdale, NJ: LEA.

Coté, N., & Goldman, S. R. (1999). Building representations of informational text: Evidence from children's think-aloud protocols. In H. Van Oostendorp & S. R. Goldman (Eds.), *The construction of mental representations during reading.* Mahwah, NJ: LEA. pp. 169–193.

Coté, N., Goldman, S. R., & Saul, E. U. (1998). Students making sense of informational text: Relations between processing and representation. *Discourse Processes,* 25, 1–53.

de León, A. G., & Carnegie Corporation of New York. (2002). *The urban high school's challenge: Ensuring literacy for every child.* New York: Carnegie Corporation of New York.

Gee, J. P. (1990). *Social linguistics and literacies: Ideology in discourses.* London: Falmer Press.

Goldman, S. R. (1997). Learning from text: Reflections on the past and suggestions for the future. *Discourse Process,* 23, 357–398.

Goldman, S. R. (2004). Cognitive aspects of constructing meaning through and across multiple texts. In N. Shuart-Faris & D. M. Bloome (Eds.), *Uses of intertextuality in classroom and educational research.* Greenwich: Information Age Publishing. pp. 313–347.

Goldman, S. R., & Bisanz, G. (2002). Toward a functional analysis of scientific genres: Implications for understanding and learning processes. In J. Otero, J. A. León, & A. C. Graesser (Eds.), *The psychology of science text comprehension.* Mahwah, NJ: LEA. pp. 19–50.

Goldman, S. R., & Bloome, D. M. (2005). Learning to construct and integrate. In A. F. Healy (Ed.), *Experimental cognitive psychology and its applications: Festschrift in honor of Lyle Bourne, Walter Kintsch, and Thomas Landauer.* Washington, DC: American Psychological Association. pp. 169–182.

Goldman, S. R., & Rakestraw, Jr., J. A. (2000). Structural aspects of constructing meaning from text. In M. L. Kamil, P. Mosenthal, P. D. Pearson, & R. Barr (Eds.), *Handbook of reading research* (Vol. 3). Mahwah, NJ: LEA. pp. 311–335.

Goldman, S. R., Lawless, K. A., Gomez, K. W., Braasch, J. L. G., MacLeod, S., & Manning, F. (2010). Literacy in the digital world: Comprehending and learning from multiple sources.

In M. G. McKeown and L. Kucan (Eds.), *Bringing reading research to life*. New York: Guilford. pp. 257–284.

Gomez, L., & Gomez, K. (2007). Reading for learning: Literacy supports for 21st century work. *Phi Delta Kappan*, 89, 224–228.

Greeno, J. G., Collins, A. M., & Resnick, L. B. (1996). Cognition and learning. In D. Berliner and R. Calfee (Eds.), *Handbook of educational psychology*. New York: Macmillan. pp. 15–41.

Guthrie, J. T., Anderson, E., Aloa, S., & Rinehart, J. (1999). Influences of concept-oriented reading instruction on strategy use and conceptual learning from text. *Elementary School Journal*, 99(4), 343–366.

Hapgood, S., Magnusson, S. J., & Palincsar, A. S. (2004). Teacher, text, and experience: A case of young children's scientific inquiry. *Journal of the Learning Sciences*, 13, 455–505.

Hartman, D. K. (1993). Intertextuality and reading: The text, the reader, the author and the context. *Linguistics in Education*, 4, 295–311.

Hynd-Shanahan, C., Holschuh, J., & Hubbard, B. (2005). Thinking like a historian: College students' reading of multiple historical documents. *Journal of Literacy Research*, 36, 141–176.

International Adult Literacy Survey (1997). Highlights from the second report of the international adult literacy survey: Literacy skills for the knowledge society. Online. Available at: http://www.nald.ca/nls/ials/introduc.htm (accessed August 15, 2002).

Janick-Buckner, D. (1997). Getting undergraduates to critically read and discuss primary literature. *Journal of College Science Teaching*, 29, 29–32.

Krajcik, J. S., Blumenfeld, P., Marx, R. W., Bass, K. M., Fredricks, J., & Soloway, E. (1998). Middle school students' initial attempts at inquiry in project-based science classrooms. *Journal of the Learning Sciences*. 7, 313–350.

Kress, G. (2003). *Literacy in the new media age*. London: Routledge.

Lave, J., & Wenger, E. (1991). *Situated learning: Legitimate peripheral participation*. Cambridge: Cambridge University Press.

Lawless, K. A., & Schrader, P. G. (2008). Where do we go now? Understanding research on navigation in complex digital environments. In J. Coiro, M. Knobel, C. Lankshear, and D. J. Leu (Eds.), *Handbook of new literacies*. Hillsdale, NJ: LEA. pp 267–296.

Lee, C. D., & Spratley, A. (2006). *Reading in the disciplines and the challenges of adolescent literacy. Report to the Carnegie Corporation of New York*. New York: Carnegie Corporation.

Lee, P. J., & Ashby, R. (2000). Progression in historical understanding among students ages 7–14. In P. Stearns, P. Seixas, & S. Wineburg (Eds.), *Knowing, teaching, and learning history*. New York: New York University Press. pp. 199–222.

Lemke, J. (1998). Multiplying meaning: Visual and verbal semiotics in scientific text. In J. R. Martin & R. Veel (Eds.), *Reading science*. London: Routledge. pp. 87–113.

Lowe, R., & Schnotz, W. (Eds.) (2008). *Learning with animation: Research implications for design*. New York: Cambridge University Press.

McNamara, D. (2004). SERT: Self-explanation reading training. *Discourse Processes*, 38, 1–30.

Magliano, J. P., & Millis, K. K. (2003). Assessing reading skill with a think-aloud procedure. *Cognition and Instruction*, 21, 251–283.

Magnusson, S. J., & Palincsar, A. (2005). Teaching to promote the development of scientific knowledge and reasoning about light at the elementary school level. In S. Donovan & J. Bransford (Eds.), *How students learn*. Washington, DC: National Academies Press. pp. 421–474.

Manning, F. H., Goldman, S. R., Ozuru, Y., Lawless, K. A., Gomez, K., & Braasch, J. L. G. (2008). Students' analysis of multiple sources for agreements and disagreements. In G. Kanselaar, V. Jonker, P. A. Kirschner, & F. J. Prins (Eds.), International perspectives in the learning sciences: Cre8ing a learning world. Proceedings of the Eighth International Conference of the Learning Sciences – ICLS 2008, (Vol. 2) Utrecht: International Conference of the Learning

Sciences (www.isls.org). pp. 19–26.). Printed proceedings printed and distributed by: Lulu (www.lulu.com).

Meltzer, J., Smith, N. C., & Clark, H. (2002). *Adolescent literacy resources: Linking research and practice*. Providence: Brown University, Northeast and Islands Regional Educational Laboratory.

Mislevy, R. J., Steinberg, L., & Almond, R. (2003). On the structure of educational assessments. *Measurement: Interdisciplinary Research and Perspective*, 1, 3–67.

Moje, E. B. (2008). Foregrounding the disciplines in secondary literacy teaching and learning: A call for change. *Journal of Adolescent Literacy*, 52, 96–107.

Moje, E. B., & O'Brien, D. G. (Eds.) (2001). *Constructions of literacy: Studies of teaching and learning in and out of secondary schools*. Mahwah, NJ: LEA.

New London Group (1996). A pedagogy of multiliteracies: Designing social futures. *Harvard Educational Review*, 66, 60–92.

Orr, L. (1986). Intertextuality and the cultural text in recent semiotics. *College English*, 48(8). 811–823.

Palincsar, A. S., & Brown, A. L. (1984). Reciprocal teaching of comprehension-fostering and comprehension-monitoring activities. *Cognition and Instruction*, 1, 117–175.

Pressley, M. (2002). Comprehension strategies instruction. In C. C. Block & M. Pressley (Eds.), *Comprehension instruction: Research based best practices*. New York: Guilford. pp. 11–27.

Rouet, J.-F., Britt, M. A., Mason, R. A., & Perfetti, C. A. (1996). Using multiple sources of evidence to reason about history. *Journal of Educational Psychology*, 88, 478–493.

Sawyer, R. K. (2006). The new science of learning. In R. K. Sawyer (Ed.), *The Cambridge handbook of the learning sciences*. New York: Cambridge University Press. pp. 1–16.

Scardamalia, M., & Bereiter, C. (1996). Engaging students in a knowledge society. *Educational Leadership*, 54(3), 6–10.

Shanahan, T., & Shanahan, C. (2008). Teaching disciplinary literacy to adolescents: Rethinking content-area literacy. *Harvard Educational Review*, 78, 40–59.

Snow, C., & Biancarosa, G. (Eds.) (2003). *Adolescent literacy and the achievement gap: What do we know and where do we go from here?* New York: Carnegie Corporation of New York.

Stevens, R., Wineburg, S., Herrenkohl, L. R., & Bell, P. (2005). Comparative understanding of school subjects: Past, present, and future. *Review of Educational Research*, 75, 125–157.

Trumbull, D. J., Bonney, R., & Grudens-Schuck, N. (2005). Developing materials to promote inquiry: Lessons learned. *Science Education*, 89, 879–900.

van den Broek, P., Risden, K., & Husebye-Hartmann, E. (1995). The role of readers' standards for coherence in the generation of inferences during reading. In R. F. Lorch Jr. & E. J. O'Brien (Eds.), *Sources of coherence in text comprehension*. Hillsdale, NJ: LEA. pp. 353–373.

VanSledright, B. (2002a). *In search of America's past: Learning to read history in elementary school*. New York: Teachers College Press.

VanSledright, B. (2002b). Confronting history's interpretive paradox while teaching fifth graders to investigate the past. *American Educational Research Journal*, 39, 1089–1115.

Wineburg, S. (1991). Historical problem solving: A study of the cognitive processes used in the evaluation of documentary and pictorial evidence. *Journal of Educational Psychology*, 83, 73–87.

Wineburg, S. (Eds.) (2001). *Historical thinking and other unnatural acts: Charting the future of teaching the past*. Philadelphia: Temple University Press.

Wolfe, M. B., & Goldman, S. R. (2005). Relationships between adolescents' text processing and reasoning. *Cognition and Instruction, 23*, 467–502.

Yarden, A., Brill, G., & Falk, H. (2001). Primary literature as a basis for a high-school biology curriculum. *Journal of Biological Education*, 35, 190–195.

4 Constraints on Learning from Expository Science Texts

Jennifer Wiley and Christopher A. Sanchez

A great deal of instruction in science occurs through reading. Even in science classes where hands-on activities, experiments, or problem-based learning approaches are becoming more popular, background information critical for integrating the activity experiences into disciplinary understanding is often presented via text. Although the genre of most instructional text, especially in science, is expository, elementary reading experiences are generally in the narrative genre, with students usually honing their basic reading skills on story-like texts. However, some time around middle school, students start to receive expository texts and are expected to learn from them (Sweet & Snow, 2003). Not surprisingly, students often have difficulty with this task. Even undergraduate students can be remarkably poor at comprehending the information presented in expository text, and have been shown to be especially inept at judging their own level of comprehension, reflecting an inability to engage in accurate monitoring (Wiley, Griffin, & Thiede, 2005). The downstream effects of poor monitoring accuracy are that readers who are studying on their own will fail to engage in re-reading of poorly learned materials, and will fail to comprehend the texts they read.

The purpose of this chapter is to explore two major constraints on learning from expository texts. The first constraint that will be discussed is that the nature of understanding or comprehending the content of an expository text is somewhat different than comprehending a narrative. In short, students need to be instructed in what it means to understand an expository text so that they can direct their attention toward the right level of processing as they read. A second constraint that will be explored builds on another essential difference between narrative and expository science texts: often, scientific texts describe how and why processes or phenomena occur. In these cases, the fundamental understanding of the content is essentially the mental representation of dynamic relations over space and time. Using an individual differences approach, the later part of the chapter examines the role that spatial abilities and working memory capacity play in comprehension and the development of such mental representations. Finally, experimental interventions that address these constraints and improve learning outcomes are discussed.

The Situation Model and Comprehension of Expository Text

The approach to expository text comprehension taken here is called a *situation model approach*, based on the work of Walter Kintsch, who delineated multiple levels of representation as one processes a text (Kintsch, 1998). The first of these levels is the *surface* level, and this level represents the exact words that are encountered. A second level, the *textbase*, represents a refinement of the actual phrasing into gist-level or macro-structural propositions. Again, this representation largely represents the information that was present in the text, with perhaps some summarization, simplification, or generalization. The final level of representation is called the *situation model*, and it is at this level that connections between the ideas in the text and ideas in prior knowledge are created. This level can also be seen as a mental model of the phenomena. This is the level of representation that captures what the text "meant" on a deep level. It includes a causal model of the phenomena, and involves the generation of inferences or connections across concepts to explain how or why phenomena happen. In other words, this representation is what we define as real comprehension of a text (Kintsch, 1998; Wiley & Myers, 2003). This assumption is based on work in the text processing literature showing how the generation and representation of text in terms of a causal model predicts comprehension measures such as reading times, recall, recognition, and inference generation (Graesser & Bertus, 1998; Trabasso & van den Broek, 1985).

A critical difference between narrative text and expository text is in the closeness of representations between the multiple levels. In narrative text, memory for the narrative (what happened to whom, and when) is largely similar to "understanding" the text. Thus, when students read narratives, they can have a reading goal of remembering what happened, and it will serve them well for most tests about the text. This could either be because representations are so similar, as suggested above, or, alternatively, because one level can be easily constructed from the other owing to the highly accessible prior knowledge about people's intentions and motivations that are the causal links required for most narrative texts. On the other hand, the discrepancy between the *textbase* and the *situation model* is much greater for expository texts. Knowing what sentences appeared in the text is not the same as understanding the process or phenomena they describe. The *situation model*, or mental model of the phenomena, is much further removed from memory for exact sentences in expository text. It requires the construction of a causal model during the reading of the text and cannot be easily constructed afterward from memory. If future comprehension tests will tap inference-level understanding, readers need to attend to their ability to make connections and not just memory (neither fluency nor retrieval cues) for text in order to judge their understanding accurately (Griffin, Wiley, & Thiede, 2008).

Constraints Due to Failure to Understand the Nature of Comprehension

The above suggests that one major constraint in the comprehension of expository text is that readers may generally fail to recognize that comprehension requires engaging in explanatory processing. In reading expository text, the reader's goal is to try to understand how and why a phenomenon occurs, and not just to try to remember the isolated ideas that were read (Wiley et al., 2005). One recent study gives direct support for this proposition. In Thiede, Griffin, Wiley, and Anderson (2010), typical college readers and at-risk college readers were given several expository texts to learn from and were asked to judge how well they would do on a test of comprehension for each text after reading. Then, they also took inference-based comprehension tests on each text. From these measures, intra-individual correlations were computed which represented how well students could predict their own actual performance on comprehension tests. The overall correlation was only .21 for typical college readers, (similar to typical levels of accuracy around .27; Thiede, Griffin, Wiley, & Redford, 2009), but the average intra-individual correlation for at-risk readers was only .08.

In addition, we asked these readers to self-report *how* they judged their level of comprehension at the end of the study. These open-ended responses were coded into four categories. The first set included reports of features of the text such as using the difficulty of the vocabulary in the texts or their length (surface cues). The second set included references to characteristics of the reader such as prior knowledge or interest about the topic (reader cues). The third set consisted of comments related to the ability to remember or recall the texts (memory cues). The final set included comments related to the ability to explain the phenomena (comprehension cues). The striking finding was that only 5 percent of readers in a college sample (and no at-risk readers) spontaneously used comprehension cues as a basis for making comprehension judgments. Most readers used surface or memory-based cues. Further, the nature of the cues used made a striking difference in metacomprehension accuracy. Readers who used only surface, reader, or memory cues had intra-individual correlations of around .15, while readers who used only comprehension cues had correlations at a level of .71. This evidence supports the two key assumptions stated above. First, readers of expository texts tend to default to monitoring their comprehension using cues based in the *surface* or *textbase* levels of representation, and this leads to poor ability to judge actual comprehension when tests tap inferences or causal connections among ideas. Second, when readers *do* use their ability to explain a text as the basis for their comprehension judgments, then students do remarkably well at judging their own understanding, presumably because such cues are linked to the quality of the *situation model.*

Finally, we note that at-risk students [college students who were enrolled in remedial reading courses because they had low American College Test (ACT) verbal scores] were even less likely than the typical readers to consider *situation model*-based cues while judging comprehension. In fact, none of the remedial readers did so spontaneously. Given that at-risk readers display similarities to younger readers, the need to support a better understanding of what it means to understand an expository text may be even more acute in these populations.

Improving Understanding of Expository Text Comprehension

Based on these results, it seems critical to explore how to get readers to have a better understanding of what it means to comprehend expository tests. A number of lines of research have been pursued to address this question using several tasks and conditions that direct a reader's attention to the *situation model.* When combined with comprehension tests that tap the creation of a causal model of the text, these interventions have been found to improve predictive accuracy on expository texts. The results of these studies are listed in Table 4.1, along with an average for no-intervention control conditions that replicates the typical finding in the literature that, without support, intra-individual correlations hover at .27.

Table 4.1 Interventions and Metacomprehension Accuracy

Intervention Condition	*Metacomprehension Accuracy (γ)*
No Intervention	
Anderson & Thiede (2008)	.20
Griffin, Wiley, & Thiede (2008) (two conditions)	.25
Thiede & Anderson (2003) (two conditions)	.28
Thiede, Anderson, & Therriault (2003)	.38
Thiede, Dunlosky, Griffin, & Wiley (2005) (seven conditions)	.28
Thiede, Griffin, Wiley, & Anderson (2010)	.21
Average across all control conditions	.27
Delayed Summary Tasks	
Thiede & Anderson (2003) (two conditions)	.61
Thiede, Griffin, Wiley, & Anderson (2010)	.64
Thiede, Griffin, Wiley, & Anderson (2010)*	.48
Anderson & Thiede (2008)	.64
Delayed Keyword Tasks	
Thiede, Anderson, & Therriault (2003)	.70
Thiede, Dunlosky, Griffin, & Wiley (2005) (five conditions)	.55
Self-explanation Tasks	
Griffin, Wiley, & Thiede (2008)	.67
Concept Mapping Tasks	
Thiede, Griffin, Wiley, & Anderson (2010)*	.67

*At risk sample.

The driving theme behind these experiments has been to get readers to have access to the right level of cues when they consider their level of comprehension. The first set of interventions showed that instituting a delay between reading texts and engaging in a generation task that prompted readers to generate keywords (delayed keyword) or summaries (delayed summary) before judging improved readers' predictions of their own level of comprehension. This result was attributed to the rapid decay that occurs in surface memory for text and the more robust nature of the *situation model* (Kintsch, 1998). Thus, when readers were asked to generate keywords or summaries after a delay, this gave them access to cues about the quality of their *situation model* for each text, which in turn improved the quality of their comprehension judgments. These effects have now been replicated across several studies (see Table 4.1).

The next wave of studies directed readers to attend to the *situation model* in a somewhat more direct fashion, by having them engage in specific reading behaviors that highlight the connections that need to be made as the text is read (i.e., self-explanation and concept mapping). Studies in this vein have used explicit reading instructions that make the purpose of reading clear, but also require the reader to engage in additional constructive activities during reading. A popular intervention to improve learning from expository texts is self-explanation (Chi, 2000; McNamara, 2004). Related approaches include asking students to write causal arguments (Wiley & Voss, 1999) or to engage in how-and-why question-asking behaviors during reading (Graesser & Bertus, 1998). Because of the emphasis of all of these interventions in asking students to engage in causal or explanatory reasoning about the phenomena as they read, these approaches should also improve students' awareness of the quality of their *situation models* and thus their comprehension monitoring ability. To test this, Griffin et al. (2008) gave students clear instructions that the tests they would be taking would be based on connections and inferences about the content of the texts. In addition, some students were prompted to self-explain during a second reading of the texts. The instructions used for this explanation instruction were based on Chi (2000), and prompted students to attempt to connect ideas and think about how and why questions. Under these conditions, students' accuracy at judging their own comprehension improved, with gamma correlations averaging at .67.

Another similar approach, particularly appropriate for at-risk or younger readers, is concept mapping. A concept map is a graphic representation of the underlying structure of the meaning of a text. Constructing concept maps can be an effective organizational strategy, which may help readers formulate the connections among concepts in a text (Weinstein & Mayer, 1986). As noted above, some previous results suggested that metacomprehension accuracy for many at-risk readers is compromised by the use of inappropriate cues based on surface features of a text (Thiede et al., 2010). Thus, supporting the use of diagnostic monitoring may be especially critical for at-risk readers. Further, because self-explanation is a process that may add extra demands on a reader, concept mapping was chosen as an alternate intervention as it has been suggested that such an approach may be particularly helpful and appropriate for low-ability readers (Nesbit & Adesope, 2006; Stensvold & Wilson, 1990). Constructing concept maps is a generative activity that shares many similarities with argumentation and explanation tasks, but because it employs the construction of external, visual representations while readers have access to the texts, it may place fewer demands on the reader than other explanation tasks. Instructing at-risk readers to construct a concept map of a text during reading should help them identify important connections and construct a *situation model* for a text. In turn, it should also increase the salience of that *situation model*-level representation, which they may then use to better monitor their comprehension of a text. Consistent with these predictions, when at-risk readers generated concept maps before judging their comprehension, gammas reached a remarkable .67.

Conclusions about Metacomprehension Constraints

Taken together, the above results are consistent with the main conjecture that readers do not really understand what it means to comprehend expository text. The studies reported above show that readers default to *surface* or *textbase* cues such as memory for the text when they attempt to monitor their understanding. Unfortunately, these cues will not be diagnostic when the goal for reading is to understand how or why phenomena occur. Additionally, monitoring comprehension on this level will result in ineffective studying behaviors and in poor learning from expository science texts.

Thus, one main obstacle that ultimately impedes comprehension and learning from expository text is a failure to understand the explanatory nature of the processing that is required to develop coherent, causal mental models of scientific phenomena. To this point, this chapter has focused primarily on how such a misconception of comprehension relates to poor monitoring accuracy, and has described several interventions that have improved accuracy specifically by putting readers in contexts that support better access to *situation model* cues for comprehension.

The ultimate goal, of course, is improved comprehension, not just improved metacomprehension for expository text. The rationale for the emphasis on metacomprehension is that exploring the conditions that support accurate comprehension monitoring should ultimately lead to better comprehension by supporting more effective study behaviors. Although there are few studies that have directly tested these downstream effects, there are some promising results already in the literature. For example, Thiede, Anderson, and Therriault (2003) gave students an opportunity to restudy whichever texts they wished after a first set of comprehension tests and before a final set. Students assigned to the delayed keyword generation condition, who on average had more accurate monitoring, also made more effective studying decisions. They chose to restudy particularly those texts that they performed poorest on, and were less likely to restudy texts that they did well on. The average first test performance for the texts they chose to restudy was .27, whereas the average first test performance for the non-selected texts was .78. This difference reveals a clear ability to discriminate the texts that were understood from those that were not. In a no-keyword generation, no-delay condition, however, students showed less of a difference in first test performance on restudied (.44) versus not selected texts (.55). Importantly, final test performance was related to differential restudy choices. Students in the delayed keyword condition improved more on final tests, indicating that the intervention, which supported better monitoring accuracy, ultimately led to better comprehension from expository text as well. Further, for other interventions such as concept mapping and self-explanation, positive effects on both metacomprehension and comprehension may be seen more immediately (Chi, 2000; McNamara, 2004; Thiede et al., 2010).

Individual Differences as Constraints on the Construction of Mental Models

Several studies have shown that readers routinely fail to spontaneously generate causal inferences or a coherent causal model while reading scientific texts. The second half of this chapter takes an individual differences approach to explore two possible sources of these failures: working memory capacity (WMC) and spatial ability.

Because the topics of such texts are usually less familiar to readers, and generating inferences requires keeping multiple ideas active at once, expository text comprehension is thought to place a load on working memory (Linderholm & van den Broek, 2002). This may be especially true for low-knowledge readers (McNamara, 2004), although all readers may need explicit support or prompting to compute inferences from expository text (cf. Singer & O'Connell, 2003). For example, Wiley and Myers (2003) have shown that causal inferences may be generated from expository text only when all necessary information is available and adjacent in the text. Putting even a sentence between a key premise and conclusion can prevent readers from generating causal inferences. Findings such as this suggest that individual differences in WMC may play an important role in who develops coherent *situation models* while reading expository texts. Such effects are largely consistent with a resource allocation view of WMC, such that readers with a higher capacity may be able to engage in more complex comprehension processes (Griffin et al., 2008).

A second way in which WMC can relate to comprehension is more consistent with a controlled attention view (Kane, Bleckley, Conway, & Engle, 2001), which posits that some individuals are better able to focus their attention than others. This approach instead suggests that high-WMC readers may be better able to focus on relevant information as they read. A demonstration consistent with this prediction is provided by a set of studies done by Sanchez and Wiley (2006). These studies demonstrated that when participants read expository text that was illustrated with irrelevant images (e.g., "seductive details") (Harp & Mayer, 1997)—images that are tangentially related to the topic, but not conceptually related to the deeper relationships contained within the text—low-WMC readers were more likely to be distracted by these irrelevant details, as seen in eye-tracking measures. This distraction, in turn, led to poorer performance on comprehension measures for low-WMC individuals. High-WMC readers were better able to focus on the relevant information, which led to the construction of higher-quality situation models and better comprehension. The results of this study suggest that, especially in cases where there is a need to ignore irrelevant information, WMC will be predictive of learning and individual differences in WMC may be seen as a constraint on expository text comprehension.

Another class of individual differences that may be particularly important for learning from expository science text is spatial ability. It is widely assumed that learning in the physical sciences requires dealing with and understanding primarily spatial phenomena (Gobert, 2000; Kozma & Russell, 2005). To the extent that comprehension of scientific phenomena requires the construction of a mental model of a process or system, then the ability to visualize or simulate the operation of that phenomenon, and integrate important temporal and spatial interactions, may be critical to the formation of a "runnable" mental model that can be used to recapitulate the physical process (Gentner & Stevens, 1983; Hegarty & Steinhoff, 1997). For example, for learners to successfully understand the geological phenomena of plate tectonics, they must identify not only the relevant conceptual units that are spatial in nature (e.g., plates, magma, plate boundaries), but also how these units interact and change over time, which must then be represented within their own internal, runnable mental model. How closely the internal representation matches the actual physical phenomenon provides a rough proxy of how well the material has been understood. Obviously, the more accurately the learner can mentally construct a representation of the phenomenon, the better the learner understands, or is capable of demonstrating understanding of the topic. Because of the inherently dynamic, spatial nature of the representation for many scientific processes, spatial abilities represent one potential constraint on comprehension.

The most commonly used measures of spatial ability are paper-and-pencil tasks taken directly from the French Reference Kit (French, Ekstrom, & Price, 1963). These tasks fall into two main classes: tasks that tap the ability to rotate objects in space (e.g., block and figure rotation tasks) and tasks that tap the ability to visualize manipulations to objects (e.g., paper folding) or the ability to reconceptualize an existing spatial representation into a revised new whole (Carroll, 1993; Pellegrino & Hunt, 1991). These two subtypes of object manipulation ability are usually highly correlated and difficult to differentiate (Stumpf & Eliot, 1995). Individual differences on measures derived from object manipulation tasks have been shown to predict performance on tasks that explicitly require visuospatial information processing, such as mechanistic reasoning tasks using drawings of physical objects such as gears or pulleys (Hegarty & Steinhoff, 1997). Individual differences in these abilities have also been found to predict the comprehension of narrative text where readers follow the actions of a character in physical space (Fincher-Kiefer & D'Agostino, 2004). Although many have posited that visuospatial abilities are key to science understanding, there is no direct evidence of a relationship with comprehension from expository text.

Similarly, some of the greatest discoveries in science, such as the double helix, the benzene ring, and theories of plate tectonics, have been attributed to the ability of great scientists to think spatially (National Research Council, 2006). However, the evidence for the link between spatial ability and performance in science is largely anecdotal and correlational, with findings that scientists and students in advanced science courses have higher spatial abilities than the general population, and some positive relations between exam performance and spatial ability (Black, 2005; Wu & Shah, 2004).

However, these studies cannot be taken as direct evidence that spatial ability leads to better comprehension in science unless one can rule out general individual differences in ability or prior knowledge as factors. Further, there are also examples of spatial ability failing to predict learning of science content, including topics in both biology (Koroghlanian & Klein, 2004) and physics (ChanLin, 2000). Although these studies span several different scientific domains and no doubt test learning in different ways, it is disconcerting that no consistent relationship has emerged between measures of spatial ability and measures of science learning, given the very visual and spatial nature of many topics within these domains (Gobert, 2000; Kozma & Russell, 2005). This recognition prompts the question as to whether or not the standard measures of spatial ability based in object-manipulation tasks are the most relevant for the kinds of representations that are needed for comprehension of some scientific topics. It is possible that the visualization and manipulation abilities captured by these tasks, which have been referred to as static spatial abilities or SSAs (Pellegrino & Hunt, 1991), are not necessarily the most critical for complex science reasoning in these domains.

Given the dynamic nature of many scientific phenomena, perhaps a spatial ability which better captures this dynamic characteristic, or what has been referred to as dynamic spatial ability (DSA), might be a better predictor of comprehension. Tests of dynamic spatial abilities attempt to capture this dynamic nature and represent the capacity to integrate spatial information across multiple instances and over time (Hunt, Pellegrino, Frick, Farr, & Alderton, 1988; Pellegrino & Hunt, 1991). While these abilities are modestly correlated, factor-analytic work has borne out their identity as separable constructs (Contreras, Colom, Hernandez, & Santacreu, 2003; Hunt et al., 1988).

The connection between DSA and science learning is slightly more tenuous at this point, as to date there have been only a handful of studies across any domain that have examined this ability, and most efforts have been dedicated to the establishment of the validity of this concept as distinct from SSA. Dynamic spatial ability has been shown to predict learning of an air-traffic control task (D'Oliveira, 2004), and has also been found to contribute unique variance over and above verbal intelligence on performance intelligence quotient (IQ) tests (Jackson, III, Vernon, & Jackson, 1993). However, given that most physical science topics involve understanding dynamic changes over time, DSA may be critical for the formation of dynamic scientific mental models.

To continue with our example from the earth sciences, when asked to understand the movements of plate tectonics and how this system produces volcanoes and earthquakes, learners need to understand not only that these plates exist, but also the underlying movements of magma as it circulates around the earth's core and causes these plate movements. In essence, the constant change and movement of magma, which causes the construction/destruction of the tectonic plates over time, is critical for understanding how, why, and where volcanoes and earthquakes occur. Further, maintaining spatial information about relations between isolated units or pairs of concepts (i.e., the intersection of two plates) can provide only a partial understanding of the entire process. In order to fully develop a deep understanding of the entire system, multiple relationships must be encoded and integrated within this time-based process, suggesting that DSA should be critical.

This indeed was found to be the case in a recent study by Sanchez and Wiley (2007). In one condition of this study, college students were asked to read a non-illustrated lesson about plate tectonics. Students were asked to read for the purposes of understanding what causes volcanic eruptions. Then, learning was assessed by having students write an essay on "What caused Mt. St. Helen's to erupt?" This essay was scored for the presence of eight main concepts from an a priori causal model of the phenomena. Consistent with the hypothesis that DSA constrains science text comprehension, the best predictor of overall learning from this lesson was DSA, with correlations of around .30 (Table 4.2). Specifically, it was found that in situations where the demands on learners to mentally animate the information to achieve understanding were highest (i.e., non-illustrated condition), DSA was especially important for generating and running this mental model. However, the presence of dynamic, external representations (e.g., animations) attenuated this relationship, while still producing high overall learning. This suggests that understanding dynamic relationships between conceptual units is necessary to form an accurate representation of the content area, and

Table 4.2 Essay Performance and Correlations between WMC, DSA, and SSA and Essay Performance by Condition in Sanchez and Wiley (2007)

	Non-illustrated (n=64)	*Static Illustrations* (n=69)	*Animations* (n =63)
Correct concepts, mean (SD)	2.08 (1.23)	2.03 (1.45)	2.54 (1.35)
Correlations			
WMC and learning	.03	.04	.24**
DSA and learning	.30*	.26*	.04
SSA and learning	−.01	.20**	.09

* $p<.05$, ** $p<.10$.

when the learner is required to generate such relationships on his or her own, DSA is important. These results suggest that building a runnable mental model or simulation of the to-be-learned phenomena is an important part of the comprehension process, and that readers with higher DSA have an easier time translating the ideas from the text into a mental simulation or internal visualization of the phenomenon.

Supporting the Creation of Runnable Mental Models in All Readers

The next question was whether the construction of such models could be supported in students with low DSA. One obvious intervention that has been used a great deal in the literature is providing readers with illustrations to support learning. However, the impact of adding visual material to text has had mixed reviews in terms of its effectiveness. On one side, the addition of static illustrations such as charts and diagrams has been used effectively to enhance science learning (see Ainsworth & Th Loizou, 2003; Mayer, Hegarty, Mayer, & Campbell, 2005). Similarly, the addition of dynamic animations or video has also been shown to sometimes provide facilitation of learning in science (see Schnotz, 2005).

However, there is also evidence that adding visual media does not always facilitate learning, and there seems to be a great deal of specificity to the contexts in which adding visualizations will be optimal. For example, in some cases, static images produce better learning than animations (Mayer et al., 2005). In other cases, static illustrations and animations have failed to enhance learning on science topics, have sometimes led to worse performance, and often seem to interact with student abilities or explicit instructional support (Geiger & Litwiller, 2005; Sanchez & Wiley, 2006; Schnotz, 2005). In a review of studies on animation, Tversky, Morrison, and Betrancourt (2002) concluded that the evidence for beneficial aspects of animation on learning were at best mixed. Similar conclusions have been reached about static graphics (Mandl & Levin, 1989).

To date, there is not a definitive explanation of when such visualizations should improve learning. Based on their review of the mixed findings, Tversky et al. (2002) proposed two general principles: *congruence* and *apprehension*. These principles simply require that information in the visualization should match the nature of the desired internal model, and should also be able to be easily apprehended and perceived (Lowe, 2004; Schnotz, 2005). Consistent with these recommendations, readers in two other conditions of Sanchez and Wiley (2007) were given either static or animated visualizations that illustrated important conceptual ideas from the text in fairly simple schematic diagrams. The static versions gave readers a visual representation of the important entities for understanding plate tectonics such as plates, layers of earth, and faults. Notably, only the animated version presented the information in a way in which essential dynamic relations and interactions between entities were obvious.

Given the discussion above, it was anticipated that providing illustrations should improve learning, but also that the type of visualization might impact learning of this science topic. In other words, animations which capture the dynamic nature of the phenomenon might lead to the best overall comprehension, while static visualizations, as they at least represent the relevant structure of the event, might produce better comprehension than the non-visual text. Further, these adjuncts may be especially important for low spatial ability learners.

The results of Sanchez and Wiley (2007) as shown in Table 4.2 demonstrate that the presence of conceptually relevant animations improved learning overall (mean = 2.54, SD = 1.35) compared with both the no-illustrations (mean = 2.08, SD = 1.23) and static illustrations (mean = 2.03, SD = 1.45) conditions and, importantly, animations also eliminated the relationship between DSA and learning. In the animated condition, only WMC was a predictor of learning. However, in the static illustrations condition, overall learning was not improved significantly over the no-illustrations condition and DSA was still a significant predictor of learning.

Results from across the three conditions in this experiment provide evidence that individual differences in spatial ability, and also WMC to some extent, impose constraints on the comprehension of expository science text. First, the inclusion of animations alongside the text led to the best performance overall, and this can be taken as evidence that representing the dynamic properties of process described in the text supported better comprehension of a topic that is an inherently dynamic phenomenon. Consistent with this interpretation, the relation between DSA and learning was eliminated in this condition. Interestingly, the addition of static illustrations did not enhance learning above the non-illustrated condition, and the relationship between DSA and learning was still strong in this situation. It appears that learners were able to represent the relevant conceptual entities from the text even without visual support. This produced similar learning effects in the no-illustrations and static illustrations conditions. Given that both of these non-dynamic conditions still placed heavy requirements on DSA, for at least this topic, a key difficulty was mentally animating the dynamic changes that occur across time.

In sum, it appears that DSA is indeed a constraint on understanding in science and is connected to how well individuals can construct mental models of a dynamic science topic. The presence of relevant conceptual animations can improve understanding and support visualization skills among readers with low DSA. However, although individual differences in DSA were eliminated in the animation condition, the role of individual differences in WMC became more apparent, consistent with previous research showing that under some conditions WMC will also act as a constraint on comprehension processes (Geiger & Litwiller, 2005). In this case, as in previous research, WMC may help readers to learn effectively from animations.

Multiple Routes to Improving Student Learning from Science Texts

The approaches discussed above have delineated two major classes of difficulties that may prevent readers from developing adequate understanding from scientific texts. The first is that readers may generally misunderstand that their goal for reading expository science texts is to develop an explanatory model of how or why phenomena happen, and that they should monitor their comprehension using that standard. The second is that some readers will have difficulty creating mental models when reading about science topics, particularly for texts about dynamic physical phenomena, due to a lack of dynamic spatial ability. However, interventions that clarify the desired nature of processing, and that support visualization and the creation of runnable mental models, are promising directions for the future which may ultimately help all students become more effective learners of science. It is also an interesting question for future research as to how the early support of expository comprehension and spatial skills in elementary school students would affect later learning in science.

Acknowledgments

A portion of the research reported here was supported by the Institute of Education Sciences, U.S. Department of Education, through Grants R305H030170 and R305B070018 to Keith Thiede, Thomas D. Griffin, and Jennifer Wiley. Additional portions of this research were supported by the APA Dissertation Research Award to the second author. The opinions expressed are those of the authors and do not represent views of these institutions.

References

Ainsworth, S., & Th Loizou, A. (2003). The effects of self-explaining when learning with text or diagrams. *Cognitive Science*, 27, 669–681.

Anderson, M. C., & Thiede, K. W. (2008). Why do delayed summaries improve metacomprehension accuracy? *Acta Psychologica*, 128, 110–118.

Black, A. A. (2005). Spatial ability and earth science conceptual understanding. *Journal of Geoscience Education*, 53(4), 402–414.

Carroll, J. B. (1993). *Human cognitive abilities: A survey of factor analytic studies*. Cambridge: Cambridge University Press.

ChanLin, L. J. (2000). Attributes of animation for learning scientific knowledge. *Journal of Instructional Psychology*, 27(4), 228–238.

Chi, M. T. H. (2000). Self-explaining expository texts: The dual processes of generating inferences and repairing mental models. In R. Glaser (Ed.), *Advances in instructional psychology* (Vol. 5). Mahwah, NJ: LEA. pp. 161–238.

Contreras, M. J., Colom, R., Hernandez, J. M., & Santacreu, J. (2003). Is static spatial performance distinguishable from dynamic spatial performance? A latent-variable analysis. *Journal of General Psychology*, 130, 277–288.

D'Oliveira, T. C. (2004). Dynamic spatial ability: An exploratory analysis and a confirmatory study. *International Journal of Aviation Psychology*, 14, 19–38.

Fincher-Kiefer, R., & D'Agostino, P. (2004). The role of visuospatial resources in generating predictive and bridging inferences. *Discourse Processes*, 37, 205–224.

French, J. W., Ekstrom, R. B., & Price, L. A. (1963). *Kit of reference tests for cognitive factors*. Princeton: Educational Testing Service.

Geiger, J. F., & Litwiller, R. M. (2005). Spatial working memory and gender differences in science. *Journal of Instructional Psychology*, 32, 49–58.

Gentner, D., & Stevens, A. L. (1983). *Mental models*. Hillsdale, NJ: LEA.

Gobert, J. D. (2000). A typology of causal models for plate tectonics. *International Journal of Science Education*, 22(9), 937–977.

Graesser, A. C., & Bertus, E. L. (1998). The construction of causal inferences while reading expository texts on science and technology. *Scientific Studies of Reading*, 2, 247–269.

Griffin, T. D., Wiley, J., & Thiede, K. W. (2008). Individual differences, rereading, and self-explanation. *Memory & Cognition*, 36, 93–103.

Harp, S. F., & Mayer, R. E. (1997). The role of interest in learning from scientific text and illustrations. *Journal of Educational Psychology*, 89, 92–102.

Hegarty, M., & Steinhoff, K. (1997). Individual differences in use of diagrams as memory in mechanical reasoning. *Learning and Individual Differences*, 9, 19–44.

Hunt, E., Pellegrino, J. W., Frick, R. W., Farr, S. A., & Alderton, D. L. (1988). The ability to reason about movement in the visual field. *Intelligence*, 12, 77–100.

Jackson, D. N., III, Vernon, P. A., & Jackson, D. N. (1993). Dynamic spatial performance and general intelligence. *Intelligence*, 17(4), 451–460.

Kane, M. J., Bleckley, K. M., Conway, A. R. A., & Engle, R. W. (2001). A controlled-attention view of working memory capacity. *Journal of Experimental Psychology: General*, 130,169–183.

Kintsch, W. (1998). *Comprehension: A paradigm for cognition*. New York: Cambridge University Press.

Koroghlanian, C., & Klein, J. D. (2004). The effect of audio and animation in multimedia instruction. *Journal of Educational Multimedia and Hypermedia*, 13, 23–45.

Kozma, R., & Russell, J. (2005). Multimedia learning of chemistry. In R. Mayer (Ed.), *Cambridge handbook of multimedia learning*. New York: Cambridge University Press. pp. 409–428.

Linderholm, T., & van den Broek, P. (2002). The effects of reading purpose and working memory capacity on the processing of expository text. *Journal of Educational Psychology*, 94, 778–84.

Lowe, R. K. (2004). Interrogation of a dynamic visualization during learning. *Learning and Instruction*, 14(3), 257–274.

Mandl, H., & Levin, J. R. (Eds.) (1989). *Knowledge acquisition from text and pictures*. Amsterdam: North-Holland.

Mayer, R. E., Hegarty, M., Mayer, S. Y., & Campbell, J. (2005). When passive media promote active learning. *Journal of Experimental Psychology: Applied*, 11, 256–265.

McNamara, D. S. (2004). SERT: Self-explanation reading training. *Discourse Processes*, 38, 1–30.

National Research Council. (2006). *Learning to think spatially*. Washington, DC: The National Academies Press.

Nesbit, J. C., & Adesope, O. O. (2006). Learning with concept and knowledge maps: A meta-analysis. *Review of Educational Research*, 76, 413–448.

Pellegrino, J. W., & Hunt, E. B. (1991). Cognitive models for understanding and assessing spatial abilities. In H. A. H. Rowe (Ed.), *Intelligence: Reconceptualization and measurement*. Mahwah, NJ: LEA. pp. 203–225.

Sanchez, C. A., & Wiley, J. (2006). An examination of the seductive details effect in terms of working memory capacity. *Memory & Cognition*, 34, 344–355.

Sanchez, C. A., & Wiley, J. (2007). Spatial abilities and learning complex science topics. *Proceedings of the 29th Annual Meeting of the Cognitive Science Society*, Nashville, TN.

Schnotz, W. (2005). An integrated model of text and picture comprehension. In R. Mayer (Ed.), *Cambridge handbook of multimedia learning*, New York: Cambridge University Press. pp. 49–70.

Singer, M., & O'Connell, G. (2003). Robust inference processes in expository text comprehension. *European Journal of Cognitive Psychology*, 15, 607–631.

Stensvold, M. S., & Wilson, J. T. (1990). The interaction of verbal ability with concept mapping in learning from a chemistry laboratory activity. *Science Education*, 74, 473–480.

Stumpf, H., & Eliot, J. (1995). Gender-related differences in spatial ability and the *k* factor of general spatial ability in a population of academically talented students. *Personality and Individual Differences*, 19, 33–45.

Sweet, A. P., & Snow, C. E. (Eds.) (2003). *Rethinking reading comprehension*. New York: Guilford Press.

Thiede, K. W., & Anderson, M. C. M. (2003). Summarizing can improve metacomprehension accuracy. *Contemporary Educational Psychology*, 28, 129–160.

Thiede, K. W., Anderson, M. C. M., & Therriault, D. (2003). Accuracy of metacognitive monitoring affects learning of texts. *Journal of Educational Psychology*, 95, 66–73.

Thiede, K. W., Dunlosky, J., Griffin, T. D., & Wiley, J. (2005). Understanding the delayed keyword effect on metacomprehension accuracy. *Journal of Experiment Psychology: Learning, Memory & Cognition*, 31, 1267–1280.

Thiede, K. W., Griffin, T. D., Wiley, J., & Anderson, M. (2010). Poor metacomprehension accuracy as a result of inappropriate cue use. *Discourse Processes*, 47, 331–362.

Thiede, K. W., Griffin, T. D., Wiley, J., & Redford, J. S. (2009). Metacognitive monitoring during and after reading. In D. J. Hacker, J. Dunlosky, & A. C. Graesser (Eds.), *Handbook of metacognition in education*. New York: Routledge.

Trabasso, T., & van den Broek, P. (1985). Causal thinking and the representation of narrative events. *Journal of Memory and Language*, 24, 612–630.

Tversky, B., Morrison, J. B., & Betrancourt, M. (2002). Animation: Can it facilitate? *International Journal of Human-Computer Studies*, 57(4), 247–262.

Weinstein, C. E., & Mayer, R. E. (1986). The teaching of learning strategies. In M. C. Wittrock (Ed.), *Handbook on research in teaching* (3rd edn). New York: Macmillan. pp. 315–327.

Wiley, J., Griffin, T., & Thiede, K. W. (2005). Putting the comprehension in metacomprehension. *Journal of General Psychology*, 132, 408–428.

Wiley, J., & Myers, J. L. (2003). Availability and accessibility of information and causal inferences from scientific text. *Discourse Processes*, 36, 109–129.

Wiley, J., & Voss, J. F. (1999). Constructing arguments from multiple sources. *Journal of Educational Psychology*, 91, 301–311.

Wu, H., & Shah, P. (2004). Exploring visuospatial thinking in chemistry learning. *Science Education*, 88(3), 465–492.

5 Two Challenges

Teaching Academic Language and Working Productively with Schools

Catherine E. Snow and Claire White

"Taking developmental science to school" is an ambiguous term. Most readers will probably think it refers to the contributions of developmental science and scientists to practice—the need for educators to embrace the contributions of developmental science. We would argue, on the other hand, that the contributions of developmental science to the improvement of educational practice have, despite the best of intentions, been modest. Thus, we interpret the phrase "taking developmental science to school" as indicating the need for developmental science and scientists to go to school, in order to learn more about children, the crucial tasks they face, the contexts of their lives, and the challenges their teachers experience. In other words, we echo here the principles that underlie the Strategic Education Research Partnership (SERP), an effort to forge dynamic and reciprocal relations between educational practitioners and researchers in the service of better teaching and learning. We use work done in the context of the SERP field site, first established in the Boston Public Schools in 2005, to demonstrate the working of some of those principles, and their relevance to one major educational challenge—preparing students to use academic language.

SERP in Action

SERP promotes educational research in a new way—a way that maximizes its relevance to schools and usability by practitioners (Donovan, Wigdor, & Snow, 2003). SERP as an organization has been effective in recruiting thoughtful, dedicated (and busy) researchers to work with it, in part because so many of those researchers have had the experiences of developing products and practices that, even when proven effective, never took hold among practitioners. The journals are full of evidence about successful interventions and research-based principles, and the publishers' catalogues are full of books that describe those principles and curricular materials that enact those interventions. Yet uptake is slow; what goes on in classrooms often seems disconnected from the knowledge base established, and even schools or districts that adopt proven practices often abandon them when new reforms come along. Thus it seems that the traditional approach, of taking the products of developmental science into schools and dropping them there, may require rethinking.

The SERP approach is one that, instead, takes developmental scientists to school, by putting them in working groups (called "design teams") that include practitioners from the field site, as well as expert practitioners with experience in other districts.

This configuration reflects the basic SERP commitment to the recognition that knowledge can move both upstream (from practice to research) and downstream (from research to practice).

Another basic SERP principle embodied in the design team configuration, as well as in the way SERP work progresses, is the commitment to work on urgent problems of practice defined by the district or school. This commitment arises from, among other sources, analysis of those many cases where researchers have worked with schools to prove the effectiveness of an intervention only to see it disappear once the "project" is completed. We argue that this reflects the fact that the intervention, though effective in improving some aspect of student or teacher performance, typically did not solve a problem the school was worried about. The traditional approach taken by researchers is to read the research literature to isolate unanswered questions—How many exposures to a word are needed to ensure learning? What is the effectiveness of direct teaching versus incidental exposure? Do dictionary definitions speed learning? These are valid and important questions, the answers to which can of course inform practice. But the developmental scientists working to answer those questions by designing a vocabulary intervention might be thought of as using the school or classroom as a test tube in which to carry out their experiments, rather than collaborating with the principal or teacher.

The collaborator lets the practitioner nominate the problem—and how to teach vocabulary might well emerge as the chosen one—and then plans to devise and test a solution to that problem informed by the researchers' knowledge base (accumulated in journals) as well as the wisdom of practice (accumulated in the practitioners' craft and ken). If the practices that emerge from collaborative problem solving are indeed a response to demands voiced by the practitioners, then they are unlikely to disappear once the study of their effectiveness is complete.

We would argue, then, that developmental scientists need to go to school to learn about the issues they should take on, and to explore the classroom conditions and constraints under which their proffered solutions will have to work. But it does not do a lot of good for them to go alone. There are myriad examples of excellent curricula and effective practices emerging from developmental research that, even if adopted by some teachers and implemented in some classrooms, end up being undermined by organizational factors: strong organizations that cannot make room for new practices; dysfunctional organizations that cannot implement new practices; or changes in organizations that lead to abandonment of effective existing practices. Excellent pieces of curriculum disappear because a new textbook is adopted, or because new teachers receive no professional development in how to use them, or because the standards that led to their development are revised, or because a new principal or superintendent introduces reforms that squeeze them out. The very best developmental research can lead to the design of educational tools that reflect what is known about children's acquisition of knowledge and conceptual change; if developmentalists are recruited to design good training for teachers to use those tools, then they are likely to be effectively implemented. But both those accomplishments can be undermined unless there is expertise at the table in how organizations work—how to help organizations take on challenges, learn from success and from failure, and move in a systematically forward direction rather than back and forth between alternative approaches.

Since 2005 we have been trying to apply these principles within the context of the SERP field site established in the Boston Public Schools (BPS). Boston, like other urban school districts, struggles with a wide array of challenges, but the urgent problem nominated by then Boston Superintendent Thomas Payzant as the first target of SERP work was adolescent literacy. Though the topic of adolescent literacy was just emerging as a target of research in 2005, there was considerable knowledge and expertise among the university researchers in the greater Boston area that could be brought to bear on it. Thus the SERP research team went to work.

We felt it was important to elicit from front-line practitioners more explicit definitions of the exact nature of the adolescent literacy problem, before offering solutions. Thus, we spent time surveying and interviewing middle school teachers, as well as observing teaching in content-area classrooms and looking at test score data, to understand the nature of the literacy challenge more precisely. That process revealed a high degree of consensus that one major problem lay in the domain of students' comprehension of academic texts—not in their capacity to read the words in the text, but in their knowledge of the meaning of many of those words, their ability to make sense of the readings required for learning, especially in science, math, and social studies, and their ability to use language to represent the text content. Thus, we decided to focus on designing and evaluating curricular materials that would help develop students' comprehension and academic language skills.

The Context for Teaching Academic Language

The broader context for teaching academic language skills in Boston was in many ways propitious. The district had already adopted "accountable talk" (Michaels, O'Connor, & Resnick, 2008) as a district-wide goal. The middle school math, social studies, and science teachers surveyed largely agreed that vocabulary and language structures were obstacles to learning for many of their students, and they also acknowledged that they had little expertise in teaching such matters. Some continued to deflect the blame for students' poorly developed academic language skills to elementary teachers or to parents, and wished to assign the responsibility for teaching vocabulary to English Language Arts (ELA) teachers, but many had heard the district's message that content-area teachers were responsible for content-area literacy. Most schools had schedules that allowed for both grade-level and departmental team meetings, so opportunities for teachers to work together on innovative practices were present. In addition, the Massachusetts Comprehensive Assessment System (MCAS), the state accountability test, was relatively challenging, required considerable reading of texts full of academic language, and included many open-ended, short-answer questions, on which BPS students performed quite poorly, relative to students in other districts or to their own performance on multiple-choice items.

In other ways though, the challenge of designing and implementing an effective academic language intervention was huge. First, in order to increase rigor and improve compliance with Massachusetts' many content standards, teachers in math, science, and social studies had been given pacing guides, which personnel in the curriculum and instruction office felt had to take priority over any academic language instruction. Furthermore, the ELA department was adopting, for the first time, a district-wide curriculum, that made many demands on teachers and was experienced as filling up their instructional time. Second, some of the middle schools in which we were working were, like middle schools in many districts, relatively low-coherence organizations with minimal levels of internal accountability (Elmore, 2004; Fuhrman & Elmore, 2004) or organizational trust (Bryk & Schneider, 2002). Interventions introduced into schools with these characteristics are unlikely to be embraced and implemented widely or with fidelity. Third, the time available for professional development to introduce and explain the intervention to whole-school teams was limited by union rules and in some schools already committed to other activities; ongoing school-site professional development activities were in most schools implemented by literacy coaches, but the coaches spent only part of a week at any school, typically worked only with ELA teachers, and were in any case not all committed to prioritizing academic language in the professional development they offered.

All these various factors had to be taken into account in thinking about the design of an academic language intervention. Fortunately, the SERP design team, which included thoughtful individuals with wide experience in intervention design, was available to jumpstart the design process, and skilled BPS practitioners were willing to brainstorm and to critique the emerging materials at every stage of design.

Word Generation

We decided to present the intervention to the district as an effort focused on vocabulary, since that was the specific skill noted as problematic by a large proportion of teachers. Various sources of information pointed to vocabulary as a promising target for enhanced and intensified instruction in Boston. In particular, teachers noted that English Language Learners (ELLs) and other students from low-education families did not know many of the words presupposed in texts. Ironically, many of the words students did not know were used in glossaries or classroom discussions to define disciplinary words teachers were trying to teach. "Photosynthesis is a process of converting solar energy into stored chemical energy" is a helpful definition only for students who know what *process, convert, solar*, and *stored* mean. Many Boston middle schoolers did not.

The vocabulary challenges for Boston students were largely located in the domain we would call "academic words," by which we mean words that are much more frequent in written text than in oral, casual, conversational language. These words are embedded in the use of language for academic purposes, and thus we felt confident that we could promote academic language generally in the context of teaching them. Promoting "academic language" is an explicit goal for BPS teachers, but a somewhat fuzzy goal given the complexity of defining what academic language is. Most practitioners and researchers would agree that academic language skills are a prerequisite to comprehending academic texts or discussions—but that is a rather circular definition. There have been many attempts to identify the features of academic language based on context and participant structures, or on linguistic features (see Snow & Uccelli, 2008, for a review). Ultimately, however, these attempts work backwards from observed differences toward definitions, without providing criteria (when is some piece of language academic enough?) or guidance to instruction. Snow and Uccelli (2008) argue that the sentence-level and text-level features of academic language should be thought of as embedded within a set of pragmatic purposes with two dimensions: self-representation and message formulation. Representing oneself as knowledgeable and authoritative, while formulating a message that is inherently complex, is key to the production of academic language. So how can middle school students learn to do that? A first step is, of course, learning to use the words that signal that authoritative self, and learning to formulate complex messages. Helping BPS teachers guide students to develop these skills was the goal we formulated for our work. That meant, then, providing models of academic language and opportunities to practice using it, at all the key levels:

1 Clause level: using academic vocabulary and the associated syntactic structures;
2 Discourse level: generating brief academic texts;
3 Message level: representing complex and sophisticated messages;
4 Self/Audience level: making explicit one's commitment to the message and willingness to argue for it by providing warrants.

We chose to address this task by developing curricular units that could engage students in discussion and writing about topics of interest to them, while giving them opportunities to learn and use academic vocabulary. In addition, in conformity to the SERP principle that real instructional change requires working on student learning, teacher learning, and organizational learning together, we designed the instructional program in such a way that teachers had to work together across the traditional content-area-defined silos in order to deliver it; in other words, we were hoping that collaboration in planning the instruction would lead to opportunities for building organizational trust, and that implementing the instruction as a grade-level team would focus the teachers on students and their learning, in addition to their own content-area responsibilities.

Local constraints had to be reflected in our design. In response to central office skepticism, we limited the teaching time to no more than 15 minutes per week in math, social studies, and science. In response to the math teachers' anxieties about the state test, we devised math activities that would also serve as MCAS preparation or practice. In response to some science teachers' lack of enthusiasm for vocabulary teaching, we provided highly structured science activities that would require little preparation time.

What vocabulary did we choose to focus on? Disciplinary, content-specific vocabulary was already being taught by BPS teachers. But none of the content-area teachers felt responsible for teaching the all-purpose academic words crucial for understanding texts and for reading the definitions offered in glossaries in those texts. Thus, we started with words that might be encountered in any of the content areas, and that occurred with reasonable frequency in academic texts. They include words for talking about thinking (hypothesize, evidence, criterion), words for classifying (vehicle, utensil, process), words that expressed nuances of communication (emphasize, affirm, negotiate), and words for expressing relationships among entities (dominate, correspond, locate). Fortunately, the work of identifying the most frequently occurring of those words had been done by Nation (1990), and helpfully made available electronically by Coxhead (1998, 2000) on the Academic Word List website (http://www.vuw.ac.nz/lals/research/awl/index.html). We selected words from the first, easier half of the Academic Word List for inclusion in the curriculum.

We designed our program, called Word Generation, in ways meant to contrast to business as usual in middle school classrooms. Whereas the texts students were typically asked to read were frequently difficult, unengaging, and disconnected from their lives, we constructed brief and engaging texts on topics of high interest. Whereas lively classroom discussion was rare, we engineered into the Word Generation activities that were opportunities for open discussion and structured debates. Whereas much school-based writing was tedious, we structured regular brief writing assignments that involved arguing for positions on which students had strong opinions and a desire for self-representation.

In brief, then, Word Generation was designed to meet goals at three levels:

1 Student level: Build knowledge of high-frequency academic words, skills at spoken and written academic discourse, and world knowledge.
2 Teacher level: Promote regular use of effective strategies usable in everyday instruction.
3 School level: Facilitate faculty collaboration across grades and across departments.

The overall strategy was one of infusion rather than displacement—a brief daily activity that could fit easily into the regular content-area instructional times. Though the instruction was focused on academic language generally, we defined it as a vocabulary program, treating vocabulary as an instructional Trojan horse or benign bacterium designed to infect the system and bring about broader change.

Downstream Effects

Word generation was also based firmly in a long history of instructional research (Beck, McKeown, & Kucan, 2002; Graves, 2006), from which several principles of instruction could be derived:

- Pick the right words.
- Present them in motivating and semantically informative contexts, not in lists.
- Provide learner-friendly definitions.
- Ensure recurrent exposures.
- Start with a context-embedded meaning and gradually expand each word's semantic mapping.
- Provide students opportunities to use the words.
- Teach word-learning strategies (morphological, inferring from context).
- Motivate "word awareness."

The materials were also designed to address the particular challenges of vocabulary learning for adolescents in urban schools. First, these students typically already possessed considerable vocabulary knowledge, but it consisted largely of the "easy words" (basic object terms, words for concrete objects, brief/monomorphemic forms, highly frequent words, minimally polysemous words, or the most frequent meaning of polysemous words). Under ideal circumstances, academic language exposure comes through reading—but BPS classrooms were full of poor and/or reluctant readers, for whom autonomous engagement with literacy was not an effective channel for word learning. We also recognized the students' need to learn content-area technical terms in order to succeed in their classes, along with the all-purpose academic words on which we were concentrating, and hoped that the instructional strategies displayed in the program might be useful to teachers focusing as well on those disciplinary words.

Details of the Word Generation Approach

In its first year of implementation, Word Generation offered 20 weeks of curricular materials, each consisting of a "launch paragraph" with guidance to the teachers about introducing the topic and the week's five target words, then activities related to the topic that used the same five words to be implemented by science, math, and social studies teachers, and a writing prompt for students' "taking a stand" essay, assigned to be completed on Friday (sample materials can be viewed at Word Generation.Org).

The first-year curriculum was fully implemented in one single, self-contained fifth-grade classroom, and the first 12 weeks of the curriculum were implemented in two different middle schools, both of which adopted a whole-school model. The students in all participating classrooms completed pre- and post-tests, and data on their performance on the state accountability exam and, when available, on a standardized test of reading were also collected.

The five target words focused on in each week were presented in the context of "journalistic" paragraphs written to display academic language features. The topics of the weekly paragraphs were selected to be of interest to adolescents, current (e.g., the sort of thing one might read about in the newspaper), and open ended, i.e., topics on which one could authentically defend a variety of opinions. Sample topics include the following: responsibility for global warming; acceptability of censorship; whether schools should enforce dress codes; whether steroid use is acceptable for athletes; whether hip-hop music should be censored; whether schools should stop selling junk food; and so on. Each paragraph was written to provide some information supporting both "sides" of the argument; in order to keep the paragraphs brief, only limited background could be provided, but additional information sources were also suggested. One of the prescribed weekly activities was a debate, in which each student participated in arguing for one of the positions on the topic of the week. These debates constituted opportunities for students to produce and hear academic reasoning and "academically productive" or accountable talk. After producing and hearing arguments in the context of the debate, students were asked to formulate their own position on the topic of the week, and argue for that in the Friday essay.

Development and Implementation

A key aspect of the rollout of Word Generation was systematic attention to guidance and feedback from practitioners. Teachers, coaches, and central office staff from BPS helped in brainstorming topics that would be of interest to adolescents. Pilot versions of the materials were critiqued by teachers, who insisted that our plan to teach 10 words a week was impractical; thus we scaled back to five. Teachers at one of the middle schools met with us to provide feedback on format and content after implementing the first five weeks of the curriculum, and to critique the activities; we incorporated this information in writing materials for subsequent weeks. All participating teachers filled out brief end-of-week reports to help us understand which activities were useful, which ones required more explanation, and how long the activities were taking. The fifth-grade teacher undertook to modify the activities so they would be appropriate for her slightly younger students, many of whom had only recently emerged from ELL status, and conducted interviews with her own students about the program. Thus, the current design of the curriculum reflects input from a wide array of practitioners.

This is not to say that either the materials or the implementation of the program were ideal in year one. Lack of time for adequate professional development, limitations on development time and funds, and other constraints left us quite humble about what we had accomplished. Nonetheless, the pre- and post-test results suggested that the students in all three implementation sites improved appreciably on the academic words taught, that many of them completed the various "wordbook" activities and the weekly essay, and that both teachers and students were enthusiastic about the program and appreciative of the learning opportunities it provided. Both the middle schools that participated in year one opted to implement a longer version of Word Generation in the subsequent year.

In year two (2007–2008), a more coherent set of curricular materials was developed. Twenty to 24 weeks of the second-year curriculum, with the theme "participating in the national conversation," were implemented in six schools. Collection of pre- and post-test data will again allow us to evaluate the program effects for participating students.

Results

Obviously, reporting of results from two schools which self-selected to participate, and in which the level of implementation varied, can only be suggestive. Both middle schools that implemented the program were serving low-income and largely minority student bodies, and both were failing to make adequate yearly progress, but the level of academic achievement in one was consistently higher than in the other, and level of teacher uptake in the two schools also differed. Nonetheless, the estimate of effect size on the words taught, based on pre- and post-tests from a total of 478 sixth- through eighth-grade students, was .44—a strong and educationally significant effect. To put the size of this effect in context, the sixth graders after 12 weeks' participation in the program scored higher on the words tested than the eighth graders had scored at the pretest. In other words, 12 weeks of instruction trumped two years of incidental learning for these academic words that are crucial to understanding academic texts.

Discussion

This small study has taught us a number of lessons about vocabulary development, about working with schools, about the needs of urban middle school students, and about the need for more rigorous research designs. We see from the effectiveness of this approach that the principles established through several decades of research on vocabulary instruction hold, but at the same time that there is much more to learn about teaching "academic words." Some traditional vocabulary teaching approaches—visual representations, study definitions—are particularly ineffective for this segment of the lexicon, where abstract and polysemous words predominate.

In this study, as in others of educational programs, ensuring effective implementation is a bigger challenge than developing programs. A key part of doing Word Generation well involves orchestrating effective classroom discussion, nurturing and guiding debates, and ensuring that the target words are used recurrently. There are documented techniques for promoting productive classroom talk, but they are not in the repertoire of all teachers, and they are not easy to teach through brief professional development sessions. Nonetheless, we feel that it is crucial to design more focus on academic discussion into future versions of Word Generation, and to keep working on the problem of how to support teachers to do a good job of it.

An unexpected discovery from this work was the degree to which the topics we chose were unknown to the students. Though many of the topics were "ripped from the headlines" (death penalty, stem cell research, genetically modified foods, gay marriage, global warming), students often had never heard of them, and in the process of the Word Generation activities they acquired information (cars produce greenhouse gases, lifetime earnings are related to years of schooling, only a minority of countries allow the death penalty) that was entirely novel to them. They also discovered, in the course of the better-run discussions and the better-structured debates, that they cared about these topics, and wanted to hone their arguments pro or con. Honing the arguments required use of academic vocabulary, but also other academic language skills of relevance to long-term academic success.

We are acutely aware of the limitations of the data currently available about the effectiveness of Word Generation. In 2007–2008, the program was expanded to a larger number of schools, and several comparable schools agreed to administer the pre- and post-test without implementing the program. We hope in the near future to launch a randomized trial in a different district. Another basic SERP principle is that innovations should be designed to respond to local conditions, but always holding in mind the possibility that they will travel to other sites. Implementing Word Generation elsewhere will reveal what aspects of the program are indeed portable, as well as whether it is effective in other settings.

References

Beck, I., McKeown, M., & Kucan, L. (2002). *Bringing words to life: Robust vocabulary instruction*. New York: Guilford Press.

Bryk, A., & Schneider, B. (2002). *Trust in schools: A core resource for improvement.* New York: Russell Sage.

Coxhead, A. (1998). *An academic word list* (English Language Institute Occasional Publication No. 18). Wellington: Victoria University of Wellington.

Coxhead, A. (2000). A new academic word list. *TESOL Quarterly*, 34, 213–238.

Donovan, S., Wigdor, A., & Snow, C. (Eds.) (2003). *Strategic Education Research Partnership.* Washington, DC: National Academies Press.

Elmore, R. (2004) *School reform from the inside out: Policy, practice and performance.* Cambridge, MA: Harvard Education Press.

Fuhrman, S., & Elmore, R. (2004). *Redesigning accountability systems for education*. New York: Teachers College Press.

Graves, M. F. (2006). *The vocabulary book: Learning and instruction.* Newark, DL: International Reading Association.

Michaels, S., O'Connor, M. C., & Resnick, L. (2008). Deliberative discourse idealized and realized: Accountable talk in the classroom and in civic life. *Studies in Philosophy and Education*, 27, 283–297.

Nation, I. (1990). *Teaching and learning vocabulary*. Boston: Heinle & Heinle Publishers.

Snow, C. & Uccelli, P. (2008). The challenge of academic language. In D. Olson & N. Torrance (Eds.), *The Cambridge handbook of literacy.* Cambridge: Cambridge University Press.

Word Generation.Org. From Strategic Education Research Partnership website. Available at: http://www.serpinstitute.org/ (accessed March 10, 2009).

6 Learning to Remember

Mothers and Teachers Talking with Children

Peter A. Ornstein, Catherine A. Haden, and Jennifer L. Coffman

Although a great deal has been learned in the last 40 years about the mnemonic skills of children of different ages, surprisingly little is known about the course of developmental changes in these abilities (Ornstein & Haden, 2001). To illustrate, age-related differences in children's use of deliberate strategies for remembering—such as rehearsal and organization—have been well documented by many researchers (Schneider & Pressley, 1997), but the development of these skills within individual children has been studied by only a few investigators (e.g., Kron-Sperl, Schneider, & Hasselhorn, 2008; Schneider, Kron, Hünnerkopf, & Krajewski, 2004).

In order to focus on development, it is necessary to supplement the cross-sectional research that has yielded valuable information about children of different ages with longitudinal methodologies in which performance within individuals is tracked over time (Ornstein & Haden, 2001). Longitudinal designs have the potential to (a) reveal information about developmental patterns of change, which often are quite different from those suggested by the cross-sectional literature (e.g., Schlagmüller & Schneider, 2002; Schneider et al., 2004); (b) generate insights into the ways in which early skill, as in talking about the past, is linked to later skill, as in the use of a deliberate strategy (Ornstein & Haden, 2001); and (c) identify factors—such as adult–child conversation—that may serve to bring about the observed developmental change (Ornstein & Haden, 2001).

For these reasons, we are committed to the use of longitudinal methods to study children's changing mnemonic skills, with the work summarized here being based primarily on two separate but related research programs in which we examine preschoolers' changing autobiographical memory skills (Haden, Ornstein, Rudek, & Cameron, 2009; Hedrick, San Souci, Haden, & Ornstein, 2009; Ornstein, Haden, & Hedrick, 2004) and elementary school-age children's developing deliberate memory skills (Coffman, Ornstein, McCall, & Curran, 2008; Ornstein, Coffman, Grammer, San Souci, & McCall, 2010). In an effort to examine both *memory* development and the *development* of memory, in each of these projects we make use of laboratory-based tasks adapted from the information-processing tradition to characterize children's skills, and we draw upon social constructivist approaches to examine social interchanges—in the form of parent–child and teacher–child conversation—that may be linked to developmental changes in mnemonic skill (Ornstein & Haden, 2001).

As such, in this chapter, we discuss our work on social factors that we feel may govern developmental changes in the encoding, retrieval, and reporting of information. We first examine linkages between adult–child social interaction in the home and children's reports of events that they have experienced, and then turn to the classroom, focusing on associations between teachers' "mnemonic style" and children's deliberate skills for remembering.

Mother–Child Conversations and the Development of Mnemonic Skills

Children begin to talk about past experiences almost as soon as they produce their first words, and their skills for reporting past events as they "reminisce" with their parents develop rapidly between two and four years of age. A central focus in the literature on parent–child conversations about the past has been on differences among parents in the "reminiscing styles" they use to structure these discussions with their young children (see Fivush, Haden, & Reese, 2006). In contrast to mothers who use a low-elaborative conversational style, those who employ a high-elaborative style ask many questions, follow in on their children's efforts to contribute information to the conversation, and continue to add new information about the event even when their children do not do so. It is clear that these reminiscing styles generalize across different types of past event discussions (e.g., holidays, museum trips, entertainment outings) and to conversations about events mothers did and did not experience with their children (Reese & Brown, 2000). Importantly, longitudinal data indicate that differences in maternal reminiscing styles are associated with later differences in children's abilities to recall personally experienced events over time. For example, Reese, Haden, and Fivush (1993) found that maternal elaborativeness during early conversations about the past with 40-month-olds is associated positively with children's recall of past experiences in later conversations at 58 and 70 months of age.

Complementing this work on reminiscing, an emerging literature illustrates the impact of conversations as events unfold on young children's remembering (Haden, Ornstein, Eckerman, & Didow, 2001; McGuigan & Salmon, 2004; Ornstein et al., 2004). For example, Haden et al. (2001) observed a substantial effect of joint mother–child conversational interaction on children's remembering. In this longitudinal study, young children took part in three specially constructed activities with their mothers: at 30 months, a camping trip; at 36 months, a birdwatching adventure; and at 42 months, the opening of an ice cream store. The children's reports of these experiences varied such that those features that were discussed jointly by mother and child were better recalled than those that were commented on only by mothers, which in turn were better recalled than those components that were not discussed at all.

Findings from prior studies thus converge to suggest that social-communicative exchanges can dramatically influence children's memory performance. Although the mechanisms underlying these effects have not been explored in detail, elaborative parent–child conversations may influence the encoding and representation of information in memory on the one hand, as well as search, retrieval, and reporting operations, on the other. For example, conversations as events unfold enable parents to direct children's attention to salient features of the activity so as to facilitate comprehension, thus impacting the ways events are encoded and represented in memory (Ornstein et al., 2004). Similar benefits of elaborative conversation may result from opportunities to reminisce with an adult about previously experienced events (Fivush et al., 2006). In addition, it seems likely that an elaborative conversational style in reminiscing may provide opportunities for children to practice searching and retrieving information from memory and experience in using narrative conventions to provide accounts of their experiences (e.g., Reese et al., 1993).

The Pathways Study

Our examination of linkages between parent–child conversational interactions during and after events and children's developing memory skills has taken place primarily in the context of a large-scale longitudinal investigation, the *Developmental Pathways to Skilled Remembering* study. In this project, we established a sample of children composed of two cohorts, one seen initially when they were 18 months of age, and the other at 36 months of age. The 110 children enrolled in the study were observed at frequent intervals for a period of three years, allowing us to focus on the multiple contributions of maternal conversational style and mother–child joint conversational engagement to the development of memory. Within the *Pathways* project, we have emphasized the development of key verbal memory skills as they are reflected in children's abilities (a) to talk about experiences in the present; (b) to discuss events in the past; and (c) to plan deliberately for future assessments of remembering.

Reminiscing about the Past

As a component of the larger *Pathways* study, Haden et al. (2009) observed 56 children in the younger cohort as they reminisced with their mothers when they were 18, 24, and 30 months old. We were particularly interested in how joint reminiscing changes over a period that is characterized by marked increases in children's recall abilities (e.g., Harley & Reese, 1999). We were motivated by questions concerning the mechanisms by which elaborativeness affects remembering, and specifically by whether contrasting types of "elaborative" conversational patterns may be differentially predictive of children's developing memory skills (see Fivush et al., 2006, for similar arguments). As a way of characterizing individual differences in reminiscing style with very young children, Haden et al. (2009) examined the extent to which mothers pose more *Wh-* questions than they make elaborative statements. Because *Wh-* questions require that children retrieve and report *who, what, where, when, why,* or *how* types of information from memory, they are more cognitively challenging than closed-ended questions that call for only a confirmation or negation of information provided by the mother. Further, because elaborative *Wh-* questions invite the child to become a partner in the conversation, they contrast with elaborative statements that essentially reflect a mother's "telling" of the tale.

At each age point—18, 24, and 30 months—the children's mothers selected different events about which to talk, ones that they had experienced with their children in the recent past. The mothers were instructed to discuss these events with their children as they naturally would, and recordings of these conversations were transcribed verbatim prior to coding. In addition to scoring the mothers' *Wh-* question elaborations (e.g., "What did you eat?"; "What did you do with Taylor?"), and statement elaborations that involved new information about the event under discussion (e.g., "We had dinner"; "You watched a Barney video"), the children's provision of new memory information was scored in terms of the number of memory elaborations that they provided (e.g., "Barney"; "We ate cookies!").

Supporting our effort to identify contrasting groups of mothers who differed in the extent to which they elicited responses from their children, we observed striking variability in the use of *Wh-* question elaborations and statement elaborations. To establish these groups, we identified 29 mothers who were above the median in their use of *Wh-* question elaborations when their children were 18 months of age. Of these 29 mothers, 22 were also found to be using as many ($n = 3$) or more ($n = 19$) elaborative *Wh-* questions as elaborative statements per event in their conversations about the past with their 18-month-olds. This "high-eliciting" group was thus composed of mothers who offered embellished details about the events under discussion, while at the same time inviting their children to participate in the creation of jointly authored personal narratives. The remaining 34 mothers—seven who were above the median for *Wh-* question elaborations and 27 who were below—all used fewer *Wh-* question elaborations than statement elaborations and were thus classified as belonging to a "low-eliciting" group who primarily told the story of the event to their children.

As shown in the top panel of Table 6.1, the high- versus low-eliciting mothers differed in their use of *Wh-* question and statement elaborations at the 18-month time point. Further inspection of the table indicates that the high- and low-eliciting groups differed in the ways in which their use of *Wh-* question elaborations and statement elaborations changed over time. All mothers increased their tendencies to pose elaborative *Wh-* questions but, over the three age points, mothers in the high-eliciting

Table 6.1 Mean Frequencies and Standard Deviations per Event for Maternal and Child Participation in the Memory Conversations as a Function of Time and Maternal Style Group

	Low-Eliciting	*High-Eliciting*
Maternal Conversational Techniques		
Wh- Question Elaborations		
18 months	1.32 (1.50)	3.50 (2.81)
24 months	2.85 (2.50)	6.77 (3.50)
30 months	4.64 (3.31)	8.53 (3.00)
Statement Elaborations		
18 months	7.98 (4.98)	2.93 (2.08)
24 months	5.39 (3.92)	2.28 (1.85)
30 months	4.64 (4.50)	3.33 (3.18)
Children's Memory Responses		
Memory Elaborations		
18 months	.09 (.19)	.57 (.64)
24 months	1.01 (1.19)	2.80 (2.95)
30 months	2.12 (1.85)	5.92 (2.44)

Adapted from Haden, Ornstein, Rudek, & Cameron (2009). © 2008 The International Society for the Study of Behavioural Development.

group consistently made greater use of *Wh-* question elaborations than did mothers in the low-eliciting group. Moreover, as can also be seen in Table 6.1, the mothers' use of statement elaborations contrasted markedly with their production of *Wh-* questions. Indeed, at 18 months, the mothers in the low-eliciting group demonstrated considerable use of statement elaborations, but their production of these statements decreased considerably over time. In fact, by 30 months, the low-eliciting mothers were not different from the high-eliciting mothers in their use of statement elaborations. In contrast, the mothers in the high-eliciting group maintained their level of use of approximately three statements per event.

Turning to the children's provision of new memory information in the event conversations, it was clear that their participation was quite limited. Indeed, more than half the sample did not provide any memory elaborations at 18 months of age. Even so, compared with children of low-eliciting mothers, children of high-eliciting mothers generated more new memory information per event. As illustrated in the lower panel of Table 6.1, these differences were present at 18 months of age, but they were magnified over time. These patterns, of course, raise the further question of the extent to which the children's early skills were predictive of the later differences in their abilities to report their past experiences. In this regard, in a regression model we found that the children's language skills at 30 months and their recall abilities at 24 months accounted for a little more than one-third of the variance in the children's provision of memory information during reminiscing at 30 months. Moreover, when maternal eliciting style was added to the regression equation, an additional increase of nearly three memory elaborations per event was predicted by having a mother in the high-eliciting style group.

Thus, consistent with the previous literature, we find that parents play a unique role in the development of children's skills for remembering personally experienced events. Moreover, the findings suggest that one mechanism by which elaborativeness may be important for remembering is that the mothers' use of open-ended elaborative questions fosters children's active participation in reminiscing. By lending their own voices to the construction of a narrative about personally experienced events, children can gain valuable practice in retrieving and reporting events in ways that can impact dramatically the development of skills for remembering.

Conversations during Events

Our examination of possible mechanisms that may underlie the impact of elaborative interactional styles on children's memory has extended to joint talk as events unfold. Across both *Pathways* cohorts, we (Hedrick, San Souci, et al., 2009) focused on 89 mother–child dyads who took part in two of the novel "adventures" in their homes that had been studied by Haden et al. (2001)—a camping event and a birdwatching adventure—when the children were 36 and 42 months of age. The children's recall of these events was assessed after delays of one day and three weeks. In order to link the degree of joint verbal engagement as the events unfolded to the children's subsequent recall, we first established two groups of mother–child dyads on the basis of the mothers' use of elaborative *Wh-* questions to which the child responded either correctly or not at all. This grouping strategy reflects the view that a richer encoding is established when children respond correctly to their mothers' questions as an event unfolds than when this is not the case. On average, at each of the two age points, the mothers posed approximately 16 *Wh*- questions during the events and the children most frequently responded to these questions with correct responses: 49 percent of the time at 36 months and 55 percent of the time at 42 months. It was also quite common for children not to respond to these queries, particularly at 36 months, when this was the case approximately one-quarter of the time.

In forming groups based on these question–response patterns, Hedrick, San Souci et al. (2009) identified those dyads in the sample who at 36 months were both at or above the median for the mother *Wh*- question–child correct response pattern, and below the median for the mother *Wh*- question–child no response pattern. A total of 36 dyads was assigned to this "high joint talk" group that was characterized by a relatively high proportion of child correct responses and low proportion of child no responses to mothers' *Wh*- questions. The remaining 53 dyads were assigned to a "low joint talk" group that was composed of pairs who were relatively low in the proportion of mother *Wh*- questions that were correctly responded to by the children, and relatively high in the proportion of mother *Wh*- questions that were not responded to at all. Although these two contrasting groups were established on the basis of the levels of joint talk at 36 months, they remained distinctly different at the 42-month assessment point.

As illustrated in Figure 6.1, in comparison with the children in low joint talk dyads, those in the high joint talk group recalled more features of the novel events (e.g., the spatula in the camping event), and produced more "event elaborations" or

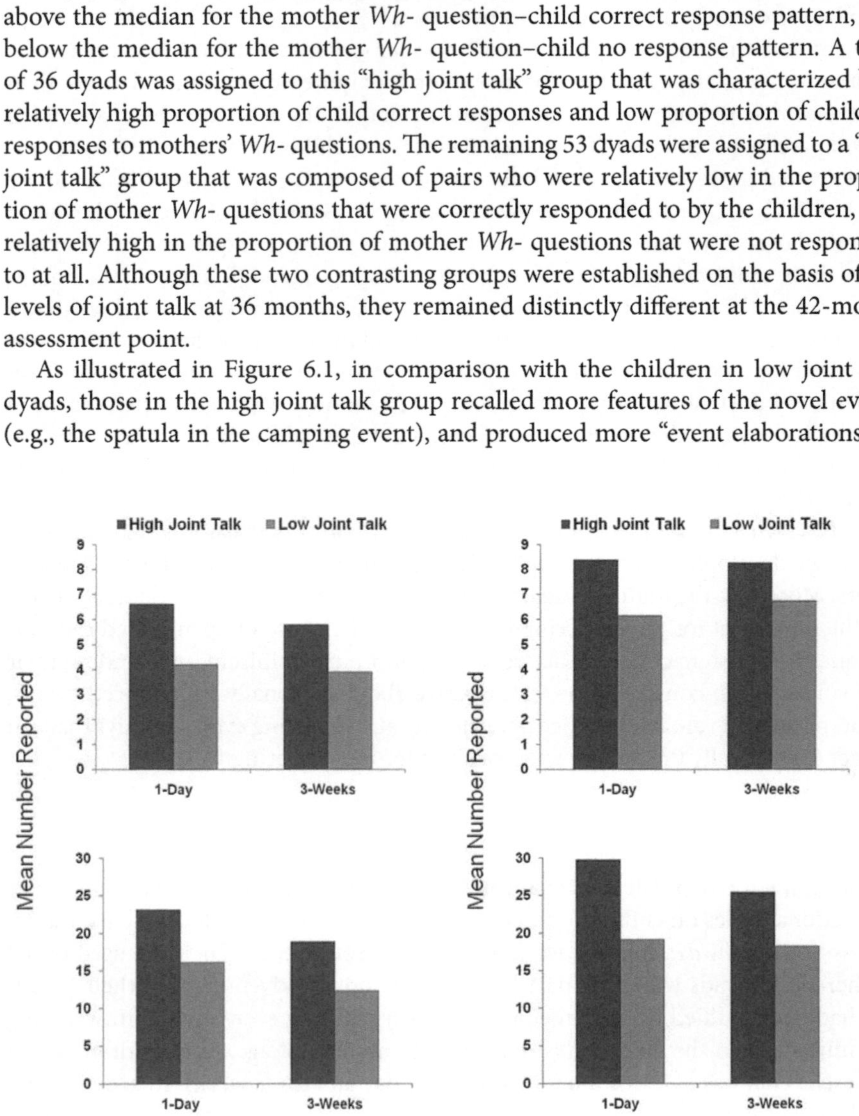

Figure 6.1 Children's Feature Recall (Upper Panels) and Feature Elaborations (Lower Panels) at the One-Day and Three-Week Delay Intervals as a Function of Joint Talk Group at the 36-Month (Left Panels) and 42-Month (Right Panels) Assessments.

informational details about the features (e.g., "My backpack was red") and the event in general (e.g., "Then we went fishing at the pond"). Moreover, inspection of Figure 6.1 reveals that these findings were robust across the one-day and three-week delay intervals at each age, and from 36 to 42 months. Nevertheless, because the children's recall of the events at one assessment was associated with their recall at other assessments, we explored further whether the children's assignment to one or the other joint talk group at 36 months uniquely predicted their 42-month recall, after accounting for any consistency in recall over time. In this regard, our finding that the children's provision of event elaborations at 42 months was predicted by their joint talk group membership is particularly noteworthy. Indeed, it adds to evidence suggesting that the strongest effects of joint talk during events may be on children's memory for the details of their experiences—for example, that a spatula was used to flip hamburgers on the grill—and not for the specific component features.

These findings extend earlier work by providing additional information concerning the contributions of joint talk as a potential mechanism by which these conversations affect children's remembering. As we see it, mothers with an elaborative style—especially those who engaged in high joint talk—are skilled in focusing their children's attention on salient aspects of an event. In this way, they are able to offer information that may affect the interpretation of a shared experience and thereby the resulting representation in memory.

Linkages to Deliberate Remembering

Although event memory and deliberate remembering are routinely treated in fairly distinct literatures, the *Pathways* study was designed to enable the simultaneous exploration of children's memory for previously experienced events as well as their performance on tasks that call for the use of strategies to support deliberate remembering. For example, we could assess children's memory for events that they experienced with their mothers (e.g., the camping "adventure" discussed above) and examine their skills on a deliberate memory task (see Baker-Ward, Ornstein, & Holden, 1984) in which they were instructed to do anything they could to remember a set of familiar but unrelated objects. The inclusion of this task enabled us to track deliberate memory performance longitudinally, to explore associations between the use of simple strategies (e.g., naming, visual examination, pointing) and recall over time, and to examine linkages between maternal elaborativeness during conversations about the past and children's strategy use.

Focusing on the children in the older *Pathways* cohort, we observed that, with increases in age, naming became a stronger predictor of the children's object recall, with r-values of .23, .47, and .54 obtained at 42, 54, and 60 months, respectively. Interestingly, given these increasing naming–recall correlations, at 54 months of age we found that maternal elaborative style during reminiscing was associated with children's use of naming during the object memory task ($r = .49$). Moreover, maternal elaborativeness at 54 months predicted children's object recall six months later, even when children's naming of objects during the study period at 60 months of age and their recall of objects at 54 months were taken into account. Although we are a long way from understanding the mechanisms by which aspects of maternal elaborative style set the stage for children's developing competencies in deliberately planning to remember, these findings are nonetheless consistent with the view that the roots of children's deliberate memory competence may be found in social interactions with adults that involve efforts to remember past experiences.

Children's Memory Strategies: The Impact of Teachers' Memory "Talk"

The findings on event memory presented here indicate clearly a linkage between mother–child conversational style and the development of preschoolers' memory skills, and even suggest that maternal elaborativeness may be associated with young children's deliberate use of a simple naming strategy and subsequent object recall. Given these relations between aspects of adult–child conversation and children's early memory skills, is it possible that conversation in the classroom might be linked to more sophisticated strategies such as rehearsal, organization, and elaboration? Admittedly, not much work has focused on factors that impact the development of these strategies, but a considerable amount of research has documented increases in children's use and effectiveness of strategies over the course of the elementary school years (Schneider & Pressley, 1997). Yet, given the importance of context for mnemonic growth in the preschool years, it seems likely that aspects of the classroom—especially teacher "talk"—would influence children's later-emerging memory skills.

A number of lines of evidence support this perspective and point to the potential impact of formal schooling on the development of memory strategies (e.g., Moely et al., 1992; Morrison, Smith, & Dow-Ehrensberger, 1995; Rogoff, 1981). Consider first the results of comparative-cultural explorations carried out in countries such as Liberia, Mexico, and Morocco of the cognitive skills of children who differed in terms of whether they had or had not attended Western-style schools. Collectively, these studies revealed that children who attended school were superior to their non-schooled peers in the memory skills that have typically been studied by Western psychologists (see Ornstein et al., 2010; Rogoff, 1981) and suggested that something in the formal school context may be related to the emergence of skills for remembering. Research by Morrison et al. (1995) complements this comparative-cultural work and suggests more precisely that the first-grade experience is particularly important for the development of children's memory skills.

Given that schooling has been identified as a potential facilitator of developmental change in mnemonic skill, what is it about the classroom context that is important for bringing about this cognitive growth? To address this issue, Moely and her colleagues (1992) identified specific instructional components that they believed were relevant for children's memory development, such as instances in which teachers provided general information about cognitive processes or gave specific strategy suggestions. Although Moely et al. observed that explicit instruction in the use of strategies was quite rare across the elementary school years, they were nonetheless able to group teachers on the basis of the amount of strategy suggestion that they provided in the course of their teaching. Importantly, even though the provision of strategy information was infrequent, it turned out to be quite important for children in the first grade. Indeed, students in the classes of first-grade teachers who were higher in their provision of strategy suggestions were more likely to spontaneously generate organizational techniques than were their peers whose teachers gave fewer strategy suggestions.

Although suggestive, these bodies of work nonetheless leave a critical developmental question unanswered: If the cross-cultural literature suggests that school is important in terms of the emergence of deliberate memory strategies, but if explicit instruction occurs infrequently, then what is it about the classroom that influences the emergence and refinement of these skills? The studies of mother–child conversational interaction and of the classroom suggest the importance of examining teacher "talk" during the course of instruction and provide the foundation for longitudinal exploration of memory in the classroom context.

The Classroom Study

To explore the classroom setting as a context for the development of skilled remembering, we launched a longitudinal investigation to examine linkages between teachers' memory "talk" and children's deliberate memory performance (e.g., Coffman et al., 2008). In this *Classroom* study, we tracked children's memory skills over the course of the elementary school years, while simultaneously making detailed observations of the language used by their teachers in the course of instruction.

Characterizing the Classroom

In the first year of the *Classroom* study, we made observations that focused on teachers' memory-relevant language in 14 first-grade classrooms. Over several visits, two research assistants observed each class for a total of one hour of whole-group instruction in each of two subject areas, language arts and mathematics. One of these observers made use of our *Taxonomy of Teacher Behaviors* (see Coffman et al., 2008; Ornstein et al., 2010) to make judgments every 30 seconds concerning the teachers' language, while the second observer simultaneously prepared a detailed narrative to capture each lesson as it unfolded.

We developed the *Taxonomy of Teacher Behaviors* to describe the nature of instruction, primarily by classifying each teacher's instructional language into four basic categories: instructional activities (e.g., providing information); "cognitive structuring activities" (encouraging children to engage the materials in ways that are known to facilitate encoding and retrieval of information); memory requests (asking students to retrieve information already acquired or to prepare for future activities); and metacognitive information (presenting or soliciting information that might facilitate performance on a range of cognitive tasks in the classroom). Moreover, the narratives of the ongoing lessons enabled judgments about the nature of the memory "demands" or goals, both implied and expressed, that the teachers were communicating to their students.

Our observations indicated that the teachers focused their efforts in the classroom largely on instructional activities, with 78.2 percent of the observational intervals containing some form of instruction. Consistent with Moely et al.'s (1992) findings, the use of metacognitive information in the context of these lessons was observed in only 9.5 percent of the intervals; moreover, only 4.9 percent of the intervals contained a suggestion for the use of a specific cognitive strategy. However, with the *Taxonomy*, we found that there actually was a considerable amount of memory "talk" in the classroom. Indeed, 52.6 percent of the intervals included some memory-related request of the children, as for example: "What do you already know about dinosaurs?" and "Remember to take your folder home." Moreover, the use of the narratives permitted a further characterization of the intervals in which the teachers made these memory-related requests, distinguishing between the relatively rare occasions (5.4 percent of the intervals) in which memory demands were expressed explicitly with words such as "remember" and the more common situations (47.2 percent) in which demands for the use of memory were implicit, without expressed prompts to remember.

Constructing a Measure of Teacher Mnemonic Style

Although these data provide an overall impression of those features of instruction that are likely important for the development of children's mnemonic skills, it is important to note that there was considerable variability across classrooms in the language used by the teachers. Moreover, this across-classroom variability—necessary for our efforts to relate the memory "talk" to which the children were exposed to actual memory performance—provided the starting point for the construction of an index of teacher "mnemonic orientation" (Coffman et al., 2008). This measure is based on a subset of component codes that focus on the extent to which teachers emphasize remembering in their classrooms and reflect the use of memory-relevant language (as captured by the *Taxonomy*) and the nature of the memory demands placed on their students (as inferred from the narratives).

These component codes that contribute to our measure of mnemonic orientation are illustrated in Table 6.2. As can also be seen in the table, there was considerable variability across classrooms in the extent to which these aspects of instruction were observed. More specifically, the top panel illustrates the clear differences in the extent to which the teachers provided *strategy suggestions* and posed *metacognitive questions*, regardless of whether or not remembering was the focus of discussion. Moreover, reflecting the importance of contextualizing requests for remembering during ongoing instruction, the bottom panel illustrates dramatic differences in the co-occurrence of *deliberate memory demands* (both expressed and implied) when paired with *instructional activities, cognitive structuring activities,* and *metacognitive information.*

This naturally occurring variability in memory-related talk allowed us to form two groups of first-grade teachers, those who were *high* and those who were *low* in their *mnemonic style* in the classroom, based on a median split of the average of the standard scores that were calculated for each of the codes. Given this identification of high- versus low-mnemonic teachers, we were then able to examine the memory performance of the children on a range of tasks, as a function of the mnemonic style of their teachers.

Linking the Classroom Context and Children's Performance

Of the 107 first graders in the longitudinal sample, 46 were taught by low-mnemonic and 61 by high-mnemonic first-grade teachers. Importantly, these groups of children did not differ on measures of basic memory capacity, as assessed by a digit span task.

Table 6.2 Component Codes in the Measurement of Teacher Memory-Relevant "Talk"

Individual Taxonomy Codes	*Examples*	*Mean Percent Occurrence (Range)*
Strategy Suggestions	Recommending that a child adopt a method or procedure for remembering or processing information	4.9% (.8–13.8%)
Metacognitive Questions	Requesting that a child provide a potential strategy, a utilized strategy, or a rationale for a strategy they have indicated using	4.9% (.8–9.6%)
Co-occurring Codes	*Definitions*	
Deliberate Memory Demands and Instructional Activities	Intervals that contain both requests for information from memory and also the presentation of instructional information by the teacher	37.6% (25.8–50.0%)
Deliberate Memory Demands and Cognitive Structuring Activities	Intervals that contain both requests for information from memory and teacher instruction that could impact the encoding and retrieval of information, such as focusing attention or organizing material	23.5% (10.0–35.4%)
Deliberate Memory Demands and Metacognitive Information	Intervals that contain both requests for information from memory and the provision or solicitation of metacognitive information	5.9% (1.3–12.1%)

Adapted with permission from Coffman, J. L., Ornstein, P. A., McCall, L. E., & Curran, P. J., *Developmental Psychology*, 44, pp. 1640–1654, published 2008 by APA.

Despite equivalence at the beginning of the first grade, we have identified linkages between the classroom context and the children's mnemonic skills at the end of the academic year (Coffman et al., 2008). More specifically, at the end of the first-grade year, the children in classes taught by teachers with the contrasting mnemonic orientations differed in their use of memory strategies and in the amount of information recalled on a range of tasks. Moreover, these differences were maintained after the first grade, when the children were taught by different teachers.

Consider, for example, the children's use of organizational grouping and recall strategies on a Sort–Recall with Organizational Task (Moely et al., 1992). In the first and second grades, students were asked to remember a set of 16 cards with line drawings that were drawn from four taxonomic categories. At the initial assessment point (the fall of grade 1), the children were presented with these materials over a series of baseline, training, and generalization trials, each of which involved opportunities to sort and then to recall the pictures. At the baseline trial, no information was provided about using the category structure as an aid to remembering. After this, the participants received such instruction on the training trial. During training, the participants were told to sort the items into categories and then to recall on the basis of these groups (i.e., to use a clustering strategy) technique. At each subsequent assessment in the winter and spring, non-instructed generalization trials were administered.

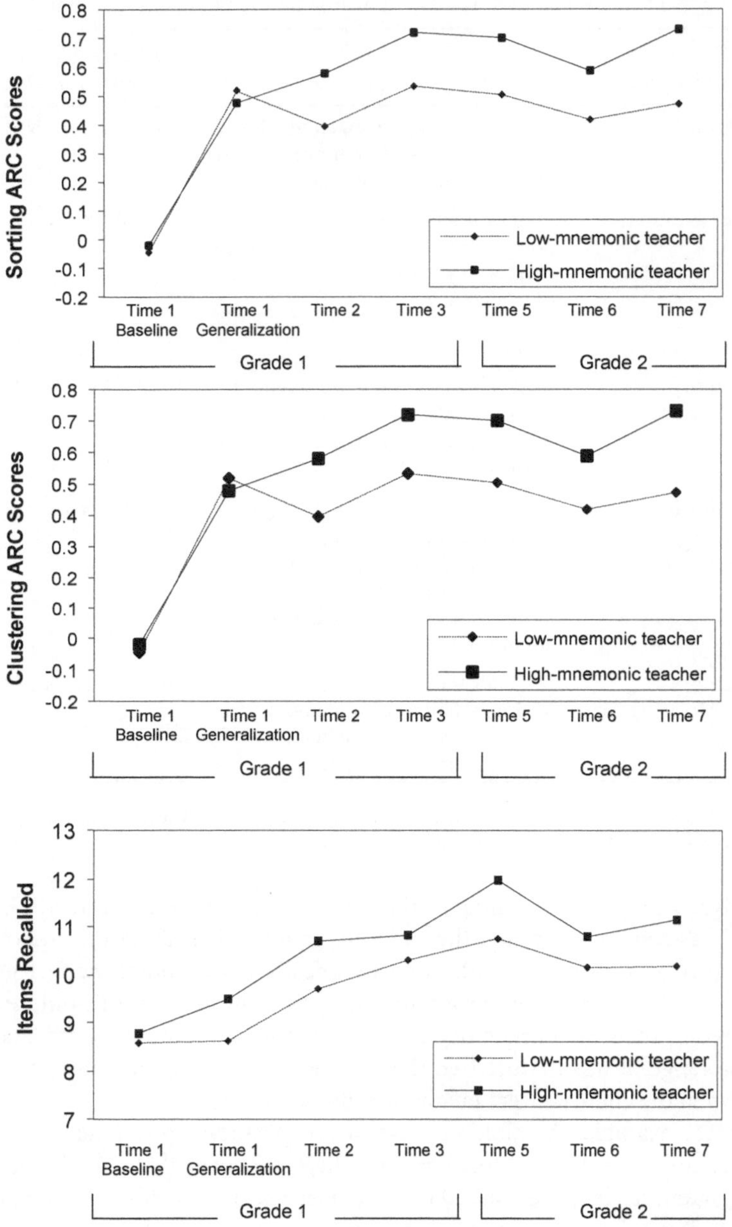

Figure 6.2 Sorting, Clustering, and Recall Scores over Grades 1 and 2 as a Function of First-Grade Teacher Mnemonic Orientation. Adapted with permission from Coffman, J. L., Ornstein, P. A., McCall, L. E., & Curran, P. J., Linking teachers' memory-relevant language and the development of children's memory skills. *Developmental Psychology*, 44, pp. 1640–1654, published 2008 by APA.

Similarly, when the children were in grade 2 (and taught by different teachers), assessments with non-instructed generalization trials were carried out in the fall, winter, and spring.

On the Sort–Recall with Organizational Training Task, the structure reflected in the children's sorts and in the order of their recall was measured with the Adjusted Ratio of Clustering (ARC) index (Roenker, Thompson, & Brown, 1971), resulting in ARC scores that could range from −1 (below chance organization) to 0 (chance) to 1 (complete categorical organization). As shown in Figure 6.2, patterns of diverging skill emerged as early as the winter of the first grade, and were maintained across grades 1 and 2, with the children assigned to high-mnemonic teachers outperforming their peers with low-mnemonic instructors. Inspection of the top and middle panels of Figure 6.2 indicates pronounced differences between the groups in the use of organizational strategies, both in sorting on the basis of meaning and in clustering in recall.

Although the initial levels of strategy use on the baseline trial at the first assessment were quite low, by the winter assessment point in grade 1 there seemed to be clear differences in the abilities of the children taught by high- versus low-mnemonic teachers to take advantage of the organizational training. Indeed, at the end of the first grade, the children in the high-mnemonic classrooms outperformed their peers in low-mnemonic classes in both sorting (mean = .58 versus .39) and clustering (mean = .72 versus .60). Moreover, the differences in sorting and clustering between these two groups of children persist across grade 2, suggesting that not only does the mnemonic orientation of the first-grade teachers matter for the memory performance of the children in their classes, but it is also important even a year later, when the children are being taught by other teachers. Further, as can be seen in the bottom panel of Figure 6.2, comparable—although less striking—patterns are seen in the children's recall performance.

These findings and others described by Ornstein et al. (2010) indicate clearly that first-grade teachers are doing something to create a context for strategy discovery and utilization, one that has implications for children's performance in later years. But just how does exposure to a *high-mnemonic* teacher influence a child's ability to acquire and use memory strategies? Two possibilities suggest themselves for future exploration. On the one hand, it may be that exposure to the memory-rich language of a high-mnemonic teacher permits children to discover strategies on their own. However, on the other hand, there could be some generalization process operating such that metacognitive and strategy information that is presented during the course of instruction in one domain (e.g., mathematics) is then modified and used in the service of memory goals.

Experimental Manipulations of Conversation and Future Directions

In the *Pathways* and *Classroom* studies we have been able to document associations between aspects of social communication—with either parents or teachers—and children's memory performance. In each of these projects, we have observed within-child developmental changes in mnemonic skill and have found suggestive evidence that aspects of adult–child conversation play an important role in bringing about the changes we have observed. These correlational findings are of considerable importance for identifying potential mediators of developmental change. Nonetheless, this work represents only the starting point for systematic developmental analyses of children's memory. Indeed, it is clearly necessary to combine longitudinal and experimental methodologies so that we can begin to make causal connections between conversation and the development of children's memory skills.

We have begun to move in this direction and have made progress in following up the *Pathways* study with experiments in which mothers and other adults who interact with children have been instructed in different types of elaborative conversation (Boland, Haden, & Ornstein, 2003; Hedrick, Haden, & Ornstein, 2009). These experiments support the view that conversational style may be linked causally to children's memory performance, and have encouraged us to launch similar experiments to extend the *Classroom* study by providing teachers with instruction in contrasting mnemonic styles. A long-term implication of these types of training studies is that it may be possible to create instructional programs—for parents and teachers—that will facilitate children's cognitive development.

Acknowledgments

The research reported in this chapter was supported by grants from NIH (HD 37114) and the National Science Foundation (BCS-0217206 and BCS-0519153).

References

Baker-Ward, L., Ornstein, P. A., & Holden, D. J. (1984). The expression of memorization in early childhood. *Journal of Experimental Child Psychology*, 37, 555–575.

Boland, A. M., Haden, C. A., & Ornstein, P. A. (2003). Boosting children's memory by training mothers in the use of an elaborative conversational style as an event unfolds. *Journal of Cognition and Development*, 4, 39–65.

Coffman, J. L., Ornstein, P. A., McCall, L. E., & Curran, P. J. (2008). Linking teachers' memory-relevant language and the development of children's memory skills. *Developmental Psychology*, 44, 1640–1654.

Fivush, R., Haden, C. A., & Reese, E. (2006). Elaborating on elaborations: Role of maternal reminiscing style in cognitive and socioemotional development. *Child Development*, 77, 1568–1588.

Haden, C. A., Ornstein, P. A., Eckerman, C. O., & Didow, S. M. (2001). Mother-child conversational interactions as events unfold: Linkages to subsequent remembering. *Child Development*, 72, 1016–1031.

Haden, C. A., Ornstein, P. A., Rudek, D. J., & Cameron, D. (2009). Reminiscing in the early years: Patterns of maternal elaborativeness and children's remembering. *International Journal of Behavioral Development*, 33, 118–130.

Harley, K., & Reese, E. (1999). Origins of autobiographical memory. *Developmental Psychology*, 35, 1338–1348.

Hedrick, A. M., Haden, C. A., & Ornstein, P. A. (2009). Elaborative talk during and after an event: Conversational style influences children's memory reports. *Journal of Cognition and Development*, 10(3), 188–209.

Hedrick, A. M., San Souci, P. P., Haden, C. A., & Ornstein, P. A. (2009). Mother-child joint conversational exchanges during events: Linkages to children's memory reports over time. *Journal of Cognition and Development*, 10(3), 143–161.

Kron-Sperl, V., Schneider, W., & Hasselhorn, M. (2008). The development and effectiveness of memory strategies in kindergarten and elementary school: Findings from the Würzburg and Göttingen longitudinal memory strategies. *Cognitive Development*, 23, 79–104.

McGuigan, F., & Salmon, K. (2004). The time to talk: The influence of the timing of adult–child talk on children's event memory. *Child Development*, 75, 669–686.

Moely, B. E., Hart, S. S., Leal, L., Santulli, K. A., Rao, N., Johnson, T., & Hamilton, L. B. (1992). The teacher's role in facilitating memory and study strategy development in the elementary school classroom. *Child Development*, 63, 653–672.

Morrison, F. J., Smith, L., & Dow-Ehrensberger, M. (1995). Education and cognitive development: A natural experiment. *Developmental Psychology*, 31, 789–799.

Ornstein, P. A., Coffman, J. L., Grammer, J. K., San Souci, P. P., & McCall, L. E. (2010). Linking the classroom context and the development of children's memory skills. In J. Meece & J. Eccles (Eds.), *The handbook of research on schools, schooling, and human development*. pp. 42–59.

Ornstein, P. A., & Haden, C. A. (2001). Memory development or the development of memory? *Current Directions in Psychological Science*, 10, 202–205.

Ornstein, P. A., Haden, C. A., & Hedrick, A. M. (2004). Learning to remember: Social-communicative exchanges and the development of children's memory skills. *Developmental Review*, 24, 374–395.

Reese, E., & Brown, M. (2000). Reminiscing and recounting in the preschool years. *Applied Cognitive Psychology*, 14, 1–17.

Reese, E., Haden, C. A., & Fivush, R. (1993). Mother–child conversations about the past: Relationships of style and memory over time. *Cognitive Development*, 8, 403–430.

Roenker, D., Thompson, C., & Brown, S. (1971). Comparison of measures for the estimation of clustering in free recall. *Psychological Bulletin*, 76, 45–48.

Rogoff, B. (1981). Schooling and the development of cognitive skills. In H. C. Triandis & A. Heron (Eds.), *Handbook of cross-cultural psychology* (Vol. 4). Boston: Allyn & Bacon. pp. 233–294.

Schlagmüller, M., & Schneider, W. (2002). The development of organizational strategies in children: Evidence from a mircogenetic longitudinal study. *Journal of Experimental Child Psychology*, 81, 298–319.

Schneider, W., Kron, V., Hünnerkopf, M., & Krajewski, K. (2004). The development of young children's memory strategies: First findings from the Würzburg Longitudinal Memory Study. *Journal of Experimental Child Psychology*, 88, 193–209.

Schneider, W., & Pressley, M. (1997). *Memory development between 2 and 20*. New York: Springer-Verlag.

Part II
Science and Learning

7 A Theory of Coherence and Complex Learning in the Physical Sciences

What Works (and What Doesn't)

Nancy L. Stein, Marc W. Hernandez, and Florencia K. Anggoro

In order for children to achieve a functional level of scientific understanding, especially in the physical sciences such as chemistry, physics, earth sciences, and astronomy, at what age do we begin formal teaching about the physical world, and what do we teach? What is our theoretical approach? Most importantly, how will this theory guide us in creating materials that foster optimal learning in the physical sciences? Finally, what types of instructional strategies should we use, especially for novices who have little or incorrect knowledge about the domain of physical science? These are the primary questions we set out to answer in our research program.

What Do We Teach and When?

The first question, "What do we teach and at what age?," is the most important and yet the most ignored by all but a few science education researchers (see Schmidt, Wang, & McKnight, 2005, for the importance of content; see Reif, 2008, for the importance of content and sequencing concepts). *What* we choose to teach children and *how* we organize this information determines over 50 percent of the variance in children's ability to learn and retain information (Stein, Anggoro, & Hernandez, 2009; Stein & Trabasso, 1992a,b). It is no surprise that American children perform poorly on international assessments of math and science, such as the Trends in International Mathematics and Science Study (TIMSS). Critical measurement units, such as the metric-based system, are rarely taught. Similarly, the active choice of omitting causal explanations that involve molecular or atomic theory when teaching concepts such as matter, weight, and density significantly impairs children's learning and retaining these concepts. Many of the difficulties children experience in science are not due to poor teaching, per se. Rather, difficulties are caused by the lack of exposure to core concepts and to causal explanations that would enable learners to retain the information presented to them.

Despite the importance of content and its organization, current approaches to science learning ignore the issue or speak to it only briefly. Even when content and organizational issues are addressed, criteria for creating optimally structured materials are missing. Criteria for evaluating content learning are also missing. Rather, the gold standard for assessing science learning has been the use of general achievement tests, such as the Iowa Test of Basic Skills (ITBS), math and science items from TIMSS, the

National Assessment of Educational Progress (NAEP), and state achievement tests such as the Illinois Standards Achievement Test (ISAT).

Although successful achievement on standardized tests *is* important, descriptions of what children have learned, the specific concepts they need to master, and the content of their misconceptions are critical for ensuring success in science learning. Teachers cannot function effectively unless they know specifically what children know, what they do not know, and the types of errors children make when thinking about scientific concepts. Given that elementary school teaching is becoming more regulated in terms of holding teachers accountable for children's progress and performance (Wyatt, 2009), we must be able to describe specifically what children are actually learning in the classroom, and we need to determine whether a clear relationship exists between the specific science concepts covered in the classroom, children's understanding of these concepts, and their performance on standardized tests.

The importance of content is not considered more seriously because the majority of researchers carrying out studies in science learning often hold one of three beliefs. The first is that existing text materials are sufficient for an accurate, long-lasting understanding of science. The second is that elementary school children cannot master the complexity and "abstractness" of science concepts. The third is that general instructional strategies can be used to teach any science content. Science researchers rarely consider or pay attention to the situated and content-based nature of learning (see Bransford, Brown, & Cocking, 1999; Brown, 1990; van den Broek & Kendeou, 2008, for the importance of the situation in producing "deep" understanding that lasts). That is, they believe that accurate understanding will result, regardless of content, as long as effective instructional strategies are used.

After carrying out a series of evidenced-based science learning studies, as well as an analysis of 15 different science texts from third grade through college, we have found serious problems with each of these beliefs. Science texts are rarely coherent across chapters, critical explanations are missing in every single text we have analyzed, and even the very best texts are generally missing 30–50 percent of the critical concepts necessary to understand and build a coherent representation of the text. Worse yet, because many textbook writers believe that elementary school children cannot understand core concepts, such as the molecular composition of matter, the most critical part of the text is purposely omitted [see the Tom Snyder Productions series (Scholastics) on evaporation and the lack of a detailed explanation about its causes and mechanisms].

Thus, we began our science learning studies with a rationale for choosing "what" is to be learned, with a theory and description of the coherence and organization of this content. We focused first on *states of water* and the molecular properties and processes that constitute each state of water (molecules in a solid versus liquid versus gas). We assessed children's pre-existing knowledge, what they learned during our study, and how their knowledge base changed as a function of instruction. We tested children in a verbal and visual mode. Creating an accurate, detailed, and situated assessment of whether and how children understood the material was essential, as it allowed a more accurate prediction of performance on learning new concepts and helped identify the requisite concepts that needed to be taught for successful mastery of the material.

Teachers cannot ensure effective learning over long periods of instruction without continual assessment of what the learner knows and how the learner's knowledge states have changed as a function of instruction. The longer the interval between assessments of learning, the more likely instruction will be ineffective. Effective knowledge assessments provide teachers with opportunities to change both the organization and content of instruction (see chapter 2 for an individualized approach to teaching reading).

Part of the problem in deciding "what" should be learned is that, until recently, learning about science has not been valued or required in elementary school. Even now, instruction in science is significantly less emphasized than instruction in reading and math. The discrepancy between science, math, and reading instruction is clearly shown in the amount of time school districts allow for science instruction. In the Chicago Public Schools (CPS), where we have conducted our science-learning studies, district administration recommends that only 120 minutes of instructional time per week be devoted to teaching science at the fourth-grade level. In comparison, CPS recommends 120 minutes for music and art, 175 minutes for social studies, 240 minutes for mathematics, and 645 minutes for language arts. Thus, science is given significantly less time (24 minutes each day) than either math (48 minutes) or language arts (129 minutes).

The teaching of science is heavily influenced by a widespread belief that children are unable to understand and retain "big or difficult ideas," especially physical properties and processes that are invisible and cannot be physically manipulated (Duschl et al., 2007; Krajcik, McNeill, & Reiser, 2008; Smith, Wiser, Anderson, & Krajcik, 2006; Smith, Wiser, Anderson, Krajcik, & Coppola, 2004). Further, some researchers (Smith, Solomon, & Carey, 2005) hold the belief that elementary school children's failure to learn certain quantitative and mathematical concepts, such as density and weight, or to comprehend the particulate nature of matter, prevents them from learning science concepts that focus on invisible properties and processes (e.g., molecular theory). This belief characterizes the status quo approach to physical science instruction in the elementary school years.

We have yet to discover explanations of matter, molecular movement, and its relationship to heat energy in any text at the fourth-grade level currently used in the CPS or any of the schools that we have described. To substantiate this claim, we examined 15 different science texts from third grade through college.[1] Four series that are used in some of the fourth- or fifth-grade classrooms in Chicago are: Tom Snyder Production's "Science Court" series, marketed by Scholastic; the "Full Option Science System" (FOSS) series, developed at the University of California's Lawrence Hall of Science; Macmillan/McGraw-Hill's "Science" series, developed in conjunction with the National Geographic Society[2]; and the "Science Companion" series, created by the Chicago Science Group (a team of STEM specialists primarily from the University of Chicago) and published by the Chicago Educational Publishing Company. In all of these series, either any mention of matter, molecular theory, thermodynamics, energy transfer, and explanations about the differences between weight and mass is missing or the topic is presented in a perfunctory fashion (see also Kali & Linn, 2009, and Schmidt, 2008, for a review on the lack of depth in elementary science texts).

How Do We Organize and Explicate Conceptual Information?

Failure to define critical concepts and include causal explanations in an elaborated, explicit fashion both destroy the causal coherence of science texts and decrease the probability that science concepts will be remembered and applied accurately (Stein & Trabasso, 1982a, 1989, 1992a; Trabasso & Bouchard, 2000, 2002). Similarly, failure to use a theory of concept learning that allows successful differentiation of core but confusable concepts impairs learning even more.

Examples of two physical science concepts that are continually confused are weight and mass. Were we to ask adults to describe the similarities and differences between weight and mass, the majority would not be able to do so. Most adults do not know the definition of mass. Even if they do, they often do not know that, for most purposes, what they consider to be weight is equivalent to mass, and that, to understand the differences between the two, they need to understand how gravity affects weight but does not affect mass. Thus, an important task that we need to accomplish is the development of a theory that will allow us to define, present, sequence, and differentiate among core concepts in physical science. The theory must also specify how the meaning of each concept is to be retained over time so that concepts are actually applied as each new situation arises.

An introduction to any of the domains of physical science entails learning a moderate but finite number of new concepts that are well specified. When learning these concepts, however, systematic efforts must be made to describe and compare concepts that are highly similar but still different in meaning. Many researchers working on reading and vocabulary acquisition argue for the repeated use of new words, especially in situations that would further differentiate the meaning of one word versus another (Beck & McKeown, 2007; Beck, McKeown, & Kucan, 2008; McCardle, Chhabra, & Kapinus, 2008). Currently, the repetition and use of core concepts are rarely tracked across situations. The impact of repetition and concept understanding across different situations needs to be studied and described.

The Importance of Measurement Concepts and Systems in Learning Science

Teaching children how different scientific concepts can be measured and quantified is also extremely important. In chemistry and physics, it is not enough to differentiate between two similar concepts, such as weight and mass, or volume and mass, and to show their similarities and differences across contexts. Advances in our understanding of physical events have occurred because we have been able to develop systems of measurement that answer the questions: How much? How heavy? How fast? How loud? How bright? That is, we have been able to use systems of measurement to determine the differing qualities of different substances.

When determining the quality of something, the unit of measure that allows for objective analysis, such as how much, how many, or how fast, is critical. These units of measurement need to be explained so that children understand the situations that led to the creation of each unit. In particular, children need to understand that units of analysis were created by people who wanted to be able to communicate quality in a more specific and agreed-upon fashion. Thus, we have units of measure for weight, in either newtons or pounds; for mass, in either grams or slugs; for volume, in either cubic centimeters or cubic inches, or in liters or gallons. These conventional measures were devised so that fellow scientists could study the same phenomena and communicate their findings to one another. The creation of a name for a unit is quite arbitrary, and often the unit is named after the scientist trying to achieve some form of quantification (e.g., the newton is named after Sir Isaac Newton), or the unit is named after a specified amount that corresponds to different parts of the body (e.g., in Portuguese, the name for inch is the same name as used for thumb). Thus, to understand quantification, children must gain a deep understanding of the origin, use, and creation of units of measurement.

Lack of knowledge about units of measure is one of the most significant obstacles to children's successful science learning. Elementary schools do not teach measurement well, and by middle school it is *assumed* that children possess an operational understanding of important measurement concepts. The reality is that children entering middle school have a poor understanding of measurement (see chapters 10 and 12), and are thus ill prepared to fully engage in science learning. The measurement of temperature, weight, pressure, light, distance, speed, and sound are all essential to gaining even a rudimentary knowledge of many physical concepts.

Further, we argue that teaching children measurement and not linking measurement to science concepts is another factor impeding children's learning. Currently, when measurement skills are taught, they are generally taught within the mathematics curriculum, rather than being integrated into the science curriculum. However, it is in the context of communicating to others that the necessity of conventionalized measures and quantification become most salient. Historically, it is in the carrying out of science that units of measure have often been created. Thus, the context in which measurement is taught, the experience of using different units of measurement, and the frequency with which one measurement concept is linked to another are important factors influencing physical science learning.

The Need for a Theory of Learning

The distinctions we make between a theory of learning and a theory of instruction are critical. A good theory of instruction focuses on *which* concepts should be learned, *how* these concepts should be organized, and *when* these concepts should be taught. To successfully address any one of these goals, however, we need a theory of learning that speaks to the ways in which learners process and remember incoming information. Such a learning theory needs to describe what is already known about the material to be taught, ways in which prior knowledge influences how incoming information is encoded, and ways in which newly learned information can be applied to novel situations (i.e., transfer of learning). Thus, a theory of learning describes what the learner knows before a task begins, the ways in which prior knowledge and the architecture of the mind constrain how the information is encoded and represented, and the ways in which new information is integrated with prior knowledge to form new cognitive structures.

As we argue throughout this chapter, texts and materials that are missing requisite causal information set up conditions for poor comprehension and poor retention

(Stein et al., 2009; Stein & Trabasso, 1982a). Worse yet, poorly structured materials often result in the generation of serious misconceptions that learners did not have beforehand. Several investigators have noted this lack of coherence in designing science materials (Kali, Linn, & Roseman, 2008; Krajcik et al., 2008; Reiser, 2004; Romance, Vitale, & Dolan, 2004; Schmidt et al., 2005; Stein & Trabasso, 1992b). All have called for a more coherent approach to designing science curricula. In order to achieve this goal, however, we need to describe the criteria by which we determine whether texts are in fact coherent. We also need a theory of learning to determine how understanding occurs and whether redesigned "coherent" units are effective. This theory will guide us in defining and measuring successful acquisition of scientific concepts.

A Theory of Complex Learning

Physical science includes many concepts, events, and mechanisms that are unfamiliar to learners, especially young children. Many of the core concepts and events of physical science are invisible to the human eye. Therefore, they need to be represented and modeled in order for a learner to form an accurate mental representation of the concept. Once we define and discover ways in which new concepts can be described and made accessible, we then need to illustrate how each of these individual concepts is embedded in and related to a larger framework of events and concepts that constitute the disciplinary knowledge of the physical sciences.

Our goal in describing a theory of complex learning is to identify the types of learning that occur in the physical sciences. We use the term "complex" because at least three different types of learning can be documented in any domain of science: concept learning, causal explanation-based learning, and argument learning (frequently called conceptual change). We discuss how these different types of learning can be integrated so that a model of science learning can be described and tested for its validity. By using a theory of complex learning, we will be in a better position to ensure that science concepts are accurately learned, retained, and used in future situations.

Concept Learning

Until now, the primary focus of coherence models has been on the description of causal relations among events and event networks (Graesser, Singer, & Trabasso, 1994; Rapp, van den Broek, McMaster, Kendeou, & Espin, 2007; Stein & Trabasso, 1982a, 1992a; Suh & Trabasso, 1993; Trabasso, Secco, & van den Broek, 1984; van den Broek & Kendeou, 2008). None of these models describes how concepts should be learned, nor do they articulate how concepts should be embedded in causal sequences. Although social and emotional understanding can be described fairly accurately with a causal explanatory model (Stein, Hernandez, & Trabasso, 2008; Stein & Trabasso, 1992a), the physical sciences and most other academic disciplines cannot. A model of concept learning (e.g., Klausmeier, 1992) is needed because physical sequences and events are contingent upon the assumption that all concepts embedded in causal sequences can be understood and identified.

Texts can communicate explicitly that two concepts are causally related. However, if learners do not understand the meaning of the concept, either they will remember the relationship only by rote memory or they will fail to retain the information for long periods of time. Unfortunately, the presence of rote learning is all too common, including in college-level general physics and chemistry courses. Students rely on rote memorization because the core concepts being taught are not described in the explicit detail necessary, the instructional time is far too short, and new concepts are not embedded in the contexts necessary for students to understand their meaning across situations.

Since the goal of science education is deep understanding of scientific theories for long-term retention and conceptual change, learning by rote memorization has been abandoned by educators at the elementary, middle-, and high-school levels, and replaced by a meaning-based approach to science learning (Chi, de Leeuw, Chiu, & LaVancher, 1994; Duschl et al., 2007). The value placed on understanding scientific knowledge is reflected in the tremendous emphasis placed on inquiry-based instruction in science [National Research Council (U.S.), 1996; Project 2061 (American Association for the Advancement of Science), 1993].

If a learner has never been exposed to the concepts under consideration, however, the text and sequence of instruction will lack coherence no matter how much effort is expended. The presence of coherence in a sequence of events is contingent upon the learner understanding and having prior knowledge about the concepts or events under consideration. If the events and concepts are not understood, then the causal relations that connect one event to another will not be inferred.

New concepts generally need multiple presentations, describing each critical dimension and the ways in which the concept is related to other concepts. Introducing a concept once or twice, even when all dimensions of the concept are described, is not enough to ensure understanding and retention. A concept needs to be presented in multiple situations so that the meaning of the concept can be inferred and checked every time the concept is used. These "rules" of concept learning are similar to the rules that regulate vocabulary acquisition in almost any domain (Beck & McKeown, 2007; Beck et al., 2008; McCardle et al., 2008).

The failure to discuss and use concepts in multiple situations is the single most frequent error that we have discovered in our analysis of current science texts. Yet the data on successful vocabulary acquisition show that the more frequently a word or concept is used, and the more situations it encompasses, the better the concept or word will be used and remembered. This variable needs exploration and testing in different scientific contexts. The meaning of a word may be grasped quite quickly in literary contexts, but physical science contexts may require much more variation and repetition to understand accurately the meaning of a concept. Illustrating to a learner which situations draw on common dimensions of two concepts and which situations draw on differences between two concepts is critical.

It should be noted that, when researchers construct ideal end-state models of physical science knowledge, they formulate these models as experts, not as novices. Thus, the resulting models often assume that learners possess knowledge that experts take for granted. Yet few learners possess this knowledge. The novice is dependent upon an explicit presentation of all features of the concept as well as all information that would disambiguate one concept from another. Our analysis of 15 different science texts revealed the overwhelming presence of poor descriptions for most of the concepts presented. Critical conceptual information (e.g., units of measure, explication of parallel processes during boiling, factors that influence rate of evaporation) was left out of every text series that we examined. Some were better than others, but even the best was poor by our standards of explicitness and unambiguous presentation.

Two researchers who actively speak to the representation of conceptual structures and their coordination with event structures are Mandler (1984) and Reif (2008). Mandler discusses conceptual structures and event representations in general, pointing out the differences and similarities in each type of structure. She also describes the conditions under which concepts and events are best learned. Reif specifically discusses concepts in the context of scientific learning, reviewing different structural representations and complexities of learning in a scientific domain. Other researchers (e.g., Keil, 2005) discuss concepts and causality, but not representations of causal structures that are involved in the learning.

The most elaborated model of concept learning comes from the work of Klausmeier (1992) and Winston (1986). In our model of complex learning, we have incorporated many of the principles laid out by Klausmeier, especially his description of the requisite knowledge that needs to be taught to a student learning a new concept. Three things are mandatory when introducing a new concept: (1) an explicit delineation of each component that defines the concept so that no critical component is left out, (2) an effort to compare and contrast any two concepts that are highly similar so that confusion errors do not occur, and (3) an explicit focus on the dimensions not included in a concept so that errors of over- or underinclusion do not occur.[3] Although science educators have not considered theories of learning when teaching new concepts, cognitive psychologists have considered explicit contrasts in conceptual learning (see Gentner & Namy, 1999, and Namy & Gentner, 2002, for the role of comparison in children's learning of conceptual structures). We believe that, in science learning, highly similar concepts especially require explicit comparison.

The dimensions of each new concept also demand their own explicit instruction. For example, when children are taught that matter has mass, weight, and volume, it is imperative that children understand the meaning of each of these three subconcepts. When we define weight as the force that the earth exerts on the mass, or more specifically the force of gravity on the mass, we need to ensure that children understand the concepts of mass, force, and gravity. These concepts need to be presented in several different contexts. More important, children need to learn how to represent and *model* these concepts in different contexts. The same holds true for the concept of energy. Thus, an analysis needs to be carried out about the requisite prior knowledge needed to understand each concept. Children receive very little instruction about matter, energy, and the different elements that make up the solar system. They are rarely taught about the similarities and differences between the composition of the materials that make up Earth and other planets in the solar system.

Causal Explanation-Based Learning

The second type of learning that occurs in science is causal understanding, especially focused on explanations of the properties and mechanisms that underlie physical properties and processes (Stein & Levine, 1989; Stein & Trabasso, 1982a, 1992b; Trabasso & Stein, 1997). Such learning is called causal explanation-based learning or explanatory coherence. Important concepts in physical science cannot be fully understood without being embedded in a physical event sequence that describes how concepts and events are causally related to one another.

It should be noted that, historically, models of causality and explanatory coherence came from descriptions of natural language understanding (Winograd, 1980), the understanding of action events (Schank & Abelson, 1977), the study of the organization and representation of narratives (Mandler & Johnson, 1977; Rumelhart & Norman, 1978; Stein & Glenn, 1979), the causal nature of narrative understanding (Stein & Trabasso, 1982b, 1992a; Trabasso et al., 1984; Trabasso & Stein, 1997), and the fact that social and emotional understanding can be described by a model of goal-directed action governed by rules of explanatory causal coherence (Liwag & Stein, 1995; Stein et al., 2008; Stein, Trabasso, & Liwag, 1993; Trabasso, Stein, & Johnson, 1981).

Models of explanatory coherence assume that specific causal structures exist in the mind and that these causal structures are used to understand the ways in which events in the universe function and are related to one another (Mandler & Johnson, 1977; Rumelhart & Norman, 1978; Schank & Abelson, 1977; Stein & Glenn, 1979; Stein & Trabasso, 1982a). Classically, these models have been used to explain how children understand narratives (N. S. Johnson & Mandler, 1980; Stein & Glenn, 1979; Trabasso et al., 1984; Trabasso & Stein, 1993), how they understand emotion and intentional action in themselves and other people (Stein & Levine, 1989), and how children and adults use cause and effect to understand proximal and distal causes, and predict future actions and events. Causal explanatory theories, however, can be used for learning in all domains. Structuring information in a temporal and causal fashion is an inherent type of cognitive representation (Carey, 2009; Mandler, 1984; Trabasso et al., 1981), and occurs in everything we do.

Thus, the study of coherence and its emphasis on causal thinking and development has been ongoing for over 30 years in multiple domains. Using models of coherence is not a new development, nor is its application new to learning and instruction. What is new is that researchers in science and mathematics instruction have discovered aspects of this theory (Linn, Lewis, Tsuchida, & Songer, 2000; Reiser, 2004; Shwartz, Weizman, Fortus, Krajcik, & Reiser, 2008) and have used coherence as a rubric for creating materials and methods of instruction that are more comprehensible.

In describing the learning that is necessary for a beginning understanding of the physical sciences, we focus first on concepts that are critical for specific physical events. We then use a causal explanation-based theory that allows us to embed concepts in causal sequences so that we can describe the causes and mechanisms that regulate and control physical events. This theory also helps us specify whether a strict temporal-causal sequence needs to be laid out during the learning of new concepts, or whether the concepts and events can be taught in different temporal sequences.

A theory of causal explanatory coherence also allows us to determine when a violation of causal sequencing occurs during instruction or when critical information has been left out of the visual or verbal description. Breaking a causal chain occurs when we fail to include explicit reasons for introducing certain topics, when we do not provide explicit descriptions that illustrate how concepts are related to one another, when we omit explanations for the physical events under consideration, and when we introduce a topic or concept at the wrong time (Stein & Ornstein, in preparation; Stein & Trabasso, 1982a, 1992a; Trabasso & Bouchard, 2000; Trabasso et al., 1984). For example, when we explain the melting of solid water or the boiling of liquid water, we discuss the role that temperature plays in regulating the changes in water and causing water to change states. We also describe that temperature measures the amount of heat energy produced by the speed and movement of molecules. Describing and making explicit the mechanisms that cause state changes in matter, as well as creating a causally coherent sequence of events to describe the phenomena, is mandatory.

Further, a theory of causal coherence guides us in determining whether or not adequate information about the causes of a state change have been given and how "deep" the explanation goes. For example, consider the question, "Why does water evaporate?" One explanation focuses on the fact that heat energy gets absorbed by water molecules. When enough heat energy is absorbed, the liquid molecules change to a gaseous state and escape into the air. Why heat energy changes water from one state to another, however, also needs to be explained. Learners need to understand that, the more energy molecules have, the faster they move. When molecules have enough heat energy, they move fast enough to break away from each other and turn into a gas, and these gaseous particles in turn become absorbed into the surrounding air. A series of "why" questions can be asked, where each question is prompted by the answer from the previous question. Explanations become more rigorous as the number of events on the causal chain for a series of "why" questions increases. When compared with a superficial explanation with one level of description, this form of questioning results in more thorough explanations and deeper understanding.

Argument Learning

Argument learning provides the basis for correcting misconceptions and errors in reasoning. Argumentative discourse requires the presentation of evidence for and against a particular belief or stance. With a model of argument, we can describe the reasons for learners' incorrect beliefs about a concept, in conjunction with assessing whether they have any correct knowledge about the concept.

When learners have misconceptions about a specific event or concept, providing only explicit feedback about the incorrectness of their beliefs rarely works (Bernas & Stein, 2001; Chinn, 2006; H. M. Johnson & Seifert, 1994; Vosniadou, 2007). What does work is providing information about the *opposite* point of view—specifically, the reasons for supporting the opposing point of view or belief. Presenting evidence against a learner's current beliefs, in conjunction with evidence supporting an opposite stance, is often very effective in increasing the conceptual knowledge of a learner and changing the learner's stance (Diakidoy, Kendeou, & Ioannides, 2003; Nussbaum & Novick, 1982; Smith, 2007; Smith, Snir, & Grosslight, 1992; Thagard, 2000; Vosniadou, 1994).

The change in stance will come, however, only when more or higher-quality reasons for the opposite stance are acquired in comparison with the reasons supporting a pre-existing position. Thus, when conceptual change does occur, it is a result of acquiring critical new information about a new position that convincingly refutes evidence for the previous position. Neither children nor adults perform well in abandoning or revising incorrect prior stances until they learn a new stance in its entirety, and compare the new stance with the old favored stance (Stein & Bernas, 1999; Thagard, 2000).

Teaching Physical Science Using a Theory of Complex Learning

We began the construction of a series of learning studies by focusing on water, its visible states (solid and liquid) and its invisible state (gas). Our goal was to teach children the three states in which matter exists, the differences between the three states, and the mechanisms that allow water to transition from one state to another. Fulfilling this goal meant that we had to teach children about invisible properties and processes, and the factors that regulated or influenced these invisible processes. We also had to teach them about complex causal mechanisms that entailed instruction beyond the simpler necessity–sufficiency causal relations that are prototypical of most studies on causal understanding. For example, in teaching children that the temperature of boiling water stays constant at 100°C, we had to describe the mechanisms that allow a boiling point to stay constant. This was particularly important since the majority of children and adults believe that adding heat to boiling water causes the temperature to continue to rise above the boiling point. Even though children and adults can verify that a thermometer stays at 100°C during boiling, most have no clue to why this occurs.

In building our science learning materials, our first task was to construct a conceptual and causal graph of the information we wanted our learners to acquire. This *end-state model* described what conceptual structure we wanted the learner to possess after completing the instructional sequence. The end-state model defined what concepts were taught and in what sequence they would be introduced. By creating our science learning material based on an end-state model, we were able to accurately assess what concepts children understood, did not understand, and needed extra help mastering.

We chose to focus on water for several reasons. First, water is unique among the elements in that it occurs in all three states naturally on the earth. Second, water is necessary for all living things. Third, the Celsius scale was created by using the freezing and melting points of water as its anchors, and then units of analysis were created to reflect a measure of the amount of heat energy in the different states of water.

The use of the Celsius scale provided us with the opportunity to teach children the idiosyncratic nature of measurement, the fact that a person created the unit of measure, and that its use depended upon whether the unit was helpful when communicating among a larger group of people. When a unit of measure is accepted and used by a group of people, we then say that the unit has become "conventionalized," which means that it carries a specific definition and meaning that everyone agrees upon, and that it is used to convey important information about how the physical world functions.

We also chose water because it is central to our understanding of weather. Water is classically introduced to children within the context of the water cycle or hydrologic cycle. Curricula use different parts of the water cycle to educate children about the importance of water. Thus, the centrality of water influenced our choice of domain. Centrality is critical to any theory of science learning. Two features define a concept as "central": the number of connections it has to other ideas, and its position in the causal hierarchy of ideas. To determine an idea's centrality, both of these factors need to be considered, and a causal graph needs to be created showing how all ideas in a text or lesson are related to one another, especially in a causal manner. Then a text needs to be written that makes all causal relationships explicit, especially if the learners are novices and have no knowledge of the central concepts.

The study of water further facilitated our long-range goals of teaching children about global warming, sources of energy, and the production of electricity. For example, nuclear power plants are significant generators of electricity, and they use massive amounts of water in all three of its states (i.e., solid, liquid, gas) to produce electricity. Thus, in choosing our concepts and themes, we first analyzed the connectedness that each concept had to other concepts and the logical dependencies that these concepts had upon one another. The number of connections a concept has to other concepts, and whether or not a strict causal order was necessary in presenting a sequence of concepts, were the two primary criteria we used in selecting concepts to be taught.[4]

We chose to focus on the molecular level and thermodynamics so that we could provide causal descriptions of the three states of water (i.e., solid, liquid, gas), based on the relationship between heat energy and the organization, speed, and movement of molecules. Molecules are the stuff of matter and the smallest, most basic part of a chemical compound. Molecules are not the smallest bits of matter, but they are the defining characteristic of all compounds. The inclusion of a molecular-level explanation results in greater coherence among concepts in the water cycle (e.g., how water transitions from a liquid to a solid or from a liquid to a gas), than when it is omitted. Describing the molecular organization and structure of the three states of water also allowed more accurate descriptions of the properties of a gas than those that acknowledge only that gas is invisible to the human eye.[5]

Given that we chose to describe the organization, speed, and movement of molecules, we needed to define and describe heat energy in order to explain its role in regulating these molecular properties. The amount of energy absorbed by molecules regulates their organization, speed, and movement. In turn, the organization, speed, and movement of molecules define the different states in which matter exists. Thus, we introduced children to the concepts of matter, molecules, energy, scales of temperature (that measure heat energy in a substance), freezing and boiling points, and the processes of melting, freezing, boiling, evaporation, and condensation.

We focused on changes or lack of changes in the shape and volume of the three states of water. We also focused on the constancy of temperature as water boils, as well as the constancy of temperature as water freezes. As a result of describing the nature of energy and its measurement on scales of temperature, we could describe in detail the relationship between heat energy and molecular speed and movement, and how this relationship regulates and defines the state in which water exists (solid, liquid, gas).

Modeling the Invisible

Although molecules are invisible, molecular properties and processes can easily be made visible by modeling them concretely, and by describing the processes in detail. When physical properties of substances are made visible, the phase changes in compounds and elements become clear. Visual models clarify and disambiguate physical events, especially in regard to the parallel and temporal properties of the events. When models are used in the proper context, experts as well as novices profit from their use (Gobert, 2004; Goldberg & Bendall, 1995; Jose & Williamson, 2005; Mayer,

Hegarty, Mayer, & Campbell, 2005; Stein et al., 2009; Tversky, Heiser, Mackenzie, Lozano, & Morrison, 2008).

The use of visual models to describe and explain invisible physical processes is very effective in teaching adults, and it is just as effective in teaching elementary school children. The clarity of a visual model as well as the provision of an accurate verbal description that maps onto the unfolding physical event provides children with an "accessible" platform from which to understand more complex scientific phenomena. Goldberg (Goldberg & Bendall, 1995), Williamson (Jose & Williamson, 2005), Mayer (Mayer, 2001; Mayer et al., 2005), Gobert (Buckley et al., 2004; Gobert, 2005a,b), and Tversky (Tversky et al., 2008) have all argued for the use of visual models, especially in situations where models convey the dynamic unfolding of physical events.

The additional questions for us are the following: In regard to the physical sciences, what information does a visual model communicate that we cannot communicate with words alone? If visual models increase performance, how much of their role is due to their static versus dynamic nature? How much is due to their conveying an explicit representation of causal sequencing and the existence of parallel processes?

Our studies tested the hypothesis that visual models would be critical for an accurate representation of molecules and their movements, given that children have little if any knowledge about molecules. The images of molecules we created were based on the particulate theory of matter. Visual models allow for the simultaneous representation of speed and movement, as well as the organization and spatial proximity of one molecule to another. Even though molecules can be very close to one another, they never touch each other, because of the attractive and repulsive electromagnetic forces of atomic nuclei and electrons. The distance between molecules and the structures in which they can be arranged, however, do vary as a function of the amount of energy molecules posses and the state of the matter. Thus, not only do we need to model molecular speed and movement, but we also need to model the spatial relationship between molecules.

By observing the speed and movement of molecules, as well as their spatial proximity to each other, in the three different states of water, learners can encode all three features at once. A verbal description, by definition, is constrained temporally by introducing one dimension at a time. If a learner has no prior knowledge of the physical event, it is extremely difficult to convey through a temporal sequence of words how three factors co-vary at the same time. Further, gaining a sense of relative molecular speed and movement in each of the three states may require a simultaneous comparison of these dynamic properties across all three states. By providing a graphic comparison of the three states, the absolute and relative rates of molecular speed movement can be encoded for each state.

Alternatively, when learners are familiar with the physical concepts, such as observable changes in shape and volume, visual models may not be necessary. Also, the elucidation of some physical events may require static rather than dynamic graphics (if at all). Thus, the questions we raise about visual models focus on the situated appropriateness of these graphics in conveying specific types of information. Graphics should be most effective when they illustrate information that cannot be easily conveyed in a verbal description, especially information that is unfamiliar and has never been seen before. Thus, our goal was to describe the conditions where graphics were necessary, and the conditions where graphics were not needed.

Science Learning Studies

The most important reason that our children do not understand science is that we have not taught them. This problem is endemic to the entire population, as most adults also were poorly educated in physical science. In fact, the majority of elementary school teachers who are currently trying to teach children science possess very limited backgrounds in physical science. Thus, in order to make real progress in educating children in the physical science, both children *and* their teachers will need to be trained. The studies we summarize in this chapter focus on fourth-grade children and their teachers learning the physical and chemical principles underlying states and state changes of water. Our learning modules taught a beginning understanding of molecular theory. In the fourth-grade study, we manipulated the presence or absence of visual models that map onto and unfold "invisible" molecular processes that cannot be seen by the human eye.

We also recorded and analyzed the entire verbal output that our learners generated. We encouraged children to ask questions about the text and visual models, to point out inconsistencies in the information provided, and ask for further explanations about the physical events that were described. Beck and colleagues (Beck & McKeown, 2007; Beck et al., 2008) have used a similar procedure to improve reading comprehension. Their procedure was designed so that students could pose questions to the author of the text during the process of reading and comprehension. They also wanted students to understand that writers can be fallible and leave information out of a text. Our request to have children think aloud and tell us what they did and did not understand during learning is similar. In addition, we wanted to identify problems in understanding so that we could correct our materials and make them more comprehensible.

We focused on a characterization of children's difficulties as well as an analysis of their learning. By developing a taxonomy of children's responses to our assessment questions, including their errors and difficulties during learning, we could discern what concepts were learned, what concepts proved difficult, and where children focused their attention during learning. We know from past studies that learners impute meaning to information in ways that instructors do not intend (Beck & McKeown, 2007; Beck et al., 2008; Stein & Trabasso, 1982a; Trabasso & Bouchard, 2000; Trabasso, Suh, Payton, & Jain, 1995). Children's questions and thoughts about incoming information predict whether or not the information is encoded accurately. Their causal inferences during instruction also predict whether a coherent structure has been formed during the presentation of new information, or whether they have difficulty because they cannot link incoming information to what they already know (Stein & Trabasso, 1982a; Trabasso & Bouchard, 2000).

We know from previous studies that spontaneous causal explanations and causal connections between concepts presented in a text or curriculum are crucial for accurate understanding (Chi et al., 1994; Ornstein & Trabasso, 1974; Siegler & Ramani, 2008; Stein & Glenn, 1979; Stein & Trabasso, 1982b; Trabasso, 2005; Trabasso et al., 1984). Those learners who seek and describe causal mechanisms better understand the novel content with which they are presented. Making an effort to explain and understand causal relationships among ideas is a necessary component of scientific reasoning. This is as true for adults as it is for children (Trabasso, 2005). Without an analysis of learners' questions, explanations, and interpretations, we cannot conclude that the materials used during instruction were coherent to the learner (Stein & Trabasso, 1992a,b; Trabasso, Stein, & Johnson, 1981). Coherence is subjective and depends upon the learner's prior knowledge and ability to use this knowledge. Unless we actually carry out a test of how science content is encoded, represented, and stored, we have no evidence for the coherence in the learner's mind or for the ease of learning the material.

Learners, no matter what age, are always active interpreters of incoming information (Bartlett, 1932; Magliano, Trabasso, & Graesser, 1999; Mandler, 1984; Stein & Trabasso, 1982a, 1992a). In the process of interpreting incoming information and making sense of new concepts, learners revise and reconstruct the input in order to relate it to what they already know. A description of the process of "meaning making" and understanding has been one of the most important contributions that cognitive science has made to the study of learning and memory. The process of meaning making affects every aspect of understanding and retention (Bartlett, 1932; Mandl, Stein, & Trabasso, 1984; McKeachie, 1995; Mandler, 1984, 1998; Rumelhart & Norman, 1978; Simon, 1995; Stein & Glenn, 1979; Stein, Ornstein, Tversky, & Brainerd, 1997; Stein & Trabasso, 1992a,b; Stein, Trabasso, & Liwag, 1993). Given the importance of prior knowledge in predicting meaning making and understanding, it is important to describe the content of prior knowledge that gets activated, as well as the content of the new structures that emerge.

Even though much of understanding is not available to conscious consideration, the products of understanding in an *on-line* situation can be described. By asking learners to explain what they understand about the concepts being introduced, and by asking them to identify what is difficult for them to understand, we can create a detailed record of how learners think they have understood the material *during* the process of learning. Asking children to identify problems during learning provides a more detailed and focused assessment than pre- or post-test measures can provide (Magliano et al., 1999; Stein & Trabasso, 1982a, 1985, 1992b; Trabasso & Bouchard, 2000, 2002; Trabasso & Magliano, 1996; Trabasso, Suh, Payton, & Jain, 1995). Children are quite good at describing what they do not understand during learning, and they are also good at identifying information that is missing from the text (Beck & McKeown, 2001; Mandl et al., 1984; Stein & Trabasso, 1985). Indeed, in our science learning studies, where we recorded the entire course of learning, children continued to ask questions and comment on the materials whether we wanted them to or not.

The ways in which we carried out on-line evaluation are different from the metacognitive procedures that Wiley uses (see chapter 4), and they are also different from most inquiry-based procedures that current textbook series use (e.g., the Tom Snyder series or the Chicago Science Group's series). Even our youngest learners identified concepts that were difficult to learn as they proceeded through a text, and they talked about what they knew and did not know. When children had the materials in front of them, they spontaneously asked questions, told us what they did not know, and asked for additional information, even when we told them we could not give them any additional information (Stein et al., 2009). Thus, asking children to become "critics" of a text is doable and *very* productive. Their ability to estimate what they knew and did not know was quite accurate, *if* they were allowed to ask questions and make corrections as the material was presented, rather than being forced to make assessments before and after reading.

The Importance of Training Teachers

In a thorough review of course requirements for elementary education majors at top universities and colleges in the United States, we found that prospective American elementary school teachers are required to take an average of only two science courses to complete their degrees. The most recent TIMSS report on the educational backgrounds of fourth-grade science teachers showed that significantly fewer teachers in the United States (12 percent) have majors or specializations in science than in other countries (37 percent) (Martin, Mullis, & Foy, 2008). The contrast is even more apparent when examining the seven countries that outperformed the United States on the 2007 TIMSS, where 46 percent of teachers had a science major or specialization. Given American elementary school teachers' lack of formal training in science, children's mediocre performance on international science assessments is not surprising. Expertise in science is not valued in the U.S. as much as it is in other countries. Thus, when elementary school teachers are asked to teach basic concepts in the physical sciences, we should not expect them to have much more knowledge than their students.

The question we asked throughout our teacher study is this: Can we teach teachers, as adult learners, enough about basic concepts in the physical sciences so that they can become good teachers to elementary school children? Can they learn the material well enough so that they can be better science instructors, and less dependent on a particular science series or science basal? Can they learn enough about the fundamental measurement concepts underlying physical science to illustrate the importance of measurement and mathematics in learning the physical sciences?

As we show from the results of our studies, the lack of science understanding is due not to age or incompetency, but rather to a lack of explicit instruction, introduced in a coherent manner, over time. The lack of explicit instruction as a cause of misunderstanding has been noted by many different researchers over the last 10 years (Klahr & Nigam, 2004; McNamara, Kintsch, Songer, & Kintsch, 1996; Mayer, 2004; Romance & Vitale, 2006; Schmidt, 2008; Schmidt et al., 2005; Taber, 2006; Trabasso & Stein, 1997), and was also noted by Cronbach (1966), Shulman (1986), and Shulman and Keislar (1966) in the 1960s.

We now describe two experimental studies in which fourth-grade children and their teachers underwent systematic instruction in the physical sciences. Over the course of eight weeks, learners were instructed in the core concepts and events required to understand three things: (1) the nature of physical states, (2) the observable as well as molecular changes that occurred as water transitioned from one physical state to another state, and (3) the ways in which heat energy regulated the organization, speed, and movement of molecules.

Our goal was to determine whether fourth-grade children could learn about observable and molecular properties of water. We hypothesized that three factors would significantly influence their learning: (1) the causally coherent nature of the learning sequence, (2) the presence or absence of graphics that modeled the observable and molecular properties of the three states of water, and (3) the static or dynamic nature of the graphics accompanying the text.

Study 1: Children's Learning about the Three States of Water

The following features characterized Study 1. First, the entire instructional sequence was presented on a computer in a one-to-one tutorial fashion, to control for the content presented. We compared children in a Control group with children who heard a tutor-read presentation of the text: (1) without graphics (No Graphics), (2) with static graphics (Static Graphics), and (3) with dynamic graphics (Dynamic Graphics). The results focused on (1) accuracy in describing the *observable* properties of water—changes or constancies in shape and volume as solid or liquid water was transferred from one container to another—(2) accuracy in describing and drawing *molecular* properties of the three states of water (i.e., the organization, speed, and movement of molecules), and (3) accuracy in *transferring* (generalizing) state knowledge to a novel substance other than water. We also measured the relationship between vocabulary skill and learning as well as the relationship between spatial ability and learning.

At pretest, children across all conditions answered accurately 56 percent of the questions about *observable properties* of water (i.e., shape and volume). At post-test, a main effect of Condition was found [$F(3, 133) = 5.78$, $p < .001$]. Children receiving tutor-read input of any type answered correctly 77 percent of the items compared with children in the Control group (54 percent of items). No effect of graphics was observed. Children in the No Graphics condition performed as well as children in the Graphics conditions. Thus, for familiar observable dimensions of shape and volume, which children have seen many times, children in all three experimental conditions improved from pre- to post-test, even those who did not see visual models.

When presented with *molecular properties* that were unfamiliar, however, visual models were critical. At pretest, fourth-grade children had no knowledge of molecules; they were able to answer only 3 percent of all questions about molecular properties of water. At post-test, a main effect of Condition was found [$F(3, 133) = 35.64$, $p < .001$]. Children in all three tutor-read conditions outperformed children in the Control condition, who could answer only 7 percent of questions accurately at post-test (see Figure 7.1). Children in the Static and Dynamic Graphics conditions significantly outperformed (83 percent correct) children in the No Graphics condition (60 percent correct). Children who saw dynamic graphics performed better (88 percent correct) than those who saw static graphics (79 percent correct), but the difference was not significant.

An analysis of children's drawings of molecular properties corroborated the verbal accuracy data. Significantly more children in the Graphics conditions (90 percent) than in the No Graphics (50 percent) or control conditions (17 percent) were able to draw distinctive molecular representations of the three states of water [$F(3, 133) = 24.29$, $p < .001$]. An analysis of the accuracy of children's drawings showed a significant main effect of Condition, $F(3, 133) = 7.03$, $p < .001$. Children in the Static Graphics condition (63 percent accurate) and Dynamic Graphics condition (54 percent accurate) outperformed children in the No Graphics condition (40 percent) and the control condition (22 percent accurate) ($p < .001$ in each case). These findings showed that presenting children with visual graphics of molecular properties during instruction led to better comprehension than presenting instruction without graphics.

For the *transfer task*, in which children identified the state of a new substance,

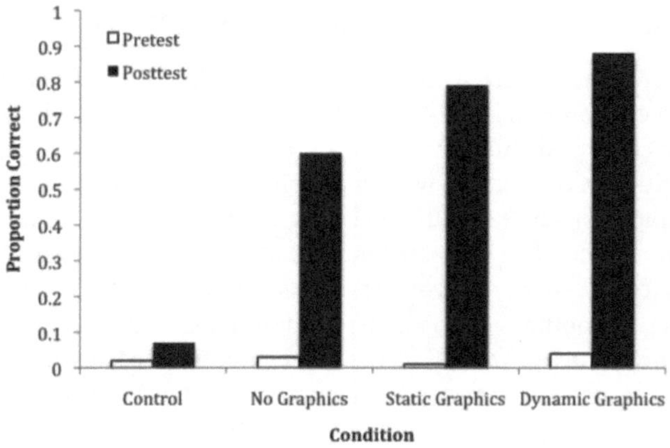

Figure 7.1 Children's Accuracy on Molecular Properties of Water.

a main effect of Condition was observed [$F(3,133)=4.14$, $p<.01$]. Children in the Graphics conditions significantly outperformed (57 percent) children in the No Graphics condition (35 percent), who significantly outperformed children in the Control condition (24 percent). The transfer data showed that using graphics during learning leads to the most successful transfer of knowledge to a new subject.

Correlations between vocabulary scores and performance on the science modules were highly significant, but only under certain conditions. The strongest correlations between vocabulary and performance were found in the No Graphics condition ($r=.65$, $p<.001$), which required children to construct their own visual models when none were presented. No significant correlations between vocabulary skill and performance were found for the Control ($r=.38$) and Graphics conditions ($r=.26$).

The only significant correlation between spatial ability and learning molecular properties was in the No Graphics condition ($r=.56$, $p<.01$), where children had to generate their own visual models to fully understand the material. Thus, when children are presented with visual models and can focus on the visual content for a guide, the relations between vocabulary and subsequent performance disappear. When critical information is missing or difficult to process, vocabulary and spatial skill become more important for learning.

Study 2: Teachers' Learning about the Three States of Water

Teachers completed the same sequence of tasks as the children, with the exception of the vocabulary and spatial tests. All teachers received the causally coherent text accompanied by dynamic graphics, and read the text to themselves. The learning modules and assessments were presented on the computer via the internet. The presentation software was programmed to control for the pacing of sessions (one session per week, comparable to the children). Teachers were also asked to write weekly logs about their learning experience.

At pretest, the mean proportion of molecular questions that teachers could answer correctly was 32 percent. That is, teachers answered 68 percent of the questions about the molecular properties of water incorrectly. In their weekly logs, almost all teachers stated that this was the first opportunity that they had received to become a learner of these particular science concepts, despite the fact that the state required them to teach about states of water. At post-test, teachers' average score on molecular content was 87 percent, identical to the average score (88 percent) of fourth-grade students in the Dynamic Graphics condition, [$F(1, 70) = .00$, ns]. Teachers' average post-test score on the observable properties of each state of water was 89 percent, which was marginally higher than fourth-grade students' average score of 79 percent, but the difference was not statistically significant ($F < 3.95$, ns). The only time teachers statistically outperformed their fourth-grade students was on the transfer task [$F(1, 70) = 16.16$, $p < .001$]. Teachers' average post-test score on this task was 93 percent, whereas fourth-grade students' average post-test score was 52 percent.

The reason for the difference between teachers and children on transfer items may have been the nature of the assessment. First, teachers could read the post-test items more than one time because all post-test items were completed on the internet and controlled by the teachers. If a teacher did not fully understand the question after the first read, she could easily read over the post-test item again or write down notes until she gained a better understanding. Children were read the item once, by a tutor, and rarely requested a second reading. Thus, the differences could be due to impoverished encoding, poorer memory for the question, or incomplete comprehension of the transfer items on the children's part, rather than an inability to generate the correct response given a verbal list of the components for each state.

Conclusions

Children successfully learned about the organization, speed, and movement of molecules in three states of water. The findings were robust in that children were post-tested four to five weeks after completing the learning modules. Under optimal conditions, in which children received tutor-read Dynamic Graphics input, children were overwhelmingly successful in learning about the molecular properties of water, attaining accuracy scores of 88 percent at post-test. The results showed that the presence or absence of visual models significantly influenced children's understanding of molecular properties. When presented with the same well-constructed texts, children who saw visual models performed significantly better (83 percent) than children who did not see visual models (60 percent). Children in the Control condition continued to perform poorly (at 7 percent), which was not surprising since molecular theory was not a part of the fourth-grade curriculum.

Of critical importance was the finding that children were able to learn about molecular properties as well as and slightly better than they learned about observable properties of water (88 percent vs. 79 percent accuracy, respectively). These results showed that focusing solely on observable-level descriptions and ignoring the molecular properties of water is a faulty criterion to use when designing curricula for fourth-grade children. Children's spatial skills were not correlated with learning when visual graphics were present. Vocabulary scores were significantly correlated with performance, but only when children had no graphics to augment their understanding of the material being read to them. Vocabulary was not significantly related to children's learning when any type of visual model was presented during the encoding for understanding.

The importance of presenting concepts in an explicit, organized, oral form must not be underestimated, in terms of both assessing children's knowledge and increasing children's conceptual understanding. Although reading and becoming literate is exceedingly important, so is using other modalities to learn about fundamental physical science concepts and theories that explain the physical world. However, it is essential that we begin to consider science as a domain of reading. In fact, one of the most important questions that can be answered about the relationship between science learning and reading focuses on the conditions in science learning that lead to best understanding when children have to get information from a written text.

We know from two of our ongoing studies that, in the initial phases of science learning, children learn better when they can listen to the material being presented and when they have visual graphics in front of them. Asking children to read with and without visual graphics decreased performance significantly, when compared with oral conditions that use graphic material. Most important, having children read the text, even with the graphics present, decreased performance more than not allowing children to have access to the graphics. Comprehension is at 84 percent when a tutor reads the text to a child versus 40 percent when the child reads the text. Comprehension without graphics in a tutor-read situation is at 60 percent. The type of input (self-read vs. tutor-read) accounted for 18 percent of the variance while the presence or absence of pictures accounted for 6 percent of the variance.

Thus, the role of reading in learning science needs careful delineation so that the optimal conditions combing science learning and gaining meaning from a text can be discovered. Most important, science learning needs to be made accessible to all children, as does being able to construct an accurate representation of text material when children do the reading. It may be that ensuring that children have a correct nonverbal representation of the text is critical to ensuring that children abstract the meaning from text in an accurate fashion. It may also be critical that explicit teaching of core concepts is mandatory before children can show progress in gaining an accurate representation of a text. In this sense, we are plotting out the conditions under which children can learn from text and conditions where they will never achieve an elaborated accurate knowledge base related to the core concepts in science.

Acknowledgments

The writing of this chapter was supported in part by NSF Grant No. 0529648 to Nancy L. Stein and by a grant from the Spencer Foundation to Nancy L. Stein.

Notes

1 The list of texts we have analyzed is available. Contact Nancy Stein at: n-stein@uchicago.edu. The original texts were analyzed in 1989–1990. A more extensive analysis was carried out in 2003–2005. We are now in the process of analyzing new science readers that have emerged since 2005, in particular the new McGraw-Hill series.

2 Macmillan/McGraw Hill no longer collaborates with the National Geographic in Macmillan/McGraw Hill's new science series, which was marketed after 2002. Many of the CPS schools, however, use an even older version of the series.
3 Clark (1993) as well as Bloom (1997; Bloom & Markson, 1998) speak to many of the issues that children face in acquiring the meaning of new words. Their work on contrastive procedures to differentiate similar concepts is especially important in concept learning tasks in science.
4 In addition to reviewing existing curricula for third grade through college, we also worked with two physicists, one chemist, and two mathematicians in choosing our content and refining the causal structure of our materials.
5 See Taber's (2006) defense of starting with the molecule rather than the atom when introducing core concepts in chemistry.

References

Bartlett, F. C. (1932). *Remembering: A study in experimental and social psychology.* Oxford: Macmillan.

Beck, I. L., & McKeown, M. G. (2001). Inviting students into the pursuit of meaning. *Educational Psychology Review*, 13(3), 225–241.

Beck, I. L., & McKeown, M. G. (2007). Increasing young low-income children's oral vocabulary repertoires through rich and focused instruction. *The Elementary School Journal*, 107(3), 251–271.

Beck, I. L., McKeown, M. G., & Kucan, L. (2008). *Creating robust vocabulary: Frequently asked questions and extended examples.* New York: Guilford Press.

Bernas, R. S., & Stein, N. L. (2001). Changing stances on abortion during case-based reasoning tasks: Who changes and under what conditions. *Discourse Processes*, 32(2–3), 177–190.

Bloom, P. (1997). Intentionality and word learning. *Trends in Cognitive Sciences*, 1(1), 9–12.

Bloom, P., & Markson, L. (1998). Capacities underlying word learning. *Trends in Cognitive Sciences*, 2(2), 67–73.

Bransford, J. D., Brown, A. L., & Cocking, R. R. (Eds.) (1999). *How people learn: Brain, mind, experience, and school.* Washington, DC: National Academies Press.

van den Broek, P., & Kendeou, P. (2008). Cognitive processes in comprehension of science text: The role of co-activation in confronting misconceptions. *Applied Cognitive Psychology*, 22, 335–351.

Brown, A. L. (1990). Domain-specific principles affect learning and transfer in children. *Cognitive Science: A Multidisciplinary Journal*, 14(1), 107–133.

Buckley, B., Gobert, J., Kindfield, A., Horwitz, P., Tinker, R., Gerlits, B., et al. (2004). Model-based teaching and learning with BioLogica™: What do they learn? How do they learn? How do we know? *Journal of Science Education and Technology*, 13(1), 23–41.

Carey, S. (2009). *The origin of concepts.* New York: Oxford University Press.

Chi, M. T. H., de Leeuw, N., Chiu, M.-H., & LaVancher, C. (1994). Eliciting self-explanations improves understanding. *Cognitive Science: A Multidisciplinary Journal*, 18(3), 439–477.

Chinn, C. A. (2006). Learning to argue. In A. M. O'Donnell, C. Hmelo-Silver, & G. Erkens (Eds.), *Collaborative learning, reasoning, and technology.* Mahwah, NJ: LEA.

Clark, E. E. (1993). *The lexicon in acquisition.* Cambridge: Cambridge University Press.

Cronbach, L. S. (1966). The logic of experiments on discovery. In L. S. Shulman & E. R. Keislar (Eds.), *Learning by discovery: A critical appraisal.* Chicago: Rand McNally. pp. 76–92.

Diakidoy, I.-A. N., Kendeou, P., & Ioannides, C. (2003). Reading about energy: The effects of text structure in science learning and conceptual change. *Contemporary Educational Psychology*, 28, 335–356.

Duschl, R. A., Schweingruber, H. A., Shouse, A. W. (Eds.), Committee on Science Learning Kindergarten through Eighth Grade, Board on Science Education, Center for Education,

Division of Behavioral and Social Sciences and Education, & National Research Council (U.S.). (2007). *Taking science to school: Learning and teaching science in grades K–8*. Washington, DC: National Academies Press.

Gentner, D., & Namy, L. (1999). Comparison in the development of categories. *Cognitive Development*, 14, 487–513.

Gobert, J. D. (2004). Collaborative discourse around students' models within a web-based inquiry science environment (WISE). Paper presented at the Winter Text Conference, Jackson Hole, WY, January 16–19, 2004.

Gobert, J. D. (2005a). The effects of different learning tasks on model-building in plate tectonics: Diagramming versus explaining. *Journal of Geoscience Education*, 53(4), pp. 444–455.

Gobert, J. D. (2005b). Leveraging technology and cognitive theory on visualization to promote students' science. In J. K. Gilbert, D. F. Treagust, J. H. v. Driel, R. Justi, & J. Gobert (Eds.), *Visualization in science education*. Netherlands: Springer-Verlag. pp. 73–90.

Goldberg, F., & Bendall, S. (1995). Making the invisible visible: A teaching/learning environment that builds on a new view of the physics learner. *American Journal of Physics*, 63(11), 978–991.

Graesser, A. C., Singer, M., & Trabasso, T. (1994). Constructing inferences during narrative text comprehension. *Psychological Review*, 101, 371–395.

Johnson, H. M., & Seifert, C. M. (1994). Sources of the continued influence effect: When misinformation in memory affects later inferences. *Journal of Experimental Psychology: Learning, Memory, and Cognition*, 20(6), 1420–1436.

Johnson, N. S., & Mandler, J. M. (1980). A tale of two structures: Underlying and surface form in stories. *Poetics*, 9, 51–86.

Jose, T. J., & Williamson, V. M. (2005). Molecular visualization in science education: An evaluation of the NSF-sponsored workshop. *Journal of Chemical Education*, 82(6), 937–943.

Kali, Y., & Linn, M. C. (2009). Designing effective visualizations for elementary school science. *Elementary School Journal*, 109(5), 181–198.

Kali, Y., Linn, M. C., & Roseman, J. E. (Eds.) (2008). *Designing coherent science education*. New York: Teachers College Press.

Keil, F. C. (2005). Explanation and understanding. *Annual Review of Psychology*, 57(1), 227–254.

Klahr, D., & Nigam, M. (2004). The equivalence of learning paths in early science instruction: Effects of direct instruction and discovery learning. *Psychological Science*, 15(10), 661–667.

Klausmeier, H. J. (1992). Concept learning and concept teaching. *Educational Psychologist*, 27(3), 267.

Krajcik, J., McNeill, K. L., & Reiser, B. J. (2008). Learning-goals-driven design model: Developing curriculum materials that align with national standards and incorporate project-based pedagogy. *Science Education*, 92(1), 1–32.

Linn, M. C., Lewis, C., Tsuchida, I., & Songer, N. B. (2000). Beyond fourth-grade science: Why do U.S. and Japanese students diverge? *Educational Researcher*, 29(3), 4–14.

Liwag, M. D., & Stein, N. L. (1995). Children's memory for emotional events: The importance of emotion-related retrieval cues. *Journal of Experimental Child Psychology*, 60(1), 2–31.

McCardle, P. D., Chhabra, V., & Kapinus, B. A. (2008). *Reading research in action: A teacher's guide for student success*. Baltimore: Brookes.

McKeachie, W. J. (1995). Learning styles can become learning strategies. *The National Teaching and Learning Forum*, 4(6), 18–23.

McNamara, D. S., Kintsch, E., Songer, N. B., & Kintsch, W. (1996). Are good texts always better? Interactions of text coherence, background knowledge, and levels of understanding in learning from text. *Cognition and Instruction*, 14(1), 1–43.

Magliano, J. P., Trabasso, T., & Graesser, A. C. (1999). Strategic processing during comprehension. *Journal of Educational Psychology*, 91(4), 615–629.

Mandl, H., Stein, N. L., & Trabasso, T. (1984). *Learning and comprehension of text*. Mahwah, NJ: LEA.

Mandler, J. M. (1984). *Stories, scripts, and scenes: aspects of schema theory*. Hillsdale, NJ: LEA.

Mandler, J. M. (1998). Representation. In W. Damon (Ed.), *Handbook of child psychology. Vol. 2. Cognition, perception, and language*. Hoboken, NJ: John Wiley & Sons Inc. pp. 255–308.

Mandler, J. M., & Johnson, N. S. (1977). Remembrance of things parsed: Story structure and recall. *Cognitive Psychology*, 9(1), 111–151.

Martin, M. O., Mullis, I. V. S., & Foy, P. (2008). *TIMSS 2007 International Science Report: Findings from IEA's Trends in Interational Mathematics and Science Study at the fourth and eighth Grades*. Chestnut Hill, MA: TIMSS & PIRLS International Study Center, Boston College.

Mayer, R. E. (2001). *Multimedia learning*. New York: Cambridge University Press.

Mayer, R. E. (2004). Should there be a three-strikes rule against pure discovery learning? The case for guided methods of instruction. *American Psychologist*, 59(1), 14–19.

Mayer, R. E., Hegarty, M., Mayer, S., & Campbell, J. (2005). When static media promote active learning: Annotated illustrations versus narrated animations in multimedia instruction. *Journal of Experimental Psychology: Applied*, 11(4), 256–265.

Namy, L. L., & Gentner, D. (2002). Making a silk purse out of two sow's ears: Young children's use of comparison in category learning. *Journal of Experimental Psychology: General*, 131, 5–15.

National Research Council (U.S.) (1996). *National Science Education Standards: Observe, interact, change, learn*. Washington, DC: National Academies Press.

Nussbaum, J., & Novick, S. (1982). Alternative frameworks, conceptual conflict and accommodation: Toward a principled teaching strategy. *Instructional Science*, 11, 183–200.

Ornstein, P. A., & Trabasso, T. (1974). To organize is to remember: The effects of instructions to organize and to recall. *Journal of Experimental Psychology*, 103(5), 1014–1018.

Project 2061 (American Association for the Advancement of Science) (1993). *Benchmarks for science literacy*. New York: Oxford University Press.

Rapp, D. N., van den Broek, P., McMaster, K. L., Kendeou, P., & Espin, C. A. (2007). Higher-order comprehension processes in struggling readers: A perspective for research and intervention. *Scientific Studies of Reading*, 11, 289–312.

Reif, F. (2008). *Applying cognitive science to education: Thinking and learning in scientific and other complex domains*. Cambridge, MA: MIT Press.

Reiser, B. J. (2004). Scaffolding complex learning: The mechanisms of structuring and problematizing student work. *Journal of the Learning Sciences*, 13(3), 273–304.

Romance, N. R., & Vitale, M. R. (2006). Science IDEAS: Making the case for integrating reading and writing in elementary science as a key element in school reform. In R. Douglas, M. P. Klentschy, & K. Worth (Eds.), *Linking science and literacy in the K–8 classroom*. Arlington, VA: NSTA.

Romance, N. R., Vitale, M. R., & Dolan, M. F. (2004). *Scientifically-based research in science education*. Washington, DC: US Department of Education.

Rumelhart, D. E., & Norman, D. A. (1978). Accretion, tuning, and restructuring: Three modes of learning. In J. W. Cotton, & R. L. Klatzky (Eds.), *Semantic factors in cognition*. Hillsdale, NJ: LEA.

Schank, R. C., & Abelson, R. P. (1977). *Scripts, plans, goals, and understanding: An inquiry into human knowledge structures*. Hillsdale, NJ: LEA.

Schmidt, W. H. (2008). Math, science education isn't working, expert says: Students are taught too much too soon. *Charleston Daily Mail*.

Schmidt, W. H., Wang, H. C., & McKnight, C. C. (2005). Curriculum coherence: An examination of US mathematics and science content standards from an international perspective. *Journal of Curriculum Studies*, 37(5), 525–559.

Shulman, L. S. (1986). Those who understand: Knowledge growth in teaching. *Educational Researcher*, 15(2), 4–14.

Shulman, L. S., & Keislar, E. R. (1966). *Learning by discovery: A critical appraisal.* Chicago: Rand McNally.

Shwartz, Y., Weizman, A., Fortus, D., Krajcik, J., & Reiser, B. (2008). The IQWST experience: Using coherence as a design principle for a middle school science curriculum. *Elementary School Journal*, 109(2), 199–219.

Siegler, R. S., & Ramani, G. B. (2008). Playing linear numerical board games promotes low-income children's numerical development. *Developmental Science*, 11(5), 655–661.

Simon, H. A. (1995). Near decomposability and complexity: How a mind resides in a brain. In H. J. Morowitz, & J. L. Singer (Eds.), *The mind, the brain, and complex adaptive systems.* Addison-Wesley. pp. 25–43.

Smith, C. L. (2007). Bootstrapping processes in the development of students' commonsense matter theories: Using analogical mappings, thought experiments, and learning to measure to promote conceptual restructuring. *Cognition and Instruction*, 25(4), 337–398.

Smith, C. L., Snir, J., & Grosslight, L. (1992). Using conceptual models to facilitate conceptual change: The case of weight-density differentiation. *Cognition and Instruction*, 9(3), 221–283.

Smith, C. L., Solomon, G. E. A., & Carey, S. (2005). Never getting to zero: Elementary school students' understanding of the infinite divisibility of number and matter. *Cognitive Psychology*, 51(2), 101–140.

Smith, C. L., Wiser, M., Anderson, C. W., & Krajcik, J. (2006). Implications of research on children's learning for standards and assessment: A proposed learning progression for matter and the atomic-molecular theory. *Measurement: Interdisciplinary Research and Perspectives*, 4(1–2), 1–98.

Smith, C. L., Wiser, M., Anderson, C. W., Krajcik, J. S., & Coppola, B. (2004). *Implications of research on children's learning for standards and assessment: Matter and the atomic molecular theory.* Washington, DC: National Research Council Committee on Test Design for K–12 Science Achievement.

Stein, N. L., Anggoro, F. K., & Hernandez, M. W. (2009). *A developmental study of physical science learning: The importance of starting early.* Unpublished manuscript, University of Chicago, IL.

Stein, N. L., & Bernas, R. (1999). The representation and early emergence of argument understanding. In P. Coirier, & J. Andriessen (Eds.), *Foundations of argumentative text processing.* Amsterdam: Amsterdam University Press. pp. 97–116.

Stein, N. L., & Glenn, C. G. (1979). An analysis of story comprehension in elementary school children. In R. O. Freedle (Ed.), *New directions in discourse processing* (Vol. 2). Norwood, NJ: Ablex, Inc. pp. 53–120.

Stein, N. L., Hernandez, M. W., & Trabasso, T. (2008). Advances in modeling emotion and thought: The importance of developmental, online, and multilevel analyses. In M. Lewis, J. M. Haviland-Jones, & L. F. Barrett (Eds.), *Handbook of emotions* (3rd ed.). New York: Guilford Press. pp. 574–586.

Stein, N. L., & Levine, L. J. (1989). The causal organisation of emotional knowledge: A developmental study. *Cognition & Emotion*, 3(4), 343–378.

Stein, N. L., & Ornstein, P. A. (in preparation). Using a theory of coherence and complex learning to study early learning in physical science. *Perspectives on Child Development.*

Stein, N. L., Ornstein, P. A., Tversky, B., & Brainerd, C. J. (1997). *Memory for everyday and emotional events.* Hillsdale, NJ: LEA.

Stein, N. L., & Trabasso, T. (1982a). What's in a story: An approach to comprehension and instruction. In R. Glaser (Ed.), *Advances in instructional psychology* (Vol. 2). Hillsdale, NJ: LEA. pp. 212–267.

Stein, N. L., & Trabasso, T. (1982b). Children's understanding of stories: A basis for moral judgment and resolution. In C. J. Brainerd, & M. Pressley (Eds.), *Verbal processes in children*. New York: Springer-Verlag. pp. 161–188.

Stein, N. L., & Trabasso, T. (1985). The search after meaning: Comprehension and comprehension monitoring. In F. Morrison, C. Lord, & D. Keating (Eds.), *Advances in applied developmental psychology* (Vol. 2). New York: Academic Press. pp. 33–58.

Stein, N. L., & Trabasso, T. (1989). Children's understanding of changing emotional states. In C. Saarni & P. L. Harris (Eds.), *Children's understanding of emotion*. New York: Cambridge University Press. pp. 50–77.

Stein, N. L., & Trabasso, T. (1992a). The organisation of emotional experience: Creating links among thinking, language, and intentional action. *Cognition & Emotion*, 6(3–4), 225–244.

Stein, N. L., & Trabasso, T. (1992b). Scientific reasoning and explanatory patterns: The effects of thinking aloud and pictorial representation. Paper presented at the American Educational Research Association, Chicago, IL.

Stein, N. L., Trabasso, T., & Liwag, M. (1993). The representation and organization of emotional experience: Unfolding the emotion episode. In M. Lewis & J. M. Haviland (Eds.), *Handbook of emotions*. New York: Guilford Press. pp. 279–300.

Suh, S., & Trabasso, T. (1993). Inferences during reading: Converging evidence from discourse analysis, talk-aloud protocols, and recognition priming. *Journal of Memory and Language*, 32(3), 279–300.

Taber, K. S. (2006). Beyond constructivism: The progressive research programme into learning science. *Studies in Science Education*, 42(1), 125–184.

Thagard, P. (2000). *Coherence in thought and action*. Cambridge: MIT Press.

Trabasso, T. (2005). The role of causal reasoning in understanding narratives. In T. Trabasso, J. Sabatini, & D. W. Massaro (Eds.), *From orthography to pedagogy: Essays in honor of Richard L. Venezky*. Mahwah, NJ: LEA. pp. 81–106.

Trabasso, T., & Bouchard, E. (2000). *Teaching children how to comprehend what they read: A review of experimental research on direct instruction of reading comprehension*. Report of the National Reading Panel. Washington, DC: National Institute of Child Health and Human Development, National Institutes of Health.

Trabasso, T., & Bouchard, E. (2002). Teaching readers how to comprehend text strategically. In C. C. Block & M. Pressley (Eds.), *Comprehension instruction: Research-based best practices*. New York: Guilford. pp. 176–200.

Trabasso, T., & Magliano, J. P. (1996). Conscious understanding during comprehension. *Discourse Processes*, 21(3), 255–287.

Trabasso, T., Secco, T., & van den Broek, P. W. (1984). Causal cohesion and story coherence. In H. Mandl, N. L. Stein, & T. Trabasso (Eds.), *Learning and comprehension of text*. Hillsdale, NJ: LEA. pp. 83–111.

Trabasso, T., & Stein, N. L. (1993). How do we represent both emotional experience and meaning [A review of Richard Lazarus's *Emotion and adaptation*, New York: Oxford University Press, 1991]. *Psychological Inquiry*, 4(4), 326–333.

Trabasso, T., & Stein, N. L. (1997). Narrating, representing, and remembering event sequences. In P. W. van den Broek, P. J. Bauer, & T. Bourg (Eds.), *Developmental spans in event comprehension and representation: Bridging fictional and actual events*. Mahwah, NJ: LEA. pp. 237–270.

Trabasso, T., Stein, N. L., & Johnson, L. R. (1981). Children's knowledge of events: A causal analysis of story structure. In G. Bower (Ed.), *Learning and motivation* (Vol. 15). New York: Academic Press. pp. 237–281.

Trabasso, T., Suh, S., Payton, P., & Jain, R. (1995). Explanatory inferences and other strategies during comprehension and their effect on recall. In R. F. Lorch Jr. & E. J. O'Brien (Eds.), *Sources of coherence in reading*. Hillsdale, NJ: LEA. pp. 219–239.

Tversky, B., Heiser, J., Mackenzie, R., Lozano, S., & Morrison, J. (2008). Enriching animations. In R. Lowe & W. Schnotz (Eds.), *Learning with animation: Research implications for design*. New York: Cambridge University Press. pp. 263–285.

Vosniadou, S. (1994). Capturing and modeling the process of conceptual change. *Learning and Instruction*, 4, 45–69.

Vosniadou, S. (2007). The conceptual change approach and its re-framing. In S. Vosniadou, A. Baltas, & X. Vamvakoussi (Eds.), *Reframing the conceptual change approach in learning and instruction*. New York: Elsevier Science. pp. 1–15.

Winograd, T. (1980). Extended inference modes in reasoning by computer systems. *Artificial Intelligence*, 13(1), 5–26.

Winston, P. (1986). Learning by augmenting rules and accumulating censors. In R. S. Michalski, J. G. Carbonell, & T. M. Mitchell (Eds.), *Machine learning: An artificial intelligence approach*. New York: Morgan Kauman. pp. 45–106.

Wyatt, K. (2009). School reform is main topic for Duncan Visit. *USA Today*, April 6, 2009.

8 Science Classrooms as Learning Labs

Rochel Gelman and Kimberly Brenneman

There is a crescendo of calls from schools, industry, and government for academic researchers to upgrade students' scientific, mathematical, and technical literacies. Advances in these fields make tremendous demands on all citizens, who are daily confronted with discussions of genetic engineering to improve the quality and quantity of food, statistical reports on the risks and benefits of pharmaceuticals, probabilistic statements about the outcome of economic policies, and so on. Jobs that require technical and scientific skills remain unfilled, owing to lack of preparation of the workforce, while jobs that do not have such requirements disappear. For these reasons, identifying ways to improve scientific and technological literacy is a critical research issue.

We have gained considerable knowledge about concept learning from controlled laboratory experiments. Still, much work remains if we are to develop a fuller knowledge base on the nature of learning about scientific and related literacies as taught in the complex environment of the classroom. In what follows, we relate our experiences using findings from carefully controlled studies of human learning as a basis for the design of classroom programs to support learning of science content and practices, as well as language and literacy. When classrooms are treated as laboratories of learning, the inputs include the interactions between students, teachers, texts, technical devices, and other prepared learning materials. This level of complexity exacerbates the challenges we face to describe student learning. Accounts of learning must acknowledge these factors as well as two premises about knowledge acquisition that interact with efforts to provide successful learning environments.

Theoretical Premises

Learners' Minds Contribute to the Definition of Relevance

Human beings, no matter how young or old, are active participants in their own learning. What one already knows influences how one interprets the data offered (R. Gelman, 1993; R. Gelman & Lucariello, 2002; Bransford, Brown, & Cocking, 1999). This presents us with a difficult circumstance: those who study learning can no longer assume full control over the definition of relevant inputs. Because of the constructivist tendencies of the mind, what the learner takes away from a learning situation can, and often does, reflect an unexpected interpretation of the input (Bartlett, 1932; Bransford et al., 1999). An example is provided by a study of how fourth graders in the US and Japan described the same section of a videotaped math lesson (Yoshida, Fernandez, & Stigler, 1993). The US students tended to talk about nonmathematical items, including what the teacher was wearing. Japanese students' comments focused on the mathematics lesson. The two groups of students interpreted the same input very differently. Why were the Japanese students better at attending to mathematically relevant novel inputs than their American counterparts? A number of variables probably contributed. Japanese teachers know more mathematics and deliver a more organized lesson (Stevenson & Stigler, 1994; Stigler & Hiebert, 1999). They also spend more time in class teaching math and encouraging their students to present and discuss different solutions to homework problems so that students are better prepared to think about material from multiple perspectives.

The description of the active learner is also supported by developmental research that tells us that young children ask information-seeking questions and work to understand the causal workings of their world (Chouinard, 2007; Schulz & Bonawitz, 2007) as well as build naïve theories about the world (R. Gelman et al., 1995; Spelke, 2000; Williams, 2000). These belief systems are built pretty much on the fly and without formal instruction. This is often good. A preschooler who is told that an unfamiliar object is an animal can infer that it moves by itself, eats, drinks, breathes, and has babies without having to be taught these facts for each animal he or she encounters. But sometimes these naïve theories generate "misconceptions" about modern scientific and mathematical concepts. Misconceptions about scientific domains can persist even when the participants in question have passed university science courses (McCloskey, 1983). Given this, each time we add sources of input for learning, from teachers, texts, parents, or information technologies, we increase the possibility of unintended interpretations by the learner. Thus it is not surprising that students often reach unintended interpretations of what they are offered in classrooms.

We must realize that the field lacks a suitable theory of environment, one that is consistent with an active theory of mind. Simply offering what we think should be learned will not do. We need principles for creating learning environments that will maximize the tendencies for learners to interpret inputs that will nurture their transfer to relevant learning paths for the topic. At the very least, teachers have to be fluent in the science and mathematics they are teaching. They also have to adopt pedagogical techniques for presenting this content. Clues for pedagogical design are related to our second premise about concept learning with understanding.

Understanding Is about Building Coherent Structure, Not Lists of Facts

Concepts do not stand alone; they exist in organized domains of knowledge that also comprise ways of using that knowledge. The language that is relevant to the concepts in a given domain is closely related to the concepts therein. The more one knows about a domain, the deeper the understanding of the language in that domain (Bransford et al., 1999; Carey, 2000). Different domains of knowledge build on different kinds of data from different sources. Likewise, the principles that organize domains vary. Scientists are concerned with explaining the world around us. Topics include the objects on, in, and surrounding the world; the states and changes of these objects; the energy conditions that influence matter; the organization and function of biological matter; the nature of inert entities; laws of motion; how entities interact with their ecological niches; as well as many others. Experiments and their related procedures provide the relevant data for explanations. Notebooks (or computer files) are used to keep track of theoretical motivations, design concerns, observed data and their organization, results of various conditions, and further hypotheses and ideas for new explorations. Different branches of science focus on particular topics, but all assume that the to-be-found laws are universal and organized, in coherent ways, and that the findings and generalizations flow from the use of experimental and modeling methods. When students learn science, all of these factors come into play and all are incorporated into the educational programs we have developed to study and foster science learning in real-world settings.

Case Studies

In what follows, we detail two projects in which we applied a domain-specific and constructivist theory of knowledge and learning to the design of classroom-based educational experiences for science. The case studies detail one program that embedded science learning opportunities into a high school English as a Second Language (ESL) curriculum and another that brought science into preschool classrooms. Each case reflects our belief that the achievement of a connectedness of concepts in the mind requires that learning experiences be conceptually linked. Because scientific vocabulary both supports and is supported by conceptual growth, learning is likely to be maximized when scientific concepts and vocabulary are taught together. When learners are given multiple opportunities to use the to-be-learned terms in meaningful contexts, they can begin to learn the actual terms that refer to the concepts they explore. Meaningful science learning contexts also incorporate the practices of science including making systematic observations, predicting, checking, comparing and contrasting, record keeping, communication, and constructing generalizations that apply across time and examples.

Some might be surprised that we treat high school students and preschoolers within the same theoretical framework regarding learning about science. After all, science and its methods are about abstract matters, and well-known developmental theorists such as Piaget (1952) and Vygotsky (1962) hold that young children lack the conceptual capacity to engage in scientific learning because they are perception bound and do not possess the requisite mental structures. However, research findings from various labs suggest that there are pockets of abstract knowledge that support young children's abstract reasoning abilities in certain domains (Bowman, Donovan, & Burns, 2001; Bransford et al., 1999; R. Gelman & Brenneman, 2004; S. Gelman, 2003; Goswami, 2002). For example, preschool children, and even toddlers, possess knowledge about some abstract differences between animate and inanimate objects and the causal conditions related to the kinds of changes and states these can assume (S. Gelman & Opfer, 2002; Massey & R. Gelman, 1988); some aspects of the life cycle in animals and plants (Hickling & S. Gelman, 1995; Rosengren, S. Gelman, Kalish, & McCormick, 1991); counting and its relation to addition and subtraction (Zur & R. Gelman, 2004); the difference between various notational systems (Brenneman, Massey, Machado, & R. Gelman, 1996); and inherent properties of materials (Au, 1994). Commitment to better science education for young children means that we acknowledge that they are active interpreters of environmental inputs and constructors of knowledge.

Our approach for preschoolers and older learners parallels current educational standards and policy that inquiry learning in the arena of science should relate the concepts, language, and processes of science (e.g., Michaels, Shouse, & Schweingruber, 2008) because they are interconnected and mutually supporting. In Table 8.1, we summarize some key components that contribute to what it means to "do and learn science."

Science into Ninth-Grade ESL

The Science into ESL program was devised for ninth-grade students attending a large public high school in the Los Angeles Unified School District (LAUSD). At that time, science was not offered to any ninth-grade ESL students in the district. Our goal was to teach ESL students enough science and related math so that they could enroll in the school's tenth-grade science and math classes taught in English. To be successful, the course had to prepare students for more advanced English *and* provide foundational conceptual and procedural knowledge about science. Because the lingua franca of science is English, we were able to use English to mitigate the fact that students' first languages varied greatly (with Spanish and Korean being the most frequent). Similarly, the choice of content was limited because it had to relate to what students would encounter in higher grades.

Table 8.1 Contributions to a Science-Relevant Education Program

Conceptual Learning Experience Considerations
Focus on the content of science
Build on what learners already know (informed by developmental research or by formative classroom assessments) and support them as they move along a conceptual learning path
Provide multiple opportunities to work with and think about a concept
Plan conceptually connected "lessons" to yield deeper understandings
Build concepts and language together
Do not settle for memorization of facts and terms
Teach Science-Relevant Processes
Careful observation skills
Predicting and checking
Comparing and contrasting: build to the notions of variable and experimentation
Use appropriate scientific language and terms; avoid metaphors
Use mathematics with meaning as part of science explorations
Communicate and document work, using other representational means (notebook, drawing, graphing, models)
Use of observation tools and measuring instruments in meaningful contexts
Be prepared to entertain new hypotheses

Given our theoretical position about the conceptual connectedness of scientific domains, the curricular material had to be organized according to underlying principles. The conceptual units were designed to be coherent and redundant, providing students with multiple opportunities to work with and think about underlying concepts and to communicate about them in English. To accomplish this, we had students work with various materials and media including written texts, the teachers' records of the class's experimental data on the blackboard, individual science notebooks, and oral reports. These multiple opportunities to work with, think about, experiment with, record, and communicate ensured that similar conceptual content was explored in multiple, varied ways. This redundancy maximizes the probability that students will attend to at least some of the offered lessons in the manner intended by the instructors. Examples that form part of the same conceptual equivalence class also make it possible for learners to compare and contrast and begin to focus on higher-order common ideas, shared terms, and relations.

To develop the content, members of the research team worked with the teachers to develop the notion of organizing principles and related concepts. We did not try to cover everything in textbooks, and instead focused on a limited number of units that interrelated language, content, and the processes of science. Dr. George Meck, who has a PhD in second-language acquisition and an extensive knowledge of science, wrote new course material to meet California's science and ESL requirements. Curriculum development meetings led to eight conceptually organized units: sun and photosynthesis; respiration; local winds, temperatures, and state; buoyancy and density; water cycle; food energy; organs and organisms; interactions and ecosystems. Each unit included an initial reading, a short pretest about vocabulary and concepts, a lab, a review, a short post-test, and journal entries about the review. Teachers also supported students' efforts to ask and answer questions.

Careful attention was given to the need to engage in formative assessments as students interacted with course materials. In addition, our commitment to monitoring whether students were on task meant that we had to probe often to determine if the experiences we designed were having the intended results. We wanted ways to obtain informative data about language and conceptual development that fit readily into classroom activities, often by taking advantage of the products students created as part of their lessons. One gauge of growth in conceptual understandings was straightforward, involving an analysis of brief pre- and post-tests on the conceptual content of each unit. We also were able to use the results of the reviews that followed each large unit of instruction. Part of the review involved the instructor and class developing a concept map that was produced on the blackboard. Then students made their own concept maps in their notebooks and wrote up to 10 sentences about these. If students wrote any sentences they received an A, and if they failed to try they received an F. Under these conditions, all students tried. Our subsequent analyses yielded a wealth of information about changes in vocabulary, grammatical complexity, and conceptual growth over the course of the semester. Once we identified the features that served this function, teachers could quickly scan students' work and then either adjust their plans or move on.

We expected that, as the semester progressed, the syntactic complexity of students' written sentences would increase as they learned more science and more English because the two are mutually reinforcing. Discussing the methods and conceptual content of science requires the use of complex grammatical structures such as embedded clauses and prepositional phrases. The former is common in scientific descriptions, such as "if you water a plant, it can grow" (Celce-Murcia & Larsen-Freeman, 1983). As conceptual understandings become more complex, so does the language required to accurately represent them. Similarly, prepositional phrases can be used to represent additional conceptual information within a single sentence.

To test our ideas about growth of language skills, we compared the syntactic complexity of students' written sentences from the first and second halves of a semester (R. Gelman, Romo, & Francis, 2002). *Simple* sentences were defined as those that consisted of a single main clause (compound sentences that were simply two main clauses connected by a conjunction were treated as separate simple sentences). *Complex* sentences included at least one main and one embedded clause. A mean proportion of complex sentences as a function of total sentences was calculated for each student, and overall analyses showed that the mean proportion of complex sentence structures increased from the first half of the course (13 percent) to the second half (31 percent) [$F(1, 19) = 38.07$, MSE = .01, $p < .001$]. In addition, the mean proportion of sentences that included prepositional phrases increased from 34 percent to 48 percent [$F(1, 19) = 14.95$, MSE = .01, $p < .001$].

Note that students were not generating perfectly grammatical sentences. In fact, the proportion of sentences that were error-free *decreased* between the two halves of the semester from 35 percent to 24 percent [$F(1, 19) = 10.03$, $MSE = .01$, $p < .01$]. This result almost certainly comes from the fact that students were generating longer sentences later in the semester. In sum, the increase in error rates reflected more mistakes in subject–verb agreement, the application of determiners, and proper use of prepositions. Importantly, the increased grammatical complexity coincided with increased conceptual complexity of students' written descriptions of the relations among the nodes of their concept maps. As shown in Table 8.2, simple definitions and descriptive statements decreased (from 47 percent to 26 percent) while the percentage of sentences that included descriptions of the specific conditions, reasons, or circumstances governing relationships between entities increased from 18 percent to 37 percent. The accuracy of students' descriptions was relatively high, averaging

Table 8.2 Average Proportion of Sentences at Each Concept Specificity Level

			First Half of Course		*Second Half of Course*
Level	*SD*	*Content Specificity Category*	p	*SD*	p
1	.13	Properties or Category Membership	.47	.14	.26
2	.1	Needs or Sources	.15	.14	.12
3	.11	Mechanics or Function	.2	.09	.25
4	.18	Conditions or Purposes	.18	.13	.37

From Gelman, R., Romo, L., & Francis, W. S. (2002). Notebooks as windows on learning: The case of a science-into-ESL program. In N. Granott & J. Parziale (Eds.), *Microdevelopment: Transition processes in development and learning*. Cambridge: Cambridge University Press. Printed p. 283.

about a 3 on a four-point scale. Correlation analyses revealed reliable, positive relationships between the complexity of students' sentences and the accuracy of their scientific understandings.

Another marker of increased conceptual growth comes from analyses of pre- and post-unit quizzes with parallel questions (R. Gelman et al., 1995). For example, prior to a unit on the sun, students were asked which color of coat (from a list of colors) would keep them warmest on a cold, sunny day. At post-test, the conceptually related item asked students to choose a color to paint a hot water tank so that the water would stay as hot as possible. Answers were scored as correct (2 points), partially correct (1), or incorrect (0). An uncorrected example of a pre-test response scored 0 is: "(I would choose the) blue jacket because the blue color is warm but also is cool for a sunny day and also is a pretty color." At post-test, this student scored a 2: "(They should paint it) black. I think than the black color because this color absorb energy and the water can make hot." Total pre-test and total post-test scores were computed for each student who took at least five pairs of tests. As shown in Figure 8.1, a paired *t*-test reveals reliable increases [$t(53) = 7.3$, $p < .0001$]. Still, there is much room for improvement. Inspection of each unit made clear that we had to revise our approach, breaking up some units and adding some transitional ones. If we had not measured learning often throughout the course, we would not have made such discoveries nor could we have designed remediation.

Despite the challenges and the variability in student learning, in general students who participated in the Science into ESL program were successful both as English language learners and as science students. Scores on the standard ESL exam were similar to those of students who completed the traditional ESL class without science content (R. Gelman et al., 1995). This reassures us that the increase found in sentences with grammatical errors reflected attempts to better describe more complex

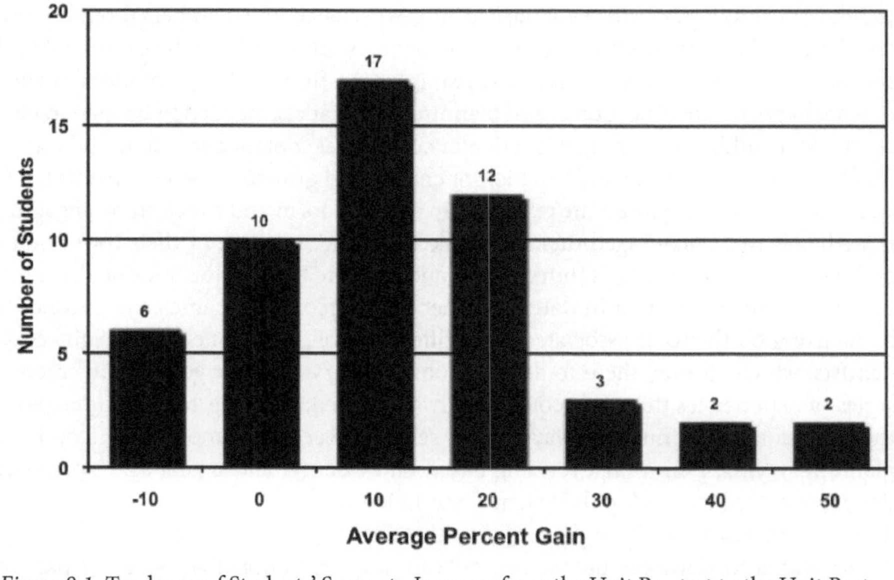

Figure 8.1 Tendency of Students' Scores to Improve from the Unit Pre-test to the Unit Post-test.

conceptual understandings using more complex syntax, rather than some real decrement in grammatical understanding. Further, 64 percent of the students who participated in the Science into ESL program moved into tenth-grade science classes taught in English. In the end, the critical goal for the school was moving students along in their second-language learning so we were all gratified that increased knowledge and skill in science did not come at the expense of English language learning. In an educational environment in which more and more needs must be met in the same instructional timeframe, it is particularly important to know that we can adequately address educational needs in both domains with one well-designed course (see also Lee, 2005, for a review of research on science education with English language learners [ELLs]).

Cognitive Science Goes to Preschool

The second educational program we present has its roots in the early 1990s, when Gay Macdonald (Executive Director of UCLA Early Care and Education) and Rochel Gelman began collaboration on a science-based preschool curriculum that would offer learning experiences that allowed children to work with and think about a particular concept for many weeks or months. This approach to science contrasts sharply with "magic science" in which children are shown an event (say a baking soda and vinegar reaction) that is exciting but unconnected to other experiences. In addition to conceptual redundancy, the Preschool Pathways to Science (PrePS) approach provides repeated opportunities to engage in the science practices listed in the lower section of Table 8.1.

From the start, the program shared many features with the Science into ESL program. Like the ESL program, PrePS was a true partnership between developmental psychologists and educators. As executive director of UCLA ECE, Macdonald set the curriculum for the school. Thus, the PrePS team, like the ESL team, was in charge of the classroom and free to develop the learning experiences the members thought were appropriate. Regular meetings occurred involving staff from both the preschool and the lab. Transcripts of these sessions reveal introspection, exchange of ideas, review of what happened in classrooms, and planning for what else needed to happen to better support children's learning of particular conceptual content and science practices. Some topics, such as form and function or change and growth, were very fruitful. For example, one class explored the relationship between form and function by engaging in activities that encouraged them to think about the structure of their own bodies and those of various animals (birds, seals) and to relate these to locomotion, by moving on land, through air, or in water. Another class explored the same general concept by focusing on the tools associated with different professions, such as firefighters or dentists, which allowed them to do their jobs. In every case, the goal was to create a series of experiences that were conceptually connected, allowing children to explore the same concept in multiple ways over a series of weeks and months. The cycle of planning, trying, examining, revising based on evidence, and trying again parallels the process used to develop the Science into ESL units.

In recent years, we have partnered with Christine Massey at the University of Pennsylvania to study the uptake of PrePS in new preschools that were not part of the program's initial development site. Differences between the original and new sites have highlighted the variables that impact the adoption of the program by teachers, which in turn impacts the types of learning opportunities offered to children. As part of the study, we have identified components of the approach that are relatively easy for teachers to adopt and those that require more intense support from the research team. We have also been able to study specific learning outcomes for children who participate in PrePS. The ultimate goal is to support teachers as they plan and offer science experiences to students that are conceptually connected and allow children to engage in science practices.

We have found that the most readily adopted aspects of PrePS involve recording and documentation. We encourage teachers to solicit children's observations and predictions and to write these on charts for classroom display. In PrePS classrooms, we have also incorporated science journals (Brenneman & Louro, 2008; R. Gelman, Brenneman, Macdonald, & Román, 2009). Children keep their own science journals to record their observations and understandings of science concepts and the related objects and events. They do so by drawing and dating their work with a date-stamp. Teachers then solicit children's descriptions and engage them in a discussion of their journal entries. These discussions provide an opportunity to probe children's understandings and possible misunderstandings. They also encourage the use of descriptive language.

As teachers record children's descriptions by writing on the journal entry or creating charts to display, they provide children with critical preliteracy experiences. The children get to see that print is used with a purpose. We suspect that the connection to literacy underlies the quick integration of charts and journals into classrooms. In fact, many of the teachers we work with already use charts (such as Know–Want to Know–Learned) to record children's ideas. None had previously used science journals, but all have begun doing so to varying degrees. Many have noted that journal entries provide rich work samples that they can use to assess and document children's progress in language, literacy, cognition, attention to detail, and so on. Because they already provide high-quality literacy experiences for students, our teacher partners easily recognize the value of science practices as another way to support the development of literacy.

Whereas incorporation of certain science practices occurs with some ease, planning conceptually connected learning activities is much more difficult. One hallmark of PrePS that we have described elsewhere (R. Gelman & Brenneman, 2004) is that it is not prescriptive. As originally designed, PRePS was a planning tool, not a set of instructions for specific activities and lessons. The ongoing coaching and collaboration that is available at UCLA to support teachers, specifically for science, is not available in most preschools. This practical issue led us to begin developing series of conceptually connected learning experiences that we could teach or that could be given to teachers for their use. To help teachers step into the program, we illustrate what we mean by "conceptually connected learning experiences that allow children to engage in authentic science practices over time" by providing clear examples. If we want to support adult learners as they begin to use the PrePS approach to preschool science teaching and learning, we must provide them many and varied experiences working with it, just as we have proposed that teachers must do for young learners working with science concepts. Lesson series have included learning experiences about senses, form and function, and growth and life cycles, among others.

One of the lesson series that we have introduced involves senses. A number of factors converged on this choice. First, as a topic addressed in many preschool classrooms, it is familiar to teachers. This allows us to build from their existing knowledge base to illustrate how to add value by incorporating science practices and knowledge into these activities. The approach goes beyond matching body parts (e.g., eyes and ears) with functions (e.g., seeing and hearing). Although this knowledge is important, preschoolers are capable of learning more, and this "more" forms a critical foundation for later science thinking skills. Specifically, children learn that senses are observation tools that they can use to learn about the objects and events in the world around them. The activity series provides children with chances to reflect on the sources of their knowledge so that they begin to understand not just *that* they know something, but *how* they know it (Massey & Roth, 2004). These developing insights are foundations for further learning about science as a way of knowing, and of coordinating and interpreting evidence (Duschl, Schweingruber, & Shouse, 2006).

Initial results from the "Using Senses as Tools for Observation" unit are encouraging,[1] as far as children's understanding of content goes. At the end of the unit, learners at our university preschool site performed comparably to a sample of kindergarteners in urban schools (Massey & Roth, 1997) on tasks that required them to determine which sense could be used to solve a discrimination problem. (Can a particular sense be used to tell which of two items is sweet, green, has a scent, and so on?) Both of the age groups represented achieved high overall scores of 84 percent (pre-K) and 85 percent (K).

Among our younger ELL students, most were unable to pass a test that required them to identify the function of each of their senses at the beginning of the school year. They had difficulty completing sentences such as "With our eyes we ________" and assessing the truth of statements such as "Do we taste with our ears?" After participating in a series of learning experiences about their senses, children in the intervention classroom were more likely (79 percent of learners vs. 27 percent) to perform well when matching senses with their functions than were children who had not participated. The results from this admittedly small sample are bolstered by recent pre- and post-intervention data from a larger sample. Among a diverse (mostly ELL) sample of four-year-olds ($n = 45$, mean age at pretest = 54.4 months, range = 48–60), 15 children were able to answer more than half of the questions at both time points, which occurred approximately three months apart. Fourteen could not do this at either time. Seventeen children moved from not passing to passing. *No* child moved in the opposite direction. Clearly, there is a positive group effect revealed by this assessment.

Introducing observation skills as part of learning about the senses is a natural marriage of science content and process skills that allows children to practice observation and to use descriptive language to communicate their observations to others. We have always begun PrePS in new settings with an emphasis on observation, prediction, and comparing and contrasting. These science practices lead naturally into experimentation. Simple experiments are incorporated into PrePS activities across content areas, providing repeated opportunities for children to engage in this critical science practice throughout the year. Test–control data suggest that repeated opportunities to engage in experimentation support children's developing ability to generate simple controlled tests (Brenneman et al., 2007). In addition to observation, prediction, compare/contrast, and experimentation, science practices in the PrePS classroom include emphases on recording and documentation and using science activities to support early mathematics skills and the development of complex language and discourse (see Table 8.1).

We turn now to examples and data that illustrate ways that PrePS supports language development. As is probably clear by now, our belief that conceptual and language development within a domain are necessarily mutually supporting leads us to provide children with the real terms for the science processes and concepts they work with. Note that these terms are provided repeatedly as part of multiple, related contexts so that vocabulary terms, while perhaps not understood completely, are not empty of meaning (see Hammer, 1999). We have related this anecdote elsewhere, but it so captures the phenomenon that we repeat it here. After being introduced to the process and the term "observation" during circle time, one of our young students approached us later in the day with a building block. He said, "It's green. It's a rectangle. I cannot *observate* any more." Clearly, the link between the term and what it meant was made. Indeed, the child was able to create a novel version of the verb, "observe," that is consistent with the rules of English.

Our focus on observation and making comparisons between objects and events also encourages the use of descriptive language. Preliminary evidence suggests that children in a PrePS classroom were reliably better at appropriately extending adjectives and adjectival phrases (such as "smells like banana," "bumpy," and "cold") than children in comparison, non-PrePS classrooms [$t(14) = 2.4$, $p < .05$], and we are currently collecting data to replicate this initial finding with a larger sample. Although our young students cannot write sentences that can be analyzed for grammatical complexity, we expect to analyze transcripts for changes over time in the syntactic complexity of the group's conversations at circle time, and of individual learners in one-on-one discussions. Children's descriptions of their journal entries could also be analyzed in this way.

Conclusion

PrePS and Science into ESL are successful applications of the science of learning to classroom learning of science. Classrooms become laboratories in which we test our approaches to integrating empirical laboratory results about learning into environments that exist to support learning. Each program has yielded measurable benefits for students. However, moving PrePS into new sites, over which our partners and we do not have control, presents new challenges. Importantly, it provides insight into situational factors that contributed to the original successes at UCLA and in the Science into ESL program in ways that were not apparent until we had a contrasting case. For instance, complete integration of a new program seems to require more than just the cooperation of school administrators. It requires that they offer continuous support to teachers as they adjust to new curricular demands and ways of thinking. Alternately, this support might come from the researchers/developers of the program who are available in classrooms to discuss, model, and plan lessons (as has occurred with PrePS in both the UCLA and NJ sites) or from a master teacher (like Meck). Success in new sites might require that there is on-site science expertise. In preschools, especially, this cannot be assumed as many teachers do not have a BA and even those who do are generally not prepared to teach domain-specific knowledge, such as science and math, to preschoolers (Isenberg, 2000). Although we have identified situational features that facilitate using basic research to develop programs for classroom practice, many of these will not be met in the typical school setting. Further, even when we operate under very favorable circumstances, we must employ a cycle of planning, implementation, and oversight as we design learning experiences that will best support learners as they build scientific understandings.

Still, these challenges must be met if we are to expand promising research-based programs beyond their development sites. Our team has begun the hard work of adapting our preschool program to make it easier to adopt in new sites, while maintaining its integrity and benefits for learners. Bridging between laboratory and classroom is a complex endeavor, but critical if we are to identify ways to upgrade students' scientific, mathematical, and technical literacies to meet the challenges and opportunities of the twenty-first century.

Acknowledgments

The work reported here and the writing of the chapter was supported by the National Science Foundation Grants (REC-0529579 and LIS 9720410) as well as research grants from UCLA and Rutgers University to the first author.

Note

1 This lesson series combines successful learning experiences developed in PrePS classrooms with adapted lessons from Massey and Roth's (2004) *Science for Developing Minds* series.

References

Au, T. K. (1994). Developing an intuitive understanding of substance kinds. *Cognitive Psychology*, 27, 71–111.

Bartlett, F. (1932). *Thinking: An experimental and social study*. New York: Basic Books.

Bowman, B. T., Donovan, M. S., & Burns, M. S. (2001). *Eager to learn: Educating our preschoolers*. Committee on Early Childhood Pedagogy, Commission on Behavioral and Social Sciences and Education. Washington, DC: National Academies Press.

Bransford, J., Brown, A., & Cocking, R. (Eds.) (1999). *How people learn: Brain, mind, experience, and school*. Washington, DC: National Academies Press.

Brenneman, K., Gelman, R., Massey, C., Roth, Z., Downs, L. E., & Nayfeld, I. (2007). Preschool pathways to science: Assessing and fostering scientific reasoning in preschoolers. Presented at the biennial meeting of the Cognitive Development Society, Santa Fe, NM, October 2007.

Brenneman, K., & Louro, I. F. (2008). Science journals in the preschool classroom. *Early Childhood Education Journal*, 36, 113–119.

Brenneman, K., Massey, C., Machado, S., & Gelman, R. (1996). Young children's plans differ for "writing" and drawing. *Cognitive Development*, 11, 397–419.

Carey, S. (2000). Science education as conceptual change. *Journal of Applied Developmental Psychology*, 21(1), 13–19.

Celce-Murcia, M., & Larsen-Freeman, D. (1983). *The grammar book: An ESL/EFL teacher's course*. Rowley, MA: Newbury House.

Chouinard, M. M. (2007). Children's questions: A mechanism for cognitive development. *Monographs of the Society for Research in Child Development*, 72, 1–112.

Duschl, R. A., Schweingruber, H. A., & Shouse, A. W. (2006). *Taking science to school: Learning and teaching science in grades K–8*. Board on Science Education, Center for Education,

Division of Behavioral and Social Sciences and Education. Washington, DC: National Academies Press.

Gelman, R. (1993). A rational-constructivist account of early learning about numbers and objects. In D. Medin (Ed.), *Learning and motivation* (Vol. 30). New York: Academic Press. pp. 61–96.

Gelman, R., & Brenneman, K. (2004). Science learning pathways for young children. *Early Childhood Research Quarterly*, 19(1), 150–158.

Gelman, R., Brenneman, K., Macdonald, G., & Román, M. (2009). *Preschool pathways to science (PrePS): Facilitating scientific ways of thinking, talking, doing and understanding*. Baltimore: Brookes.

Gelman, R., & Lucariello, J. (2002). Role of learning in cognitive development. In H. Pashler & C. R. Gallistel (Eds.), *Stevens' handbook of experimental psychology (3rd edn), Vol. 3: Learning, motivation, and emotion*. New York: Wiley. pp. 395–443.

Gelman, R., Meck, G., Romo, L., Meck, B., Francis, W., & Fritz, C. O. (1995). Integrating science concepts into intermediate English as a second language (ESL) instruction. In R. F. Macias & R. Garcia-Ramos (Eds.), *Changing schools for changing students: An anthology of research on minorities, schools, and society*. Santa Barbara: University of California Linguistic Minority Research Institute. pp. 181–203.

Gelman, R., Romo, L., & Francis, W. S. (2002). Notebooks as windows on learning: The case of a Science into ESL program. In N. Granott & J. Parziale (Eds.), *Microdevelopment: Transition processes in development and learning*. Cambridge: Cambridge University Press.

Gelman, S. A. (2003). *The essential child: Origins of essentialism in everyday thought*. New York: Oxford University Press.

Gelman, S., & Opfer, J. (2002). Development of the animate–inanimate distinction. In U. Goswami (Ed.), *Blackwell handbook of childhood cognitive development*. Malden: Blackwell. pp. 151–166.

Goswami, U. (Ed.) (2002). *Blackwell handbook of childhood cognitive development*. Malden: Blackwell.

Hammer, D. (1999). Physics for first graders? *Science Education*, 83, 797–799.

Hickling, A., & Gelman, S. (1995). How does your garden grow? Evidence of an early conception of plants as biological kinds. *Child Development*, 66, 856–876.

Isenberg, J. P. (2000). The state of the art in early childhood professional preparation. In D. Horm-Wingerd & M. Hyson (Eds.), *New teachers for a new century: The future of early childhood professional preparation*. Washington, DC: U.S. Department of Education. pp. 17–58.

Lee, O. (2005). Science education with English language learners: Synthesis and research agenda. *Review of Educational Research*, 75, 491–530.

Massey, C., & Gelman, R. (1988). Preschoolers' ability to decide whether a photographed unfamiliar object can move itself. *Developmental Psychology*, 24(3), 307–317.

Massey, C., & Roth, Z. (1997). Feeling colors and seeing tastes: Kindergarteners' learning about sensory modalities and knowledge acquisition. Presented at the biennial meeting of the Society for Research in Child Development, Washington, DC, April 2007.

Massey, C., & Roth, Z. (2004). *Science for developing minds series: A science curriculum for kindergarten and first grade*. A fully evaluated four volume NSF-funded curriculum series. Philadelphia: Edventures.

McCloskey, M. (1983). Naïve theories of motion. In D. Gentner & A. L. Stevens (Eds.), *Mental models*. Hillsdale, NJ: LEA. pp. 299–324.

Michaels, S., Shouse, A. W., & Schweingruber, H. A. (2008). *Ready, set, science! Putting research to work in K–8 science classrooms*. Board on Science Education, Center for Education, Division of Behavioral and Social Sciences and Education. Washington, DC: The National Academies Press.

Piaget, J. (1952). *The origins of intelligence in children*. Oxford: International Universities Press.

Rosengren, K. S., Gelman, S. A., Kalish, C. W., & McCormick, M. (1991). As time goes by: Children's early understanding of growth in animals. *Child Development*, 62, 1302–1320.

Schulz, L. E., & Bonawitz, E. B. (2007). Serious fun: Preschoolers engage in more exploratory play when evidence is confounded. *Developmental Psychology*, 43, 1045–1050.

Spelke, E. (2000). Core knowledge. *American Psychologist*, 55, 1233–1243.

Stevenson, H., & Stigler, J. W. (1994). *The learning gap: Why our schools are failing and what we can learn from Japanese and Chinese education*. New York: Simon & Schuster.

Stigler, J. W., & Hiebert, J. (1999). *The teaching gap: Best ideas from the world's teachers for education in the classroom*. New York: The Free Press.

Vygotsky, L. S. (1962). The development of scientific concepts in childhood. In L. S. Vygotsky, E. Hanfmann, & G. Vakar (Eds.), *Thought and language: Studies in communication*. Cambridge: MIT Press. pp. 82–118.

Williams, E. M. (2000). Causal reasoning by children and adults about the trajectory, context and animacy of a moving object. Los Angeles: UCLA, unpublished doctoral dissertation.

Yoshida, M., Fernandez, C., & Stigler, J. (1993). Japanese and American students' differential recognition memory for teachers' statements during a mathematics lesson. *Journal of Educational Psychology*, 85(4), 610–617.

Zur, O., & Gelman, R. (2004). Young children can add and subtract by predicting and checking. *Early Childhood Research Quarterly*, 19, 121–137.

9 A Research-Based Instructional Model for Integrating Meaningful Learning in Elementary Science and Reading Comprehension

Implications for Policy and Practice

Nancy R. Romance and Michael R. Vitale

An emerging trend in education is the attempt to dynamically link ongoing research initiatives for advancing the quality of K–12 teaching and learning with the more generally evolving process of systemic school reform [e.g., No Child Left Behind (NCLB), Bush, 2001]. In advocating an operational strategy that integrates and applies paradigmatically different interdisciplinary research perspectives (e.g., Bransford, Brown, Cocking, Donovan, & Pellegrino, 2000) to the persistent problems of science education reform, this chapter is designed to raise the awareness of educational practitioners, researchers, and policy developers regarding the mechanisms associated with advancing (and the potential for) in-depth meaningful learning in science as a critical element in furthering school reform efforts, in general, and science education at the elementary level, in particular.

The broader perspective presented in the chapter departs from current elementary education reform initiatives that emphasize the improvement of achievement outcomes in literacy (e.g., reading comprehension, writing) as ends in themselves rather than as the means for furthering meaningful learning in content domains (e.g., science). Also addressed is how reform has neglected issues associated with aligning meaningful learning outcomes in science and literacy within a conceptually coherent and well-articulated curricular structure in science in a manner consistent with advancements in interdisciplinary research having implications for enhancing the quality of student learning. Finally, the chapter advocates how such systemic interdisciplinary science education initiatives are necessary for producing student achievement outcomes in both science and literacy (Romance & Vitale, 2007).

The chapter offers a set of interdisciplinary perspectives and evidence about how a researched-based instructional model, Science IDEAS (and several related initiatives), implemented across multiple years provides a framework relevant to identifying key research and policy issues associated with achieving meaningful, in-depth science learning in K–5 classrooms in a manner that also furthers the literacy development of all students. Such interdisciplinary views, all relevant to the three aspects of science education (student learning, teaching, research), have the potential to accelerate meaningful learning in science.

Trends and Issues Relating to Reform in Reading and Science

Despite ongoing reform initiatives, national and international reports including NAEP 2005 Trial Urban Science (Lutkus, Lauko, & Brockway, 2006), NAEP 2007

Reading (Lee, Grigg, & Donahue, 2007), TIMSS 2007 Science (Gonzales et al., 2008), and NCEE/RA 2008 Reading First (Gamse et al., 2008) have found the quality of student achievement in science and literacy in the United States to be continuing a systemic problem. For example, while the 2005 Trial Urban District Assessment (Lutkus et al., 2006) noted a slight increase in fourth-grade science scores, eighth-grade scores remained flat and twelfth-grade scores actually declined. The 2007 TIMSS (Gonzales et al., 2008) indicated that science achievement in grades 4 and 8 was not measurably different from scores attained in 1995 and that only 15 percent and 10 percent of fourth and eighth graders, respectively, scored at or above the advanced international benchmark in science. It is not surprising, then, that upon reaching high school many students—representing all socio-economic status (SES) strata—do not have sufficient prior knowledge in the form of the conceptual understanding necessary to perform successfully in secondary science courses. Further, researchers have suggested such poor student achievement trends in science are logically related to the lack of instructional time devoted to in-depth science teaching in elementary schools (see Dillon, 2006; Jones et al., 1999; Klentschy & Molina-De La Torre, 2004), a key issue for successful reform in science (Hirsch, 1996; Vitale, Romance, & Klentschy, 2006), and, in a related fashion, to reading comprehension (Chall, 1985; Guthrie & Ozgungor, 2002).

Within a school accountability framework, the predominant reform strategy (see Weiss, 2006) has been to increase the time allocated to basal reading programs by reducing the instructional time allocated to science, especially for at-risk students most dependent upon school to learn. Such reform strategies, however, have not resulted in the desired outcomes in student reading achievement. For example, although the 2007 NAEP (Lee et al., 2007) reported an increase in fourth-grade reading achievement at or above the basic level, only 33 percent and 8 percent scored at the proficient or advanced levels, respectively. And, similar to science results, middle school achievement in reading remained flat at the basic level, with only 3 percent scoring at the advanced levels in reading. More recently, focusing on young students, the large-scale Reading First Impact Study (Gamse et al., 2008) found no significant achievement in third-grade general reading comprehension after three years of implementation of an early literacy initiative.

Science and literacy researchers have addressed aspects of these reform concerns involving the interdependency of science and literacy. For example, Duke and Pearson (2002) noted little involvement in either "doing" science or reading informational text at the primary level and documented teachers' erroneous belief that science comprehension must wait until students become proficient decoders in reading. However, emphasis on K–2 instructional interventions that emphasize the development of meaningful knowledge in science is becoming consistent with emerging literacy trends (Palmer & Stewart, 2003) that emphasize the use of informational text for developing background knowledge and comprehension proficiency at the primary levels (see also Holliday, 2004; Klentschy & Molina-De La Torre, 2004; Ogle & Blachowicz, 2002; Gould, Weeks, & Evans, 2003, for related views).

A related problematic approach for linking science and literacy is the use of basal reader selections for science learning. As Smith (2001) found, these materials have a narrow and fragmented focus on science concepts. Additionally, such use of basal readers in place of science curricular resources and the consequent lack of time devoted to meaningful science instruction are further exacerbated by elementary teachers' lack of science content knowledge, a fact well documented in the literature (e.g., Weiss, 2006).

The lack of emphasis in linking content-area reading comprehension to science at the elementary level effectively withholds opportunities for meaningful science learning and literacy development (i.e., in-depth reading comprehension proficiency) from K–5 students. The negative effect of such curricular decisions is further magnified when struggling (at-risk) learners subsequently are enrolled in middle- and high-school science courses and is more likely a major contributor to the "Black–Hispanic–White" test gap in science and reading comprehension than at-risk SES status (e.g., Lutkus et al., 2006). Although the short-term pressures of NCLB accountability mandates might be difficult for elementary schools to overcome, of even greater importance are the long-term curricular implications that serve as barriers for preparation of students for middle- and high-school science courses and general content-area reading comprehension that ultimately become manifest at the high-school level (Lee et al., 2007; Lutkus et al., 2006; Snow, 2002).

Linking Consensus Research Perspectives to Meaningful Learning in Science

Current interdisciplinary research related to meaningful learning as summarized in the National Academy Press report, *How People Learn* (Bransford et al., 2000), provides a foundation as to why and how early conceptual understanding in content domains, such as science, establishes the prior knowledge and eventual organizational knowledge structure necessary to support all future content-area learning and literacy development (e.g., reading comprehension as a form of understanding, coherent writing). In their overview, Bransford et al. summarized studies of experts and expertise as a unifying framework for understanding meaningful learning. Experts, in comparison with novices, demonstrate a highly developed organization of knowledge that emphasizes an in-depth understanding of concepts in their discipline that, in turn, they are able to access efficiently and apply with automaticity. Although the instructional implications of such perspectives (discussed below) are highly supportive of the importance of building student conceptual understanding in science, these same implications are in direct conflict with present trends in elementary education that advocate emphasis on narrative, non-content reading and an overemphasis on test preparation skills (e.g., Hart & Risley, 2003; Hirsch, 1996, 2003; Walsh, 2003).

In considering domain expertise as a foundation for in-depth learning, the notion of knowledge-based instruction provides a methodological perspective for approaching curriculum and instruction in a conceptually coherent fashion. More specifically, cognitive scientists involved in the development of intelligent tutoring systems (e.g., Kearsley, 1987; Luger, 2008) have noted that the distinguishing characteristic of knowledge-based instruction models is that all aspects of instruction, including (a) the determination of learning sequences, (b) the selection of teaching methods, (c) the specific activities required of learners, and (d) the evaluative assessment of student learning success, are related explicitly to an overall design representing the logical structure of the concepts within the subject-matter discipline to be taught. In this regard, the emphasis by Bransford et al. (2000) on expertise is consistent with and amplifies the importance of an explicit curricular focus on core concept relationships, the enhancement of prior knowledge, and the development of conceptual understanding and use of knowledge in application tasks as being of paramount importance for meaningful learning to occur (see also Schmidt et al., 2001).

The preceding also emphasizes the extensive role of varied experiences (i.e., cumulative practice) that focus on conceptual knowledge to be learned. These conceptual relationships become critical to the development of the different aspects of automaticity associated with expert mastery in any discipline (see Anderson, 1992, 1993). In related research, Sidman (1994) and others (e.g., Artzen & Holth, 1997; Dougher & Markham, 1994) have explored the conditions under which extensive practice to automaticity focusing on one subset of concept relationships can result in additional subsets of relationships being learned without explicit instruction. In these studies, the additional relationships were not taught, but rather were implied by the original set of relationships that was taught (i.e., formed equivalence relationships). In other work, Niedelman (1992) and Anderson (1996) have offered interpretations of research issues relating to transfer of learning that are consistent with a knowledge-based approach to learning. Considered together, these findings represent a set of perspectives on what constitutes meaningful learning (in science) that must be strategically linked to the use of age-appropriate instructional interventions in order to engender meaningful learning in science. The active development of such in-depth conceptual understanding serves as a foundation (e.g., Carnine, 1991; Glaser, 1984; Kintsch, 1998; Vitale & Romance, 2000) for the use of existing knowledge in the acquisition and communication of new knowledge as well as scientific literacy and general comprehension.

Representative Research Demonstrating the Importance of Science Instruction in Elementary (K–5) Settings

Building on the preceding perspectives, a major emphasis of a sound K–5 science curriculum is that the science knowledge being taught offers a meaningful context through which students are able to experience learning more about what is being learned in a cumulative fashion that enhances their capacity for understanding and in-depth learning (i.e., comprehension). Such science conceptual knowledge deals with everyday events that students experience, enabling them to (a) link together different events they observe, (b) anticipate the occurrence of events (or manipulate conditions to produce outcomes), and (c) make meaningful interpretations of events that occur, all of which are key elements of meaningful understanding in science (Vitale & Romance, 2000; Vitale, Romance, & Dolan, 2006).

Representative Research in Grades K–3

Early childhood researchers (Conezio & French, 2002; French, 2004; Smith, 2001) reported that science learning and early literacy development resulted from curricular approaches in which science experiences provide rich learning contexts. Gelman and Brenneman (2004) demonstrated how a preschool science program rich with guided hands-on activities served as the basis for instruction that supports early subject-matter learning in young children. Smith's (2001) work with three- to six-year-olds described how active learning in science is naturally motivating if topics are approached with sufficient depth and time, a position emphasized in the 1995 "National Science Education Standards" (see Rakow & Bell, 1998). Further, their analyses of the curricular trends of competing nations, Schmidt et al. (2001) noted that high-achieving nations had a conceptually coherent, meaningfully sequenced, and well-articulated science curriculum for all students. Finally, Ginsburg and Golbeck (2004) suggested that both developmental researchers as well as practitioners should be critically open to the possibilities of unexpected competence in young children in learning science (e.g., Asoko, 2002; Newton, 2001; Revelle et al., 2002; Sandall, 2003).

Representative Research in Grades 3–5

The importance of building cumulative student background knowledge in science has been demonstrated repeatedly by the extensive work of Guthrie and his colleagues (e.g., Guthrie, Wigfield, & Perencevich, 2004; Guthrie & Ozgungor, 2002) as enhancing student reading comprehension of upper elementary students. Armbruster and Osborn (2001) summarized numerous research findings that demonstrated positive student achievement in reading comprehension resulting from integrating science with reading/language arts. Others (Beane, 1995; Ellis, 2001; Hirsch, 1996, 2001; Schug & Cross, 1998; Yore, 2000) also have presented findings in support of interventions in which curriculum content serves as a powerful framework for building background knowledge and increased proficiency in reading comprehension.

The Science IDEAS Instructional Model

As a cognitive science-oriented model, Science IDEAS in grades 3–5 exemplifies an in-depth, instructional approach (e.g., Mintzes, Wandersee, & Novak, 1998) that emphasizes students learning more about what is being learned in a meaningful fashion. The model is designed to prepare teachers for instruction that engenders student in-depth understanding of both science concepts and the nature of science that is consistent with national science standards (e.g., AAAS, 1993; NRC, 1996) and articulated across grade levels. The architecture of the model involves using a conceptually coherent framework of concepts (see Figure 9.1) for sequencing different types of classroom activities (e.g., hands-on, reading, concept-mapping, journaling/writing). This approach is consistent with recommendations (e.g., Donovan, Bransford, & Pellegrino, 1999; Romance & Vitale, 2006; Vitale & Romance, 2006a) that also provide the means for a curricular-embedded approach to assessment (e.g., Pellegrino, Chudowsky, & Glaser, 2001; Vitale et al., 2006).

Implementation of the Science IDEAS model (see Figure 9.1) involves teacher construction of propositional concept maps representing the conceptual structure of the science concepts to be taught. This serves as the framework for identifying, organizing, and sequencing all instructional activities and assessments. As a result, Science IDEAS requires comprehensive professional development that focuses on increasing teacher science understanding and providing support through teacher leaders (e.g., King & Newmann, 2001).

Science IDEAS amplifies the importance of focusing all aspects of instruction on the cumulative development of age-appropriate student mastery of core concept

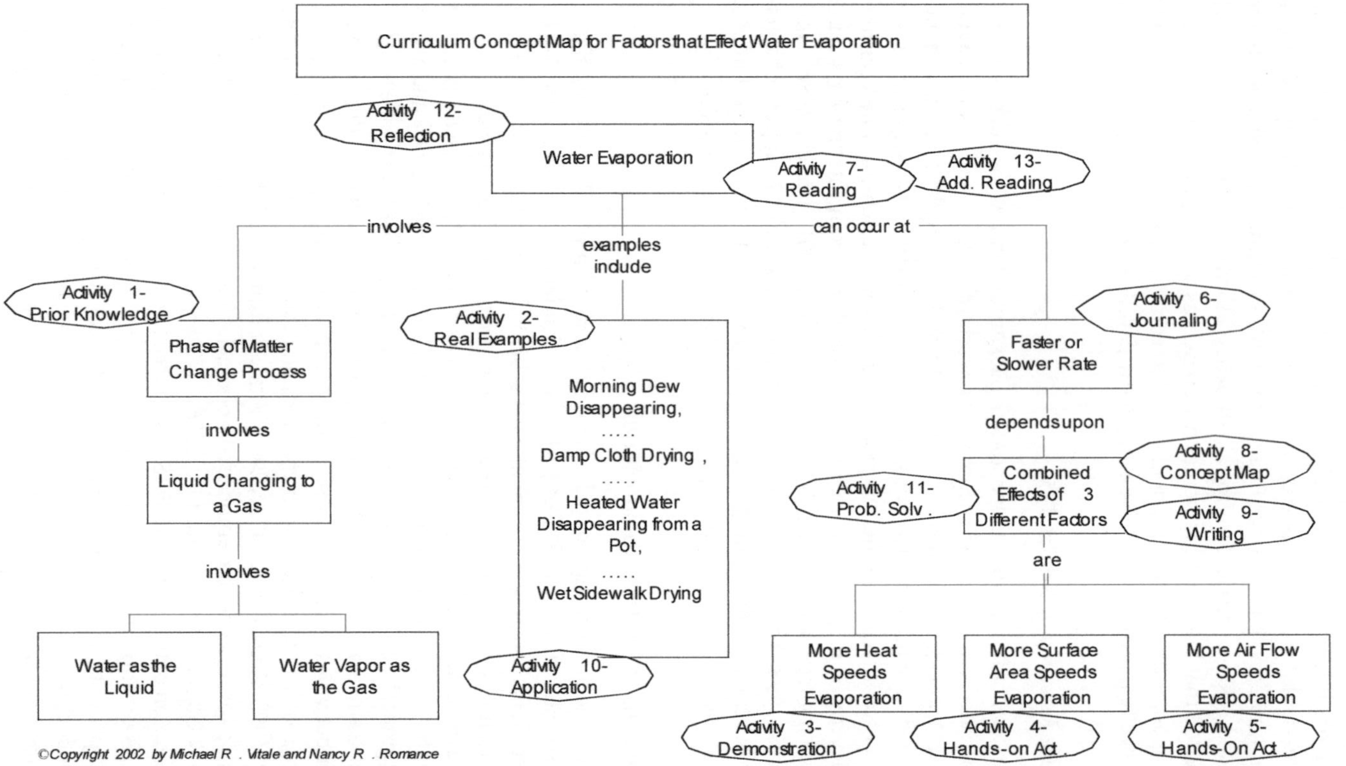

Figure 9.1 Simplified Illustration of a Propositional Curriculum Concept Map Used as a Guide by Grade 4 *Science IDEAS* Teachers to Plan a Sequence of Knowledge-Based Instruction Activities. Adapted from "Implementing an in-depth expanded science model in elementary schools: Multi-year findings, research issues, and policy implications" by N. Romance & M. Vitale, *International Journal of Science Education*, 23(4), 373–404 (2001). Reprinted by permission of the publisher Taylor & Francis Ltd. (http://www.tandf.co.uk/journals).

relationships within physical, earth, and life sciences consistent with learning progression methodology (e.g., Duschl, Schweingruber, & Shouse, 2007). Science IDEAS involves daily two-hour blocks of time, which replace regular reading/language arts instruction across grades 3–5, and consists of multiday science lessons emphasizing cumulative learning experiences. In referencing Figure 9.1, when teaching core concept relationships, teachers may use a variety of instructional approaches (e.g., hands-on science experiments, reading text/trade/internet science materials, writing about science, science projects, maintaining science journals, propositional concept mapping) focused on enhancing conceptual understanding (Hapgood, Magnusson, & Palincsar, 2004; Romance & Vitale, 1992, 2001).

The Science IDEAS model emphasizes the use of student-constructed science journals for archiving all lessons and activities, posing questions, and communicating what has been learned in varied formats (e.g., charts, graphs, summaries and conclusions, questions, illustrations) for linking new information with prior knowledge as a natural part of science learning (Hapgood et al., 2004; Harlen, 1988, 2001; Rivard, 1994). This approach also is consistent with recent research (e.g., Klentschy & Molina-De La Torre, 2004; Magnusson & Palincsar, 2006; Palincsar & Magnusson, 2001) demonstrating how the integration of hands-on science activities (first-hand investigations) with reading and writing (second-hand investigations), rather than hands-on science alone, can result in increased student achievement outcomes in science and literacy.

Evidence in Support of the Effectiveness of the Science IDEAS Model

Overall Research Design

The proposition that replicability of research findings in diverse settings is the goal of all scientific enterprises (e.g., Sidman, 1960) provides a framework for interpreting the multi-year findings associated with the Science IDEAS model. These multi-year findings also are consistent with the concept of "patch" experiments and the associated implications for external validity outlined by Stanley and Campbell (1963). The following sections overview student achievement outcomes associated with implementation of the Science IDEAS model reported in the literature and other professional outlets from 1992 through 2007.

Pattern of Research Evidence: 1992–2001

The research studies completed from 1992 to 2001 consisted of a series of year-long studies conducted in authentic school settings. In the first study (Romance & Vitale, 1992), three grade 4 classrooms in an average-performing school implemented the Science IDEAS model. The achievement measures were Iowa Test of Basic Skills (ITBS) Reading and Miller Analogies Test (MAT) Science subtests. Results showed that Science IDEAS students outperformed comparison students by approximately one year's grade equivalent (GE) in science achievement (+ .93 GE) and one-third of a GE in reading achievement (+ .33 GE). In the second study, conducted the following school year, Science IDEAS was again implemented with the same three teachers/

classrooms in grade 4. The results of this second-year replication obtained similar levels of achievement effects, with Science IDEAS students outperforming comparison students by + 1.5 GE in science and + .41 GE in reading (Romance & Vitale, 2001).

In the third and fourth studies, which followed (Romance & Vitale, 2001), the robustness of the model was tested by (a) increasing the number of participating teachers/schools, (b) broadening the grade levels to grades 4 and 5, and (c) enhancing the diversity of participants by focusing on district-identified at-risk students. The results of the year 3 study (Romance & Vitale, 2001) showed that the low-SES, minority at-risk Science IDEAS students in grade 5 significantly outperformed comparable controls by + 2.3 GE in science and by + .51 GE in reading over a five-month (vs. school year) intervention. In contrast with the grade 5 findings, no significant treatment effect was found for the younger grade 4 at-risk students. However, in a supplementary study, the levels of achievement growth for the original grade 4 classrooms studied were comparable to those obtained originally in years 1 and 2.

In the fourth study, the number of participating schools and teachers was increased to 15 school sites and 35 classroom teachers. The results of the fourth study showed that Science IDEAS students displayed greater overall achievement on both science (+ 1.11 GE) and reading (+ .37 GE). As in year 3, no interactions were found between student demographics and treatment, indicating that Science IDEAS was effective consistently across grade levels (grade 4 and grade 5) and with both regular and at-risk students.

Pattern of Research Evidence: 2004–2007

All of the preceding studies (1992–2001) focused on individual teachers/classrooms located in a variety of different school sites. However, beginning with 2002, the Science IDEAS research framework was composed of two different initiatives. The primary initiative (Romance & Vitale, 2008) involved implementing Science IDEAS on a schoolwide basis in grades 3, 4, and 5 in an increasing number of participating schools (from two to 13 over the multi-year project). The increasing number of such schoolwide interventions provided a framework for the study of issues relating to scale-up of the Science IDEAS model through a project supported by the National Science Foundation. The second initiative consisted of two small-scale studies embedded within the overall scale-up project that explored extrapolations of the Science IDEAS model to grades K–2 (Vitale & Romance, 2007a) and as a setting for reading comprehension strategy effectiveness (Vitale & Romance, 2006b).

This section overviews the effect of Science IDEAS on student achievement in science and reading (Romance & Vitale, 2008). Figure 9.2 shows the adjusted GE means for grade 4 and 5 Science IDEAS and Basal Reading classrooms during the 2003–2004 school year. After statistically equating students for differences on the preceding year's state-administered Florida's Comprehensive Assessment Test (FCAT) Reading achievement, Science IDEAS students displayed significantly higher ITBS achievement on reading and science.

Figure 9.3 shows the effect of Science IDEAS on student achievement in new and continuing project schools during the 2004–2005 school year. After statistically equating students for differences on the preceding year's state-administered FCAT Reading achievement, Science IDEAS students in schools with three years' experience ($n = 4$)

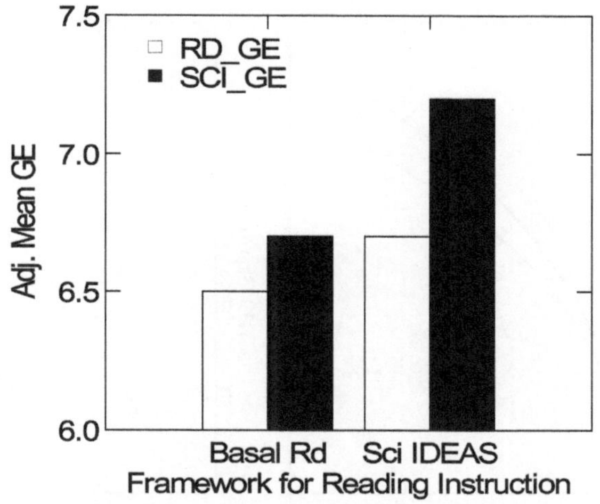

Figure 9.2 Adjusted Grade-Equivalent Means on ITBS Reading and Science for Science IDEAS and Comparison (Basal) Students for 2003–2004.

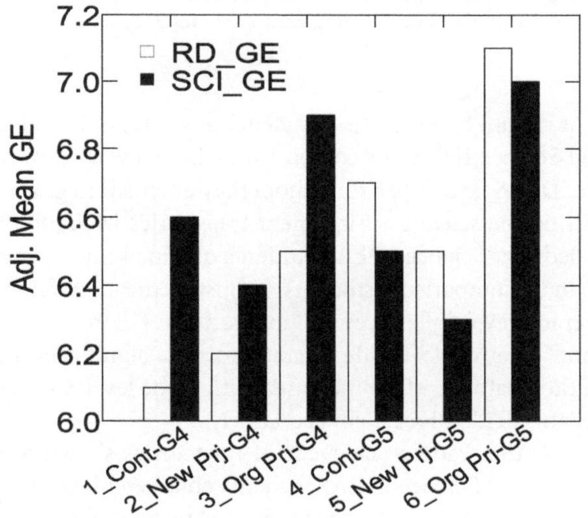

Figure 9.3 Adjusted Grade-Equivalent Means on ITBS Reading and Science for Continuing and New Science IDEAS and Comparison (Basal) Students.

displayed significantly higher ITBS achievement than Basal Reading schools on both reading and science. However, at the same time, results for Science IDEAS schools in their initial year ($n = 4$) varied, suggesting that more than one year for implementation experience is required before the Science IDEAS model is implemented with effectiveness.

Figure 9.4 shows the cross-sectional effect of Science IDEAS across grades 3–8 on ITBS science and reading achievement across 13 participating and 12 comparison schools in 2006–2007. Both groups of schools were comparable demographically (approximately 60 percent minority, 45 percent free/reduced lunch). In interpreting

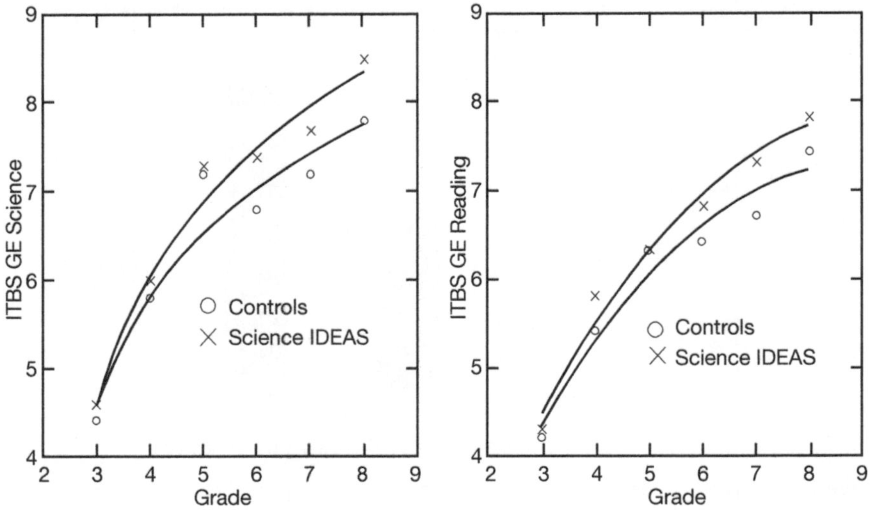

Figure 9.4 2006–2007 ITBS Achievement Trajectories for Science IDEAS and Control Schools in Science and Reading across Grades 3–8.

these figures, it should be noted that students in grades 6, 7, and 8 (who had previously attended Science IDEAS or comparison schools) were expressed as extensions of the Science IDEAS or comparison school they attended in grade 5.

In interpreting the science achievement trajectories in Figure 9.4, linear models analysis revealed that Science IDEAS students obtained higher overall ITBS science achievement than comparison students (adjusted mean difference = + .38 GE in Science with grade-level differences ranging from + .1 GE to + .7 GE). Both Treatment main effect and Treatment × Grade interaction were significant, indicating that the magnitude of the treatment effect increased with grade level. Covariates were Gender and At-Risk Status (Title I Free/Reduced Lunch).

In interpreting the reading achievement trajectories shown in Figure 9.4, linear models analysis found Science IDEAS students obtained higher overall ITBS reading achievement than comparison students (adjusted mean difference = + .32 GE in reading with grade-level differences ranging from .0 GE to + .6 GE). Whereas the overall Treatment main effect was significant, the Treatment × Grade interaction was not. Covariates were Gender and At-Risk Status (Title I Free/Reduced Lunch).

Other results of the analyses were that (a) the treatment effect was consistent across at-risk and non-at-risk students for both ITBS science and reading, and (b) girls outperformed boys on ITBS reading (there was no gender effect on science).

Elaborative Science IDEAS Mini-Studies in K–2 and Grade 5

The second initiative consisted of two mini-studies that explored extrapolations of the Science IDEAS model to grades K–2 and as a setting for reading comprehension strategy effectiveness. The objective of the K–2 study (Vitale & Romance, 2007a) was to adapt the grade 3–5 Science IDEAS model to grades K–2 in two Science IDEAS schools (vs. two comparison schools). In grades K–2, teachers incorporated a daily 45-minute science instruction block while continuing their daily basal reading instruction. Results found an overall main effect in favor of Science IDEAS students on both ITBS science (+ .28 GE) and reading (+ .42 GE). However, for ITBS reading, a significant Treatment × Grade interaction was found. Subsequent simple effects analysis showed a significant difference in grade 2 of + .72 GE on ITBS reading, but no effect in grade 1. Other results showed a significant effect of White vs. Nonwhite (+ .38 GE), but no Treatment × Ethnicity interaction.

The grade 5 study (Vitale & Romance, 2006b) explored whether research-validated reading comprehension strategies (see Vitale & Romance, 2007b) would be differentially effective in the cumulative meaningful learning setting established by Science IDEAS in comparison with basal reading instruction that emphasized narrative reading. After a seven-week intervention in which reading comprehension strategies were implemented in both Science IDEAS and basal reading classrooms, a 2 × 2 factorial design (with prior state-administered FCAT reading as a covariate) was used and the results showed that Science IDEAS students performed significantly better than basal students on both ITBS science (+ .38 GE) and reading (+ .34 GE). Although the main effect of reading comprehension strategy use was not significant, the Instructional Setting × Strategy Use interaction was significant [i.e., use of the reading comprehension strategy by Science IDEAS student improved their overall performance in both science (+ .17 GE) and reading (+ .53 GE), but Strategy Use had no effect in basal classrooms].

Summary of Science IDEAS Research Findings

The major conclusion based on the multi-year findings is that Science IDEAS has been shown to be effective in accelerating student achievement outcomes in science and reading comprehension in grades 3, 4, and 5. Further, the magnitude of the effects expressed in grade equivalents on nationally normed tests [ITBS, Scholastic Aptitude Test (SAT), MAT] was educationally meaningful. Based on these studies, Science IDEAS can be considered as a more effective replacement for basal reading programs that currently dominate instruction across grades 3–5. Another key finding was that the impact of the effects of Science IDEAS in grades 3, 4, and 5 was transferable to grades 6, 7, and 8. As a result, the Science IDEAS model offers major implications for curricular policy at the elementary level (Vitale et al., 2006). Other findings show the feasibility of adapting the model for grades K–2 and its effectiveness with regular and at-risk students. Overall the Science IDEAS model is suggestive of changes in curricular policy for linking science and literacy in elementary schools (Romance & Vitale, 2006).

Future Directions and Implications for Systemic Educational Reform

The evidence presented in this chapter suggests that the instructional time allocated to traditional reading instruction represents a misdirected curricular commitment to reform that has resulted in minimal policy emphasis on content area instruction (see Gamse et al., 2008; Hirsch, 1996). The "opportunity cost" of allocating instructional time to basal reading programs in present school reform initiatives denies students the benefits of interacting with the very forms of content-oriented instruction and reading materials that are necessary for success in middle- and high-school courses and the development of a potentially transferable proficiency in content-area reading comprehension. The implications for school reform are: (a) the preparation of students for successful, meaningful learning in middle and high school should be considered a major reform goal on which minimal progress has yet to be met, (b) practitioners should be encouraged to disregard the misconception that considers reading as a curriculum contact area in grades 3–8, and (c) the replacement of academically oriented science (and social studies) instruction that emphasizes meaningful learning with the non-content-oriented "literature" materials common to "basal reading curricula" is a major barrier that must be overcome if educational reform is to be successful.

Acknowledgments

Preparation of this chapter was supported by IES Project R305G04089 and National Science Foundation/IERI Project REC 0228353.

References

American Association for the Advancement of Science. (1993). *Benchmarks for science literacy.* New York: Oxford University Press.

Anderson, J. R. (1992). Automaticity and the ACT theory. *American Journal of Psychology*, 105, 15–180.

Anderson, J. R. (1993). Problem solving and learning. *American Psychologist*, 48, 35–44.

Anderson, J. R. (1996). ACT: A simple theory of complex cognition. *American Psychologist*, 51, 335–365.

Armbruster, B. B., & Osborn, J. H. (2001). *Reading instruction and assessment: Understanding IRA standards.* New York: Wiley.

Artzen, E., & Holth, P. (1997). Probability of stimulus equivalence as a function of training design. *Psychological Record*, 47, 309–320.

Asoko, H. (2002). Developing conceptual understanding in primary science. *Cambridge Journal of Education*, 32(2), 153–164.

Beane, J. A. (1995). Curriculum integration and the disciplines of knowledge. *Phi Delta Kappan*, 76, 646–622.

Bransford, J. D., Brown, A. L., Cocking, R. R., Donovan, S., & Pellegrino, J. W. (Eds.) (2000). *How people learn: Brain, mind, experience, and school* (Expanded edition). Washington, DC: National Academies Press.

Bush, G. W. (2001). *No child left behind.* Washington, DC: Educational Resources Information Center.

Carnine, D. (1991). Curricular interventions for teaching higher order thinking to all students: Introduction to a special series. *Journal of Learning Disabilities*, 24(5), 261–269.

Chall, J. S. (1985). Afterword. In R. C. Anderson et al. (Eds.), *Becoming a nation of readers: The report of the commission on reading.* Washington, DC: National Institute of Education. pp. 123–124.

Conezio, K., & French, L. (2002). Science in the preschool classroom: Capitalizing on children's fascination with the everyday world to foster language and literacy development. *Young Children*, 57(5), 12–18.

Dillon, S. (March 26, 2006). Schools push back subjects to push reading and math. *New York Times*. http://nytimes.com/2006/03/26/education/26child.html?pagewanted=1&_r=1.

Donovan, M. S., Bransford, J. D., & Pellegrino, J. W. (Eds.) (1999). *How people learn: Bridging research and practice*. Washington, DC: National Academies Press.

Dougher, M. J., & Markham, M. R. (1994). Stimulus equivalence, functional equivalence and the transfer of function. In S. C. Hays, L. J. Hays, M. Santo, & O. Koichi (Eds.), *Behavior analysis of language and cognition*. Reno: Context Press. pp. 71–90.

Duke, N., & Pearson, P. D. (2002). Effective practices for developing reading comprehension. In A. E. Farstrup & S. J. Samuels (Eds.), *What research has to say about reading instruction*. Newark: International Reading Association. pp. 205–242

Duschl, R. A., Schweingruber, H. A., & Shouse, A. W. (2007). *Taking science to school: Learning and teaching science in grades K–8*. Washington, DC: National Academies Press.

Ellis, A. K. (2001). *Research on educational innovations*. Larchmont, NY: Eye on Education.

French, L. (2004). Science as the center of a coherent, integrated early childhood curriculum. *Early Childhood Research Quarterly*, 19, 138–149.

Gamse, B. C., Jacob, R. T., Horst, M., Boulay, B., Unlu, F., Bozzi, L., et al. (2008). *Reading first impact study final report*. (NCES 2009-4038). Washington, DC: National Center for Education Evaluation and Regional Assistance, Institute of Education Sciences, U.S. Department of Education.

Gelman, R., & Brenneman, K. (2004). Science learning pathways for young children. *Early Childhood Research Quarterly*, 19, 150–158.

Ginsburg, H. P. & Golbeck, S. L. (2004). Thoughts on the future of research on mathematics and science learning and education. *Early Childhood Research Quarterly*, 19, 190–200.

Glaser, R. (1984). Education and thinking: The role of knowledge. *American Psychologist*, 39(2), 93–104.

Gonzales, P., Williams, T., Jocelyn, L., Roey, S., Kastberg, D., & Brenwald, S. (2008). *Highlights from TIMSS 2007: Mathematics and science achievement of U.S. fourth- and eighth-grade students in an international context*. (NCES 2009-001). Washington, DC: U.S. Department of Education.

Gould, C. J., Weeks, V., & Evans, S. (2003). Science starts early. *Gifted Child Today Magazine*, 26(3), 38–41.

Guthrie, J. T., & Ozgungor, S. (2002). Instructional contexts for reading engagement. In C. C. Block & M. Pressley (Eds.), *Comprehension instruction: Research-based best practices*. New York: Guilford Press. pp. 275–288.

Guthrie, J. T., Wigfield, A., & Perencevich, K. C. (Eds.) (2004). *Motivating reading comprehension: Concept-oriented reading instruction*. Mahwah, NJ: LEA.

Hapgood, S., Magnusson, S. J., & Palincsar, A. S. (2004). Teacher, text, and experience: A case of young children's scientific inquiry. *The Journal of the Learning Sciences*, 13(4), 455–505.

Harlen, W. (1988). *The teaching of science*. London: David Fulton.

Harlen, W. (2001). *Primary science, taking the plunge* (2nd edn.). Portsmouth: Heinemann.

Hart, B. H., & Risley, T. R. (2003). The early catastrophe: The 30 million word gap. *American Educator*, 27(1), 4–9.

Hirsch, E. D. (1996). *Schools we need: And why we don't have them*. New York: Doubleday.

Hirsch, E. D. (2001). Seeking breadth and depth in the curriculum. *Educational Leadership*, 59(2), 21–25.

Hirsch, E. D. (2003). Reading comprehension requires knowledge—of words and the world: Scientific insights into the fourth-grade slump and stagnant reading comprehension. *American Educator*, 27(1), 10–29.

Holliday, W. G. (2004). Choosing science textbooks: Connecting science research to common sense. In W. Saul (Ed.), *Crossing borders in literacy and science instruction*. Newark: International Reading Association and NSTA Press. pp. 383–394.

Jones, M. G., Jones, B. D., Hardin, B., Chapman, L., Yarbrough, T., & Davis, M. (1999). The impact of high-stakes testing on teachers and students in North Carolina. *Phi Delta Kappan*, 81, 199–203.

Kearsley, G. P. (Ed.) (1987). *Artificial intelligence and instruction: Applications and methods*. New York: Addison-Wesley.

King, M. B., & Newmann, F. M. (2001). Building school capacity through professional development: Conceptual and empirical considerations. *International Journal of Education Management*, 15(2), 86–93.

Kintsch, W. (1998). *Comprehension: A paradigm for cognition*. Cambridge: Cambridge University Press.

Klentschy, M. P., & Molina-De La Torre, E. (2004). Students' science notebooks and the inquiry process. In E. W. Saul (Ed.), *Crossing borders in literacy and science instruction: Perspectives on theory and practice*. Newark, DE: International Reading Association. pp. 340–354.

Lee, J., Grigg, W., & Donahue, P. (2007). *The nation's report card: Reading 2007. National assessment of educational progress at grades 4 and 8*. (NCES 2007-496). Jessup: National Center for Education Statistics.

Luger, G. F. (2008). *Artificial intelligence: Structures and strategies for complex problem-solving*. Reading: Addison Wesley.

Lutkus, A. D., Lauko, M. A., & Brockway, D. M. (2006). *The nation's report card: Trial urban district assessment science 2005* (NCES 2007-453). U.S. Department of Education, National Center for Education Statistics. Washington, DC: U.S. Government Printing Office.

Magnusson, S. J., & Palincsar, A. S. (2006). Teaching and learning inquiry-based science in the elementary school. In J. Bransford & S. Donovan (Eds.), *Visions of teaching subject matter guided by the principles of how people learn*. Washington, DC: National Academies Press.

Mintzes, J. J., Wandersee, J. H., & Novak, J. D. (1998). *Teaching science for understanding: A human constructivist view*. Englewood Cliffs, NJ: Academic Press.

National Research Council. (1996). *National science education standards*. Washington, DC: National Academies Press.

Newton, L. D. (2001). Teaching for understanding in primary science. *Evaluation and Research in Education*, 15(3), 143–153.

Niedelman, M. (1992). Problem solving and transfer. In D. Carnine & E. J. Kameenui (Eds.), *Higher order thinking*. Austin: Pro-Ed.

Ogle, D., & Blachowicz, C. L. Z. (2002). Beyond literature circles: Helping students comprehend informational texts. In C. C. Block & M. Pressley (Eds.), *Comprehension instruction*. New York: Guilford Press. pp. 247–258.

Palincsar, A. S., & Magnusson, S. J. (2001). The interplay of first-hand and second-hand investigations to model and support the development of scientific knowledge and reasoning. In S. M. Carver & D. Klahr (Eds.), *Cognition and instruction: Twenty-five years of progress*. Mahwah, NJ: LEA.

Palmer, R. G., & Stewart, R. A. (2003). Nonfiction trade book use in primary grades. *Reading Teacher*, 57(1), 38–48.

Pellegrino, J. W., Chudowsky, N., & Glaser, R. (Eds.) (2001). *Knowing what students know*. Washington, DC: National Academies Press.

Rakow, S. J., & Bell, M. J. (1998). Science and young children: The message from the National Science Education Standards. *Childhood Education*, 74(3), 164–167.

Revelle, G., Druin, A., Platner, M., Bederson, B., Hourcade, J. P., & Sherman, L. (2002). A visual search tool for early elementary science students. *Journal of Science Education and Technology*, 11(1), 49–57.

Rivard, L. (1994). A review of writing to learn in science: Implications for practice and research. *Journal of Research in Science Teaching*, 31(9), 969–983.

Romance, N. R., & Vitale, M. R. (1992). A curriculum strategy that expands time for in-depth elementary science instruction by using science-based reading strategies: Effects of a year-long study in grade 4. *Journal of Research in Science Teaching*, 29, 545–554.

Romance, N. R., & Vitale, M. R. (2001). Implementing an in-depth expanded science model in elementary schools: Multi-year findings, research issues, and policy implications. *International Journal of Science Education*, 23, 373–404.

Romance, N. R., & Vitale, M. R. (2006). Making the case for elementary science as a key element in school reform: Implications for changing curricular policy. In R. Douglas, M. Klentschy, & K. Worth (Eds.), *Linking science and literacy in the K–8 classroom*. Washington, DC: National Science Teachers Association. pp. 391–405.

Romance, N. R., & Vitale, M. R. (2007). Elements for bringing a research-validated intervention to scale: Implications for leadership in educational reform. Paper presented at the Annual Meeting of the American Educational Research Association, New York, NY, April 11, 2007.

Romance, N. R., & Vitale, M. R. (2008). Science IDEAS: A knowledge-based model for accelerating reading/literacy through in-depth science learning. Paper presented at the Annual Meeting of the American Educational Research Association, New York, NY, March 25, 2008.

Sandall, B. R. (2003). Elementary science: Where are we now? *Journal of Elementary Science Education*, 15(2), 13–30.

Schmidt, W. H., McKnight, C. C., Houang, R. T., Wang, H. C., Wiley, D. E., Cogan, L. S., et al. (2001). *Why schools matter: A cross-national comparison of curriculum and learning*. San Francisco: Jossey-Bass.

Schug, M. C., & Cross, B. (1998). The dark side of curriculum integration. *Social Studies*, 89, 54–57.

Sidman, M. (1960). *Tactics of scientific research*. New York: Basic Books.

Sidman, M. (1994). *Stimulus equivalence*. Boston: Author's Cooperative.

Smith, A. (2001). Early childhood—a wonderful time for science learning. *Investigating: Australian Primary & Junior Science Journal*, 17(2), 18–21.

Snow, C. E. (2002). *Reading for understanding: Toward a research and development program in reading comprehension*. Santa Monica, CA: RAND.

Stanley, J., & Campbell, D. (1963). Experimental and quasi-experimental designs for research on teaching. In N. Gage (Ed.), *Handbook of research on teaching*. Chicago: Rand-McNally. pp. 171–246.

Vitale, M. R., & Romance, N. R. (2000). Portfolios in science assessment: A knowledge-based model for classroom practice. In J. J. Mintzes, J. H. Wandersee, & J. D. Novak (Eds.), *Assessing science understanding: A human constructivist view*. San Diego: Academic Press. pp. 168–197.

Vitale, M. R., & Romance, N. R. (2006a). Research in science education: An interdisciplinary perspective. In J. Rhoton and P. Shane (Eds.), *Teaching science in the 21st century*. Arlington, VA: NSTA Press. pp. 329–351.

Vitale, M. R., & Romance, N. R. (2006b). Effects of embedding knowledge-focused reading comprehension strategies in content-area vs. narrative instruction in grade 5: Findings and research implications. Paper presented at the Annual Meeting of the American Educational Research Association, San Francisco, CA, April 10, 2006.

Vitale, M. R., & Romance, N. R. (2007a). Adaptation of a knowledge-based instructional intervention to accelerate student learning in science and early literacy in grades 1–2. Paper presented at the Annual Meeting of the American Educational Research Association, Chicago, IL, April 11, 2007.

Vitale, M. R., & Romance, N. R. (2007b). A knowledge-based framework for unifying content-area reading comprehension and reading comprehension strategies. In D. McNamara (Ed.), *Reading Comprehension Strategies*. Mahwah, NJ: LEA. pp. 73–104.

Vitale, M. R., Romance, N. R., & Dolan, F. (2006). A knowledge-based framework for the classroom assessment of student science understanding. In M. McMahon, P. Simmons, R. Sommers, D. DeBaets, & F. Crawley (Eds.), *Assessment in science: Practical experiences and education research*. Arlington, VA: NSTA Press. pp. 1–14.

Vitale, M. R., Romance, N. R., & Klentschy, M. (2006). Improving school reform by changing curriculum policy toward content-area instruction in elementary schools. Paper presented at the Annual Meeting of the American Educational Research Association, San Francisco, CA, April 9, 2006.

Walsh, K. (2003). Lost opportunity. *American Educator*, 27(1), 24–27.

Weiss, I. (2006). Professional development and strategic leadership to support effective integration of science and literacy. In R. Douglas, M. Klentschy, & K. Worth (Eds.), *Linking science and literacy in the K–8 classroom*. Washington, DC: National Science Teachers Association. pp. 359–372.

Yore, L. (2000). Enhancing science literacy for all students with embedded reading instruction and writing-to-learn activities. *Journal of Deaf Students and Deaf Education*, 5, 105–122.

10 Children's Cognitive Algebra and Intuitive Physics as Foundations of Early Learning in the Sciences

Friedrich Wilkening

If one wants to teach science to children of all ages, it seems vital to know their current knowledge about basic facts and laws of nature. This chapter will focus on children from about four to ten years old. Amazing knowledge structures will be revealed, not only pertaining to specific domains of intuitive physics and mathematics but also to scientific concepts in general. In particular, children's concepts will be shown to be governed by a cognitive algebra, pointing to the capabilities of coordinating variables and function thinking, both being essential for grasping the quantitative side of scientific concepts.

These findings arose from developmental applications of information integration theory (Anderson, 1981, 1996), a broad theory of cognition covering almost any field of psychology. Their implications have not readily been accepted within developmental psychology, possibly because of conflicts with several traditional claims about young children's cognitive capabilities (e.g., Piaget, 1970; Sander, 1932). From the present point of view, these traditional claims can be seen as myths. Because of their ostensible plausibility, they have been particularly resistant to change. First, we will see how three orthodox claims about cognitive development have been overcome within the framework of information integration theory and by its methodology. Second, we will go beyond the myths and focus on children's intuitive physics and explore if and how this can serve as a foundation for early learning in the sciences.

Children's Cognitive Capabilities Contrasted with Three Traditional Views

Three orthodox claims will be considered. If true, each of them would highly constrain young children's learning in the sciences. The claims are that (1) young children's cognitive structures are strictly one-dimensional and thus not ready for the integration of information, (2) the holistic, non-analytic mode of information processing has an ontogenetic primacy and is the natural one in young childhood, and (3) cognitive development can best be characterized by a logical sequence of pure conceptual structures.

Multidimensional Thinking, Integration Capacity, and Cognitive Algebra

The traditional assertion that young children's thinking is one-dimensional up to the age of about six to seven years goes back to Piaget's (1970) theory of cognitive development and has been maintained in more recent variants (Case, 1992; Siegler, 1998). Piaget's theory can be interpreted in various ways, ranging from young children having limitation in their information processing to children only being capable of processing one dimension of a stimulus in the extreme (centration). If centration were a true characteristic of young children's thinking, the question of information integration would of course be irrelevant for this age.

During the past 30 years, developmental applications of information integration theory have found a large amount of evidence to the contrary. In fact, the integration of multiple dimensions to form an overall judgment is at the core of the theory and has been studied in various domains. Most importantly in the present context, children as young as five years of age have been found not only to take into account more than one dimension but also to integrate the information according to meaningful algebraic rules. Developmental differences rarely occurred, but typically entailed the nature of integration rather than the amount of information integrated, particularly in situations in which the mathematical rule describing the objective world required a nonadditive integration such as multiplication. Both behaviors, needless to say, speak against the myth of one-dimensional cognition.

Figure 10.1 shows just three examples of many findings. The left panel refers to judgments of rectangular area (Wilkening, 1979). Children were shown chocolate bars, varied in a 4 × 4 factorial width × height design, and had to judge for each bar how long a row of the single pieces would be, if joined together. The data in Figure 10.1 show the judgments of one individual child, five years old. The graphs for the other children in that age group were essentially similar, and the overall pattern averaged over all five-year-olds looked even more systematic, clearly indicating an adding rule (Wilkening, 1979).

The center panel of Figure 10.1 shows data from a developmental study on time quantification by Wilkening, Levin, and Druyan (1987). Children of different ages judged the overall duration of two successive events, varied in a factorial design. The mean judgments of the group of six-year-olds are presented as an example here. It can clearly be seen that the judgment pattern is in virtually perfect agreement with the adding rule—in this case, the objectively correct one. There was no sign of *centration* in any individual child.

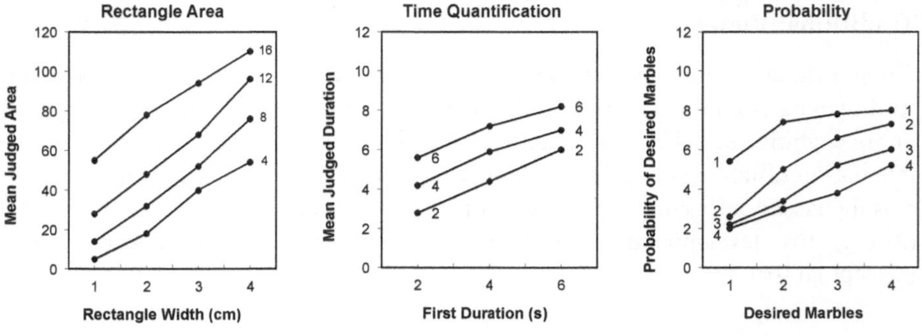

Figure 10.1 Children's Judgments Patterns in Three Different Experiments (see text for details). Curve parameters are: rectangle height in cm (left panel), duration in s (center panel), and number of undesired marbles (right panel).

The right-hand panel of Figure 10.1 shows data from a probability experiment by Wilkening and Anderson (1991). Children were shown a plate of marbles, containing one to four marbles of both a desired color and an undesired color. The task involved estimating probability based on the child's degree of happiness at picking a desired color in a blind draw. In effect, the children had to perform proportional reasoning. According to Piaget, this is impossible below the stage of formal operations, approximately 14 years old. The data in Figure 10.1 show the judgments of eight-year-olds. They mirror the pattern for the mathematically correct probability ratio rule almost perfectly. In any case, they speak against the myth of one-dimensional thinking (Falk & Wilkening, 1998).

Then why has the myth of one-dimensional thinking persisted? The answer lies in the reliance on the use of choice-task methodology, virtually without exception. In a choice task, the child is typically confronted with two stimuli, and the child has to decide which of the two has a greater value on a specified variable. Such tasks do not seem to be suited to detect unexpected integration rules (Wilkening & Anderson, 1982). In fact, neither Piaget's theory nor most of the more modern computational models of cognitive development have ever considered and tested the possibility of non-normative integration rules, such as the adding rule instead of a normative multiplying one. Furthermore, a major problem of choice tasks is that they tend to elicit cursory short-cut responses that are based on one aspect or one dimension only, masking children's cognitive capabilities (Anderson & Wilkening, 1991). This will be elaborated upon in the next section.

Analytic Information Processing and Separability of Stimulus Dimensions

The assertion that young children's perception and cognition is dominated by a holistic, non-analytic mode of processing can be traced back to the German *Ganzheitspsychologie* (holistic psychology) of the first half of the past century (e.g., Sander, 1932). It was taken up in a methodologically more sophisticated framework in the context of the *separability hypothesis* almost 50 years later (Shepp, 1978; Smith & Kemler, 1978). This idea can range in impact from suggesting maturation from predominantly holistic to an analytic mode to claiming a complete lack of access to the dimensional structure of multidimensional stimuli for young children. It should be noted that this idea posits the contrary of what had been said by Piaget about young children's *centration* tendencies. Centration in this sense can be seen as the prototype of an analytic mode of information processing. If a child focuses on one stimulus dimension only, he or she must have filtered it out of the whole array, that is, seen it as separable from the context. This is exactly what was meant by an analytic processing by the proponents of the separability hypothesis. Thus, we have the strange situation that there are two contradictory statements about fundamental characteristics of young children's cognition, a puzzle that existed largely unnoticed between the disconnected literatures and remained unsolved for quite a long time.

To investigate this progression from holistic to analytic, Garner's (1974) distinction between separable and integral stimuli was used, specifically the restricted classification task. In the prototypical example, the child is shown three squares. Two are identical in size, but differ considerably in brightness: (A) a very light grey and (B) almost black. The third square (C) is a little smaller in size and also differs in brightness. However, C is almost as light as A in brightness. The assumption is that the stimulus triad is constructed such that A and C are closer in overall similarity than A and B. However, A and B do share an identical value on the size dimension.

When preschool age children were asked the critical question, "Which two most go together?", the percentage of their classifications by identity (A and B) was found to not be significantly higher than chance. In contrast, older children and adults had AB classifications that were significantly above chance. The same conclusion was drawn from the data of many similar experiments. All conclusions were based on the assumption that children do not have access to the dimensional structure of the stimuli and thus cannot understand that two dimensions of the stimuli can be independently varied. Children then base their classifications on overall similarity. The problems with this conclusion are so obvious (see Wilkening & Lange, 1989, for details) that it remains a mystery why these studies could enter the major developmental journals for more than a decade. One reason might have been that the wrong conclusions drawn from the data were in congruence with the myth of the naturally holistic young child.

To mention just one problem with the restricted classification paradigm and the interpretations derived from it, consider the possibility that young children grouped A and C together because they judged "overall similarity" by following a so-called city-block metric (Garner, 1974). This would mean that, in fact, they added the dissimilarities existing on both dimensions, size and brightness, and, because the sum of differences for A and C was subjectively smaller than that for A and B, the children chose the overall-similarity alternative. Garner himself would have interpreted such a behavior as a clear case of separable, analytic processing. According to his logic, an overall-similarity classification based on an additive rule as mentioned would have to be taken as a clear indicator of analytic processing, to the same extent as an identity classification. Thus, children's classification behavior in the cited studies may have been as analytic as that of adults. This possibility was obviously not considered by the proponents of the separability hypothesis in their attempt to extend Garner's theory to issues of cognitive development.

To test this possibility and to shed more light on the problem, Wilkening and Lange (1989) designed an experiment employing principles of information integration methodology by varying the stimulus dimensions in a factorial design and introducing a fine-graded rating scale, thus enabling a more direct assessment of children's processing mode. Size and brightness were again used as stimulus dimensions, different levels on each characterizing the belly of a dwarf, presented as upright ellipses. Two schematic pictures served as end anchors, a dwarf with either: (a) a thin, light belly or (b) a thick, dark belly. Children were told that the lower end anchor showed how the dwarf normally looks and the upper showed how he looked after consuming a bag full of magic candies. For the main experiment, the child was told that the dwarf had gotten curious and wanted to try out the effects of other, smaller amounts of candies. To this end, all 3×4 size–brightness combinations of the factorial design were presented in random order, and for each stimulus the child's task was to "guess" how many candies (from a rating scale of 1 to 20) the dwarf might have consumed.

The data plots of the mean ratings of each of the three age groups investigated in that study were roughly parallel, like those shown in the leftmost and center panels of

Figure 10.1, with no significant interaction in any group, thus indicating the use of an adding rule over the entire age range from five years to adulthood. No developmental trend as to the integration rule can be seen, an impression that was corroborated by analyses of the individual data patterns.

It appears, thus, that there is no acceptable evidence for the claim of a developmental trend from holistic to analytic processing, implying that young children have no access to the dimensional structure of stimuli that are separable for adults. The data obtained in developmental applications of information integration theory revealed the contrary. They detected the additive-type rule that the proponents of the separability hypothesis would have required for the assessment of children's analytic processing but were unable to find with their methods.

Adaptive Thinking and Multiple Knowledge Representations

For almost a century now, mainstream research in the field of cognitive development has been built upon the assumption that the changes occurring in the course of development can be best understood as a logical sequence of conceptual structures, from primitive ones to the highest forms. A corollary of this assumption is the belief that such logical sequences will show up in developmental studies—provided that the concepts at each stage of development will be revealed and diagnosed in their pure, uncontaminated form. Again, this assertion goes back to Piaget's seminal theory and can still be found in the more recent post-Piagetian variants. This assumption makes claims from a developmental sequence of pure concepts within domains to potentially across all domains. In any case, a child's knowledge within a domain should be characterized by only one conceptual structure in any phase of development, given that the knowledge is assessed in an adequate way, in tasks freed from all collateral demands.

Albert Einstein's question for Jean Piaget illustrates this traditional view: Which concept develops first in children: time or speed? Obviously, the idea behind the question was that a child can have only one time and/or one speed concept at a certain point of development, and that it is the task of the researcher to diagnose that single, pure concept. Of course, Piaget could easily take up this idea. The answer he presented two decades later, based on several studies employing his choice methodology was: The speed concept comes first in the course of development, years before the time concept emerges. Siegler and Richards (1979), in a more modern variant of the choice methodology derived from the information-processing approach, arrived at essentially the same conclusion.

Experiments using integration theory methodology arrived at dramatically different conclusions about children's concepts of time, speed, and their interrelations (Wilkening, 1981). In particular, it was found that the conceptual structures in the different tasks were not at all reversible, in sharp contrast to Piaget's notions. Even more importantly in the present context, remarkable knowledge dissociations appeared in each age group, especially in the young children. If, for instance, the task suggested the use of an eye-movement strategy for integrating the information about time and speed, children as young as five years of age produced judgment patterns that were in virtually perfect agreement with the normative multiplying rules. If such a psychomotor action was prevented, the children fell back on adding rules—deviating from the normative rule but still evidencing a much higher knowledge than conceded for this age in Piagetian and information-processing theories.

Krist, Fieberg, and Wilkening (1993) elaborated on these findings by investigating children's and adults' intuitive physics about trajectories of moving objects. In effect, the participants had to estimate the speed a tennis ball had to have at the end of an elevated horizontal ramp to hit a target on the floor. Height of the ramp above the floor and horizontal target distance from the ramp were varied in a 4×3 factorial design. Estimates could be given either on a speedometer-like rating scale (judgment condition) or by actually producing the speed by pushing the ball on the ramp (action condition). In both conditions, no feedback was given; in the latter, this was prevented by hiding the ball's downward trajectory with a curtain.

The two conditions yielded quite different results. Whereas the data patterns obtained in the action condition were in almost perfect agreement with the normative multiplying rule in all age groups, including children as young as five years, this was not generally true for the judgment condition, particularly in the younger groups. Figure 10.2 shows a comparison of particular interest in the present context. Both data patterns come from the same subgroup of participants, containing individuals from all age groups. The left panel shows the speed productions, which mirror the physically correct multiplying rule. The right panel shows the speed ratings, which exhibit a dramatically different pattern. Most remarkable is that the levels of the height dimension changed places in the judgment condition. Instead of the correct inverse relation between release height and speed, a direct relation emerged here. These participants appear to have judged according to what can be termed a false height heuristic: the higher the ball's release, the higher the speed. None of these individuals acted according to this heuristic when producing the speed with a part of their own body.

These data provide impressive evidence that people, particularly children, can have conceptual knowledge on different levels at the same time and in the same domain. Analogous findings were obtained in more recent studies by Huber, Krist,

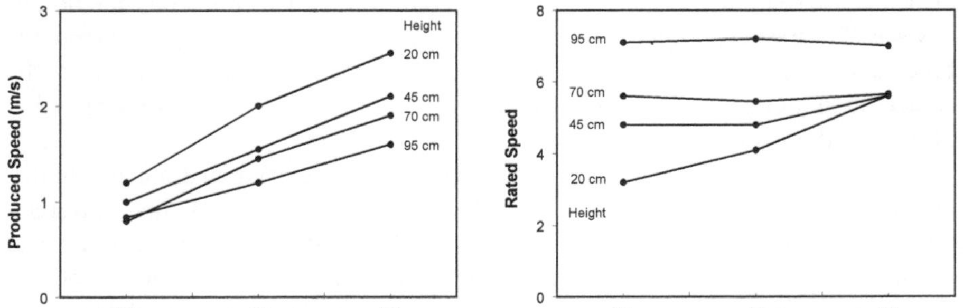

Figure 10.2 Mean Produced Speed in the Action Condition (Left Panel) and Rated Speed in the Judgment Condition (Right Panel) of a Straight Throw. The different data patterns shown in both panels stem from the same group of participants, predominantly children, and point to a striking knowledge dissociation.

and Wilkening (2003) and by Wilkening and Martin (2004). Each knowledge level deserves investigation and is interesting in its own right. It seems arbitrary, if not meaningless, to decide that one of the different levels (from implicit to explicit expressions) represents the pure concept. Hence, the search for pure concepts, in an attempt to find empirical support for the myth of a logical developmental sequence of conceptual structures, seems seriously misleading.

Beyond the Myths: Children's Embodied Knowledge as a Springboard for Science Learning

The studies just cited have shown that what appears as a misconception in explicit, verbalized knowledge can coexist with virtually perfect behavioral responses in implicit, embodied knowledge. The coexistence seems to be a peaceful one—as if both forms of knowledge did not know of each other's existence in the same child. In the following, we will focus on children's implicit knowledge, while exploring the potential transfer to explicit knowledge—that of primary interest in the school.

Adjusting Speed

Whereas young children have been found to have remarkable knowledge about the functional relations that exist in the time–speed–distance triad (Wilkening, 1982), adults exhibit striking misconceptions in the same field. These misconceptions already appear in basic tasks, which—from a normative point of view—do not seem more complicated than those used in the studies with young children. The following problem, here presented in a numerical format for the sake of convenience, provides an example.

If you planned to drive at a constant speed of 60 mph for a certain distance, but for the first half of the total distance you drove 45 mph, how fast would you need to drive to arrive at the time initially planned? The standard answer is 75 mph, typically given with a high level of confidence. Even people with a good education in physics hold the belief that a speed reduction on the first half of a distance can be compensated for by an increase of the same amount (15 mph) on the second half. The necessary speed would be 90 mph, not 75 mph, for arriving at the planned time to compensate for the initial time loss and achieve the average speed of 60 mph over the whole distance. That the standard answer cannot be true should become evident in a thought experiment. Imagine you could drive only 30 mph instead of the planned 60 mph on the first half. Then you would reach the halfway point exactly at the point in time you wanted to be at the endpoint of your trip—and no speed in the world would compensate for your loss of time over the first half of the distance.

People's difficulties with this task in the present format seem in large part due to an inappropriate focus on distance and to ignorance or at least an undervaluation of the time variable. They seem to intermix time with distance. Interestingly, the wrong standard answer would be perfectly right if the two times for the two speeds were the same, that is, if the driver would have had the chance to increase the speed after half of the time that was initially anticipated for the total trip, which of course was elapsed much before he reached the half-way point in distance. Alternatively, non-linear thinking can lead to the correct solution, recognizing that adding the speed differences is not generally adequate. Seen in this way, the cognitive demands for giving the right response are relatively high and it may not come as a surprise that even adults fail in this task.

What would happen if we presented the same problem in another format, with a response mode better suited for tapping people's implicit knowledge? Would children's thinking then be in line with the laws of physics, allowing them to distinguish between speed change at half-time versus half-distance?

Wilkening and Martin (2004) introduced the following task to investigate these questions: two toy cars on parallel tracks. While one car moved with a constant speed, the other car started at one of three slower speeds or the same speed as a control. Exactly at the midpoint of the total distance the first car entered a tunnel until the end of the trip. The question was how fast the second car had to go on the remaining part of the trip so that both cars would be at the finish line at the same time. It was made clear to the child that the speed change could be initiated either (a) when the second car reached the midpoint of the distance (half-distance) or (b) when the first car entered the tunnel (half-time). The first condition was termed the nonlinear one, because the correct solution required speed increases following a nonlinear function. The second condition was termed the linear one, because the absolute amount of difference in speed of the slower car had to be added to that of the faster one for the second section of the trip (in line with the general misconception for the nonlinear condition). Children could respond in two different modes: on a speedometer-like rating scale (judgment condition) or by actually producing the speed by pushing the second car on its track (action condition).

In the judgment condition, the ratings followed the linearity principle, that is, were in line with the common misconception in all age groups and in both problem cases, linear and nonlinear. First signs of nonlinearity appeared in the speed production condition in the 10-year-olds. That is, when allowed to act only on the speeds, at least some of the children began to differentiate between both problem cases and seemed to have overcome the misconception, which one would have to infer from their judgments.

Huber et al. (2003) elaborated on these findings by presenting the task on a computer in a virtual world with the relevant information. The speed of the car presented on the screen could be produced manually using a force-feedback device simulating a haptic environment (Haptic Phantom) or again be judged on a graphic rating scale. The virtual setting allowed an easy variation of the tunnel length, such that the car with the constant speed was always visible until the point in time at which the other car could change its speed. This prolongation of time in which the constant-speed car could be seen apparently simplified the task and helped children in mentally simulating the speed of that car when it was in the tunnel, while they produced the speed of their car with the goal of closing the gap.

Figure 10.3 shows the speed productions of the children, 10 years old, in both conditions, linear and nonlinear. The data are in impressive congruence with the shape of the normative functions. If the speed of the initially slower car could be increased at the midpoint in total time, children graded the increases linearly. If the speed could be increased at the midpoint in distance, children graded the increases nonlinearly, as required by the physical law. The data thus exhibit a clear-cut dissociation of knowledge for the judgment and action condition. It deserves mention that such dissociation was not found for the judgment condition. There, the ratings in the nonlinear condition essentially followed the same linear trend as those for the linear condition, thus exhibiting the common misconception manifesting itself at the explicit level.

Tilting Glasses

From an early age, children have many everyday experiences with tilting glasses filled with liquids. From these experiences the following question can be asked to tap their intuitive geometry: If there are two drinking glasses of equal height but different diameters, each filled with water to the same level, which one can be tilted more without spilling any of the water—the narrow or the wide glass? Schwartz and Black (1999) introduced the water-tilting task to investigate adults' mental simulation of actions. They found that adults solve the problem by using dynamic imagery of the whole tilting movement, rather than by imagining the final static picture with its relevant geometrical features. In a recent experiment, we adopted the tilting task for use with children to investigate their action knowledge in this domain (Frick, Daum, Wilson, & Wilkening, 2009).

Participants of different ages, five years to adulthood, were shown cylindrical glasses of two different inner diameters, 25 or 65 mm, and told to imagine that they would be filled with water up to two different heights, 15 or 45 mm below the rim. These heights were shown in a reference glass that actually contained water but could

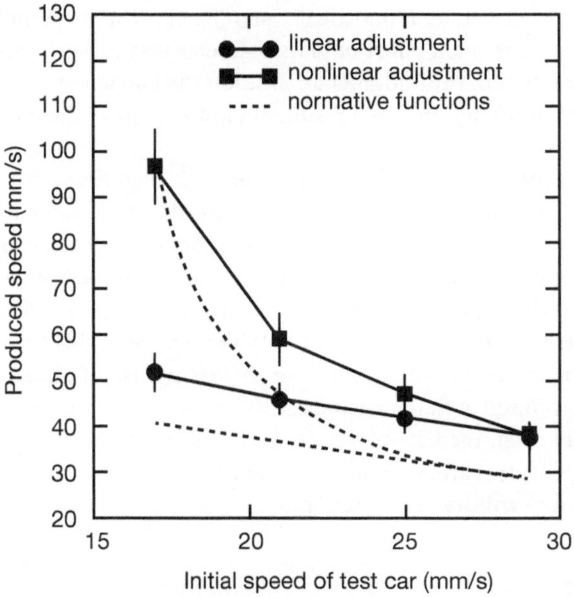

Figure 10.3 Children's Mean Speed Productions for the Second Part of the Trip as a Function of the Initial Speed on the First Part. Adapted from Huber, S., Krist, H., & Wilkening, F. (2003). Judgment and action knowledge in speed adjustment tasks: Experiments in a virtual environment. *Developmental Science*, 6, 197–210. Blackwell Publishing.

not be moved throughout the experiment. For each of the four glasses of the 2×2 diameter×height design, the question was how far it could maximally be tilted so that none of the water would be spilled. In the action task, the children could actually grasp the glass and tilt it. They were allowed to fine-tune and adjust their tilts as often and as long as they liked; the final position was registered as main data. In the judgment task, the children were asked to indicate the presumed tilt by positioning a pointer fixed to a vertical board at the desired angle from the vertical line.

According to the laws of geometry, the narrower glass can be tilted more than the wider one if filled up to the same height. And, of course, a glass with a low water level can be tilted more than a glass with a higher water level. As can be seen from the data in Figure 10.4 (upper panels), children as well as adults exhibited this knowledge in the action condition. For each age group, the actual angles of tilt were higher for the narrower than for the wider glasses, the differences being significant in each single group. In comparison, the other effect appearing in the data may seem less surprising—each age group recognized the impact of the water level factor on the angle of tilting. At least for the youngest age group, however, the significance of the main effects of both factors is not trivial. Apparently, even the five-year-olds integrated both pieces of information—in contrast to traditional claims about the limitations of information-processing capacity at this age.

The data obtained in the judgment condition were quite different, as shown in Figure 10.4 (lower panels). The vertical separation of the data lines, indicating knowledge about the effect of the glass diameter, is virtually absent. A slight effect, if any, can be seen for the data of the nine-year-olds, but it goes in the opposite direction. However, this effect was not statistically significant. The same holds for the three other age groups; the main effect of glass diameter was insignificant in each case. It seems fair to say that in this judgment condition the participants did not have access to their action knowledge and, as a result, did not recognize the relevance of this factor at all ages.

Additional experiments by Frick et al. (2009) showed that perceiving the motion and being able to control it in some way are not the decisive factors for the action knowledge found in the young children. When the participants could tilt the glasses by means of a remote-control device only, a clear age trend was found, in contrast to the action and judgment conditions just mentioned. Controlling the movement of the glasses from the distance did not help the five-year-olds to access their action knowledge. For the older participants, an increasing use of glass diameter as a relevant factor was found with increasing age. It appears, thus, that the younger children are more reliant on immediate motor control and feedback, and that the coupling of psychomotor processes and mental representations of the external world is especially intertwined before children enter school.

Rotating Hands

The conclusions just made are strongly supported by data from experiments on mental imagery development by Funk, Brugger, and Wilkening (2005). Elaborating on previous research on mental rotation, Funk et al. presented photographs of hands to children five and six years old and asked them to decide as quickly as possible if the hand shown was a left or a right hand. The visually presented hands were left

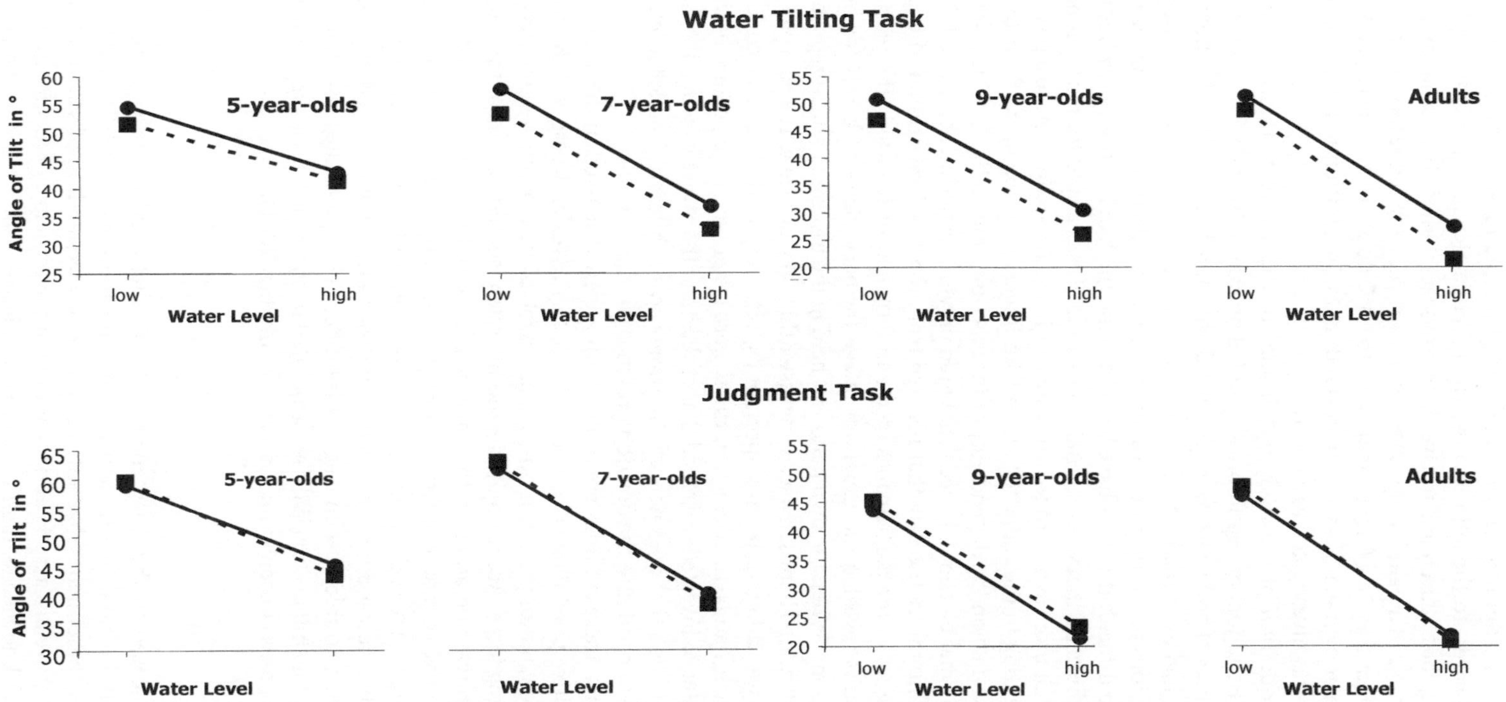

Figure 10.4 Mean Produced Angles of Tilt for Each Age Group in the Glass Tilting Task (Solid Lines = Narrow Glass, Dashed Lines = Wide Glass). Adapted from, Frick, A., Daum, M. M., Wilson, M., & Wilkening, F. (2009). Effects of action on children's and adults' mental imagery, *Journal of Experimental Child Psychology*, 104, 34–51. Copyright 2009, with permission from Elsevier.

and right hands in palm or back view in different orientations: the fingers pointing upwards, downwards, to the left or to the right. While viewing the photographs of the single hands, the children had to give their responses with their own hands either in a palms-down or in a palms-up posture, each time pressing the response key with the same hand that they saw (right–right, left–left). While giving responses, children could not see their own hands as a cloth covered them. In addition to the children, a group of adults was investigated as a control.

The crucial question of the study was if and to which extent children's kinetic imagery—as a useful tool of cognition—is guided by motor processes. It was for this reason that pictures of body parts were presented, the same body parts with which the motor response had to be made. There were many indications in the data pointing to strong contributions of motor processes. In particular, children's reaction times were longer the more difficult it would have been to bring the hand of their own body to the position of the hand that was externally presented. For example, for most people it is physically much more awkward to bring their right hand in palm view to a position with fingers pointing to the right (90°) than with fingers pointing upwards or to the left (270°). These data on the dependency of the reaction time on the specific rotation angle replicated what had already been shown in previous experiments with adults.

Most compelling were the data that resulted from the novel variation of the task, varying the posture of the participant's own hand: regular or inverted. The reaction times of the adults revealed the usual advantage for back views of hands for both response conditions. When the own hand was held in the inverted palms-up posture, the adults still identified hands in back view faster than in palm view. The data pattern of the children was different in an important aspect. Despite the fact that children could not see their hands, their reaction times were faster when the posture of their own hand (regular or inverted) matched that of the hand they were viewing. It is hard to find an explanation that does not refer in some way to children's implicit attempts to bring their own hand into the position presented in the stimulus.

Taken together, these findings are in line with a view discussed in the recent literature in cognitive psychology and neuroscience (e.g., Wexler, Kosslyn, & Berthoz, 1998). In this view, motor processes not only are involved in dynamic mental imagery but are the driving force. Accordingly, the motor system should be seen as the engine driving the cognitive operations, rather than just as an output system. For children, the effects indicative of motor processes were generally higher than those found for adults. Children's motor and cognitive processes seem to be tightly linked early in development. This view is a modern version of Piaget's notion but expanded beyond the age of two years to at least school age. This modern view postulates essentially the same notion: that all intellectual skills are grounded in and supported by motor activity, even at high levels (Rosenbaum, Carlson, & Gilmore, 2001).

Conclusions

The data summarized in this chapter have shown that children have an amazing knowledge about the basic laws of nature, at least on an implicit level. Their intuitive physics contains components that are vital for the understanding of scientific concepts in general as well as a basic understanding of the principle of factorial interaction. This kind of knowledge, highly adaptive and largely isomorphic to the laws of the objective world, is quite different from what would have been predicted by traditional developmental theories. In the studies cited here, this kind of knowledge was revealed by nonverbal methods, typically through children's actions. Assessed thus, it appears to be already there when a child enters school.

It does not seem wise, then, for a teacher to ignore these implicit competencies that are already there or to adhere to myths about fundamental limitations in children's cognitive capabilities. Children's intuitive physics is full of these components, and the educational challenge is to transfer the basic forms of implicit knowledge into other, more explicit formats—particularly verbalization. Systematic research on how such implementations for the classroom could work is still lacking; this remains a grand task for the future.

References

Anderson, N. H. (1981). *Foundations of information integration theory.* New York: Academic Press.

Anderson, N. H. (1996). *A functional theory of cognition.* Mahwah, NJ: LEA.

Anderson, N. H., & Wilkening, F. (1991). Adaptive thinking in intuitive physics. In N. H. Anderson (Ed.), *Contributions to information integration theory. Vol. 3: Developmental.* Hillsdale, NJ: LEA. pp. 1–42.

Case, R. (1992). *The mind's staircase: Exploring the conceptual underpinnings of children's thought and knowledge.* Hillsdale, NJ: LEA.

Falk, R., & Wilkening, F. (1998). Children's construction of fair chances: Adjusting probabilities. *Developmental Psychology*, 34, 1340–1357.

Frick, A., Daum, M. M., Wilson, M., & Wilkening, F. (2009). Effects of action on children's and adults' mental imagery. *Journal of Experimental Child Psychology*, 104, 34–51.

Funk, M., Brugger, P., & Wilkening, F. (2005). Motor processes in children's imagery: The case of mental rotation of hands. *Developmental Science*, 8, 402–408.

Garner, W. R. (1974). *The processing of information and structure.* Hillsdale, NJ: LEA.

Huber, S., Krist, H., & Wilkening, F. (2003). Judgment and action knowledge in speed adjustment tasks: Experiments in a virtual environment. *Developmental Science*, 6, 197–210.

Krist, H., Fieberg, E. L., & Wilkening, F. (1993). Intuitive physics in action and judgment: The development of knowledge about projectile motion. *Journal of Experimental Psychology: Learning, Memory, and Cognition*, 19, 952–966.

Piaget, J. (1970). Piaget's theory. In P. H. Mussen (Ed.), *Carmichael's manual of child psychology, Vol. 1.* New York: Wiley. pp. 703–732.

Rosenbaum, D. A., Carlson, R. A., & Gilmore, R. O. (2001). Acquisition of intellectual and perceptual-motor skills. *Annual Review of Psychology*, 52, 453–470.

Sander, F. (1932). Funktionale Struktur, Elebnisganzheit und Gestalt. *Archiv für die gesamte Psychologie*, 85, 23–260.

Schwartz, D. L., & Black, T. (1999). Inferences through imagined actions: Knowing by simulated doing. *Journal of Experimental Psychology: Learning, Memory, and Cognition*, 25, 116–136.

Shepp, B. E. (1978). From perceived similarity to dimensional structure: A new hypothesis about perceptual development. In E. Rosch & B. B. Lloyd (Eds.), *Cognition and categorization.* Hillsdale, NJ: LEA. pp. 135–167.

Siegler, R. S. (1998). *Children's thinking* (3rd edn.). Upper Saddle River, NJ: Prentice-Hall.

Siegler, R. S., & Richards, D. D. (1979). Development of time, speed, and distance concepts. *Developmental Psychology*, 15, 288–298.

Smith, L. B., & Kemler, D. G. (1978). Levels of experienced similarity in children and adults. *Cognitive Psychology*, 10, 502–532.

Wexler, M., Kosslyn, S. M., & Berthoz, A. (1998). Motor process in mental rotation. *Cognition*, 68, 77–94.

Wilkening, F. (1979). Combining of stimulus dimensions in children's and adults' judgments of area: An information integration analysis. *Developmental Psychology*, 15, 25–33.

Wilkening, F. (1981). Integrating velocity, time, and distance information: A developmental study. *Cognitive Psychology*, 13, 231–247.

Wilkening, F. (1982). Children's knowledge about time, distance, and velocity interrelations. In W. J. Friedman (Ed.), *The developmental psychology of time*. New York: Academic Press. pp. 87–112.

Wilkening, F., & Anderson, N. H. (1982). Comparison of two rule-assessment methodologies for studying cognitive development and knowledge structure. *Psychological Bulletin*, 92, 215–237.

Wilkening, F., & Anderson, N. H. (1991). Representation and diagnosis of knowledge structures in developmental psychology. In N. H. Anderson (Ed.), *Contributions to information integration theory. Vol. 3: Developmental*. Hillsdale, NJ: LEA. pp. 45–80.

Wilkening, F., & Lange, K. (1989). When is children's perception holistic? Goals and styles in processing multidimensional stimuli. In T. Globerson & T. Zelniker (Eds.), *Cognitive development and cognitive style*. Norwood: Ablex. pp. 141–171.

Wilkening, F., Levin, I., & Druyan, S. (1987). Children's counting strategies for time quantification and integration. *Developmental Psychology*, 23, 823–831.

Wilkening, F., & Martin, C. (2004). How to speed up to be in time: Action–judgment dissociations in children and adults. *Swiss Journal of Psychology*, 63, 17–29.

11 Learning Newtonian Physics with Conversational Agents and Interactive Simulations

Arthur C. Graesser, Don Franceschetti, Barry Gholson, and Scotty Craig

Newtonian physics is an area of science that is part of the curriculum in middle and high schools throughout the country. It is easy to set up demonstrations of physics principles that are easy to observe and provocative to talk about. Students drop objects of different size or weight and wonder whether they hit the pavement at the same time. They observe how objects float in water, how objects collide, and how moving objects land at particular locations. Some of what the students see is counterintuitive, which ideally stimulates them to ask questions and perform mini-experiments to answer the questions. Students discuss what they see and learn with their peers and teachers.

Consider, for example, the following conceptual physics question:

> If a lightweight car and a massive truck have a head-on collision, upon which vehicle is the impact force greater? Which vehicle undergoes the greater change in its motion, and why?

When students answer this question, one hopes that they would communicate principles (*P*) such as 1 and 2 below. However, they might be misled by the two misconceptions (*M*).

P1: The magnitudes of the forces exerted by A and B on each other are equal.
P2: If A exerts a force on B, then B exerts a force on A in the opposite direction.
M1: A lighter/smaller object exerts no force on a heavier/larger object.
M2: Heavier objects accelerate faster for the same force than lighter objects.

The hope is that they will acquire more correct principles and fewer misconceptions over the course of learning. Research on physics learning has shown, however, that it is very difficult to correct many of the misconceptions, particularly those that are entrenched in a student's everyday experiences (Chi, 2005; diSessa, 1993; Hunt & Minstrell, 1996; Ploetzner & VanLehn, 1997). For example, M1 would appear to be confirmed perceptually when a child throws a rubber ball against a wall.

The tight correspondence of Newtonian physics principles to everyday experience has both advantages and liabilities. The advantages lie in (a) the natural correspondence between the constructs of physics and everyday actions, events, and perceptions, (b) the ease of setting up demonstrations to illustrate many physics principles, and (c) the comparatively low density of jargon to memorize. The disadvantages lie in (a) the clash between some mental models that are grounded in everyday experience and the proper mental models of Newtonian physics and (b) the difficulty of correcting mental models that are entrenched in everyday experiences.

It is illuminating to explore the mental models of students by asking them to generate explanations while solving conceptual physics problems (Gentner & Stevens, 1983; VanLehn et al., 2007), such as the example above or the question below:

> When a car without headrests on the seats is struck from behind, the passengers often suffer neck injuries. Why do passengers get neck injuries in this situation? Explain why.

The nature of students' mental models is often manifested by asking them to draw sketches of the processes. Many college students believe that a rear-end collision will directly push the head of the victim forward. Their mental model is that the head goes forward much like a billiard ball goes forward when hit from behind. They may have a memory of a person's head going through the windshield from a movie or personal experience. However, this is flawed reasoning. The head first goes backwards after the impact of the collision by virtue of the forces underlying Newton's laws, which explains whiplash in accidents. The head subsequently goes forward after recoil. Knowledgeable students identify the initial stage of the head going back when asked to draw pictures, but students with shallow understanding miss this step. It is not until the student can reason with abstract vectors, forces, and Newton's laws that a correct answer emerges. The generation of verbal explanations and pictures is an excellent task for diagnosing misconceptions.

Methods of Training Newtonian Physics

It is beyond the scope of this chapter to summarize all of the traditional and inventive ways that Newtonian physics has been taught. Instead, this section briefly addresses some of the typical learning environments, including their benefits and potential liabilities. Subsequent sections focus on our own work involving human and computer tutoring. Most of our research on learning Newtonian physics has been on college students, but we have collected some data from ninth and eleventh graders. Our expertise in early child development and elementary school education is limited, however, so our remarks on physics research would only be speculative at the younger ages and grade levels. It is a very important research question whether the findings from middle school and above could be extrapolated to a child in the second grade or even younger (Stein, Hernandez, & Gamez, 2007). The learning of science at young ages is a foundational question for this edited volume so we will offer some remarks in the final section of this chapter.

At this point we will turn to some of the frequent pedagogical methods and learning environments, including their salient advantages and limitations.

Reading Textbooks

Textbooks are routinely assigned to students in physics courses so one would hope that this conventional approach would promote learning. However, students normally have low background knowledge of science so it is a struggle for students to comprehend the text and stay motivated as they struggle (Otero, Leon, & Graesser, 2002; Vitale & Romance, 2007). They no doubt acquire shallow knowledge from reading, such as definitions of terms, facts, lists of properties, and historical details. However, it is more difficult for students to acquire deep knowledge, such as causal explanations, dynamic processes, and complex mechanisms. There is indeed evidence that adults fail to acquire deep knowledge from reading textbooks. This can be seen in our assessment of deep learning with a popular science textbook (*Conceptual Physics*, Hewitt, 1998). Posttests revealed no significant difference between a Do Nothing and a Textbook Reading condition, while various tutoring conditions showed significantly higher levels of deep learning (VanLehn et al., 2007). The counterintuitive claim here is that students get shallow knowledge but not deep knowledge from reading physics texts, so it is not sufficient to merely assign a textbook and expect students to learn physics very deeply.

Teachers Lecturing

Students presumably learn well from the lectures of award-winning teachers (Hunt & Minstrell, 1996). However, it is hardly a secret that most physics science teachers in schools are not adequately trained in science and they vary substantially in pedagogical methods. One of the drawbacks to lecturing environments is that the training is not tailored to individual students. In contrast, individualized instruction is one of the hallmarks of computer-based training and more advanced computer learning environments (Woolf, 2009).

Discovery Learning Environments

Students would no doubt be fascinated with a museum of physics artifacts that allow them to discover physics principles on their own. However, available evidence is that little is learned unless there is an instructor or tutor who assigns activities and mediates learning (Klahr, 2002). Researchers have carefully created inquiry-learning environments that stimulate students to ask questions, generate hypotheses, and plan activities for experimenting in pursuit of answers (Goldman, Duschl, Ellenbogen, Williams, & Tzou, 2003; White & Frederiksen, 1998). However, it is extremely difficult to design such environments to guarantee effective inquiry and discovery. Students need guidance.

Human Tutoring

Human tutoring is an effective method of teaching science topics, so this process has been investigated in depth (Chi, Siler, Jeong, Yamauchi, & Hausmann, 2001; Graesser, Person, & Magliano, 1995; VanLehn et al., 2007). Meta-analyses show learning gains from human tutors that vary in expertise of .42 sigma (effect size in standard deviation units) compared with classroom controls and other suitable controls (Cohen, Kulik, & Kulik, 1982). More will be said about tutoring later in the chapter.

Animation

Computers can be designed to present animations that exhibit physics principles through event sequences that unfold over time. However, the animations run a number of risks: being difficult to understand, being transient, moving too quickly, presenting distracting material, placing demands on working memory, and depicting processes in a fashion other than what the learner would otherwise actively construct (Hegarty, 2004). Therefore, animations have failed to improve learning in a large percentage of systematic studies (Ainsworth, 2008; Tversky, Morrison, & Betrancourt, 2002).

Multimedia

Physics lessons can be delivered in different presentation modes (verbal, pictorial), sensory modalities (auditory, visual), and delivery media (text, video, simulations). The impact of different forms of multimedia has been extensively investigated by Mayer and his colleagues (Mayer, 2005). Meta-analyses by Dodds and Fletcher (2004) report an effect size of .50 sigma for multimedia learning, whereas that in the meta-analyses reported by Mayer is considerably higher, more like 1.00. Mayer has documented and empirically confirmed a number of psychological principles that predict when different forms of multimedia will facilitate learning: multimedia, modality, spatial and temporal contiguity, coherence, and redundancy. Multimedia presentation allows for conceptually richer and deeper representations; however, it is important that there is not too large a cognitive load placed on the learner (Kalyuga, Chandler, & Sweller, 1999).

Interactive Simulation

Interactive simulation is expected to enhance learning because the learner can actively control input parameters and observe the results on the system. The learner can slow down animations to inspect the process in detail, zoom in on important subcomponents during the course of a simulation, observe the system from multiple viewpoints, and systematically relate inputs to outputs (Kozma, 2000). Interactive simulation has indeed shown a positive impact on science learning in several studies but others have shown no gains compared with various control conditions (Jackson, Olney, Graesser, & Kim, 2006; van der Meij & de Jong, 2006). These learning environments may have complex content and human–computer interfaces that are unfamiliar to learners, causing difficulties in getting started, managing the interface, and strategically interacting with the simulation to advance learning.

Our research group has investigated interactive simulation in microworlds that captured principles of Newtonian physics (Jackson et al., 2006). We developed an interactive simulation world with people, objects, and the spatial setting associated with the conceptual problems that were illustrated earlier. The college students could manipulate parameters of the situation (e.g., mass of objects, speed of objects, distance between objects) and then ask the system to simulate what will happen. The results showed only a small, nonsignificant increase in learning compared with a conversation-based computer tutoring system called AutoTutor, which will be discussed later. The good news is that there were substantial gains for those students who ran several simulations for a problem. However, most students ran only one simulation, which is entirely inadequate for tracing the impact of variables on system behavior.

For example, in the collision problems, a student might want to compare the impact of the mass of the vehicles and air resistance on the displacement of vehicles in the collision. In order to discover that the mass of vehicles is important, but air resistance is not, it is necessary to run four simulations that cross large versus small mass with high versus zero air resistance (see Klahr, 2002). Students rarely performed such a systematic comparison. Examples such as these illustrate that students need to be trained on how to effectively use interactive simulations.

Intelligent Tutoring Systems

Intelligent tutoring systems (ITSs) track the knowledge states of learners in fine detail and adaptively respond with activities that are sensitive to these knowledge states. The processes of tracking knowledge (user modeling) and adaptively responding to the learner incorporate computational models in artificial intelligence and cognitive science, such as production systems, case-based reasoning, Bayes networks, theorem proving, and constraint satisfaction algorithms. Successful systems have been developed for Newtonian physics, such as Andes and Why/Atlas (VanLehn et al., 2002, 2007). These systems show impressive learning gains (.50 to 1.00 sigma), particularly for deeper levels of comprehension.

Games

The game industry has captured the imagination of the current and next generations of students. Serious teenage gamers play games over 20 hours per week. There are many types of games that have the potential of incorporating physics principles. The challenge of combining entertainment and pedagogical content is the foundational question of serious games (Gee, 2003; O'Neil, Wainess, & Baker, 2005; Ritterfeld, Cody, & Vorderer, 2009). Presumably, the success of a game can be attributed to such factors as feedback, progress markers, engaging content, fantasy, competition, challenge, uncertainty, curiosity, control, and other factors that involve cognition, emotions, motivation, and art. The technical and psychological components of games have been analyzed at considerable depth, but there has been very little research on the impact of these components on learning gains, engagement, and usability (Malone & Lepper, 1987; O'Neil et al., 2005; Virvou, Katsionis, & Manos, 2005).

Animated Conversational Agents

Animated conversational agents play a central role in some of the recent advanced learning environments (Atkinson, 2002; Baylor & Kim, 2005; Graesser, Chipman, Haynes, & Olney, 2005; McNamara, Levinstein, & Boonthum, 2004; Moreno & Mayer, 2005; Reeves & Nass, 1996). These agents interact with students and help them learn either by modeling good pedagogy or by holding a conversation. The agents may take on different roles: mentors, tutors, peers, players in multiparty games, or avatars in the virtual world. The students communicate with the agents through speech, keyboard, gesture, touch panel screen, or conventional input channels. In turn, the agents express themselves with speech, facial expression, gesture, posture, and other embodied actions. Intelligent agents with speech recognition essentially hold a face-to-face, mixed-initiative dialogue with the student, just as humans do (Graesser, Jackson, & McDaniel, 2007; Johnson & Beal, 2005). Single agents model individuals with different knowledge, personalities, physical features, and styles. Ensembles of agents model social interaction.

From the standpoint of learning, there are at least three fundamental reasons why these agents would be effective in facilitating knowledge construction. First, it is well documented that one-to-one tutoring is one of the most effective methods of helping students learn, as discussed earlier. Second, computer-generated agents can consistently and reliably apply tutoring strategies, unlike teachers and human tutors. Third, agents can demonstrate (i.e., model) most learning activities and strategies that involve interactions between people or interactions between people and external media. Both single agents and ensembles of agents can be carefully choreographed to mimic virtually any activity or social situation: curiosity, inquiry learning, negotiation, interrogation, arguments, empathetic support, helping, and so on.

Two Example Learning Environments with Agents: AutoTutor and iDRIVE

This section describes two computer learning environments with agents that we believe have some potential for helping young children learn Newtonian physics. The two systems (AutoTutor and iDRIVE) have already proven to be effective for college students, as well as middle and high school students. Although they are untested in young children, we believe they hold some promise because an agent can simulate face-to-face tutorial conversations with the learner, or two agents can model good conversations with each other.

The distinctive characteristic of both is their spotlight on conversation in scaffolding learning. As discussed earlier, it is well documented in meta-analyses that one-on-one human tutoring is superior to classroom instruction (Cohen et al., 1982; Graesser, D'Mello, & Person, 2009). Moreover, Graesser et al.'s (1995) in-depth analysis of human tutoring resulted in the fundamental conclusion that there is something about conversational discourse that explains why human tutoring is superior to classroom instruction. AutoTutor and iDRIVE are instantiations of conversational tutoring. Aside from the spotlight on conversation, the two systems can be blended with many other promising pedagogical approaches discussed in the previous section.

AutoTutor

AutoTutor is an intelligent tutoring system that helps students learn through tutorial dialogue in language (Graesser, Chipman et al., 2005; Graesser et al., 2004; VanLehn et al., 2007). AutoTutor's physics dialogues are organized around conceptual physics problems that require reasoning and explanations in the answers. These questions require the learner to construct approximately three to seven sentences in an ideal answer and to exhibit reasoning in natural language. It takes a conversation to answer each one of these questions, typically 30 to 100 conversational turns between AutoTutor and the student.

Table 11.1 illustrates AutoTutor in a conversation that was extracted from an actual tutoring session. This session was with a relatively verbose, knowledgeable college student so it is comparatively short. Lower-ability students would take many more conversational turns before a good answer would be constructed. When students are asked these challenging questions, their initial answers are typically only one or two sentences in length. However, one or two sentences provide insufficient information to adequately answer the question so tutorial dialogue is needed to flesh out a complete answer. AutoTutor engages the student in a mixed-initiative dialogue that draws out more of what the student knows and assists the student in the construction of an improved answer.

AutoTutor can be viewed as proof that a computer tutor can manage a reasonably smooth and pedagogically effective conversation. It is beyond the scope of this chapter to discuss the theoretical and computational mechanisms of AutoTutor, but some

Table 11.1 Example Dialogue with AutoTutor

TUTOR:	Now let's move on. A lightweight car and a massive truck have a head-on collision. On which vehicle is the impact force greater? Which vehicle undergoes the greater change in its motion? Explain why.
STUDENT:	The lightweight vehicle experiences the greater impact and the greater change in motion because the mass of the lighter vehicle is less, therefore having less inertia to resist the change in motion. The momentum of the truck is greater than that of the car.
TUTOR:	Right! What does Newton's third law of motion say about the forces that bodies exert on each other?
STUDENT:	The forces must be equal in magnitude and opposite in reaction.
TUTOR:	Yes, correct! Can you add to that?
STUDENT:	The forces must be equal and opposite reactions acting on opposite bodies.
TUTOR:	Ya! We've almost covered it, but there's one part I think is still missing. I am sure you can get this. During an interaction, each body exerts an equal but opposite force on the other __?
STUDENT:	Body.
TUTOR:	Yes, right! On the other body. So when two balls, one plastic and the other iron, hit each other, does the heavier or neither ball exert the greater impact force on the other?
STUDENT:	Neither.
TUTOR:	Yes, correct! Neither, the forces are equal in magnitude. Let's take a moment for general physics questions. Ask me one and I'll do my best to answer it.
STUDENT:	No.
TUTOR:	Alright. To recap what we have just discussed, please give me your complete answer to the original question, as presented at the top of the screen.
STUDENT:	Although the lighter car experiences a greater change in motion, both vehicles exert an equal impact force, but opposite in direction. The larger object has a larger mass, and therefore a larger inertial resistance than that of the car.
TUTOR:	That is right! Putting everything together, um, according to Newton's third law of motion, if one body exerts a force on the other body then the other body must exert an equal and opposite force on the first body. Therefore, the car and the truck both experience the same impact force. By Newton's second law, the car will experience the greater change in its motion because its mass is much smaller. This question is finished so let's go on to the next question.

highlights are in order. The structure of the dialogue in both AutoTutor and human tutoring (Chi et al., 2001; Graesser et al., 1995; VanLehn et al., 2007) can be segregated into three structural components: (1) expectation- and misconception-tailored dialogue, (2) a five-step dialogue frame, and (3) composition of a conversational turn. These three levels can be automated and produce respectable tutorial dialogue.

Expectation- and Misconception-Tailored Dialogue

Both AutoTutor and human tutors typically have a list of expectations (anticipated good answers) and anticipated misconceptions associated with each main question. For example, expectations E1 and E2 (corresponding to principles P1 and P2) and misconceptions M1 and M2 are relevant to the physics problem presented earlier. AutoTutor guides the student in articulating the expectations through a number of dialogue moves: *pumps*, *hints*, and *prompts* to fill in missing words. Hints and prompts are carefully selected by AutoTutor to produce content in the answers that fill in missing content words, phrases, and propositions. As the learner expresses information over many turns, the list of expectations is eventually covered and the main question is scored as answered. Complete coverage of the answer requires AutoTutor to have a pool of hints and prompts available to extract all of the content words, phrases, and propositions in each expectation. AutoTutor adaptively selects those hints and prompts that fill missing constituents and thereby achieves pattern completion.

AutoTutor is dynamically adaptive to the learner in other ways than coaching him or her to articulate expectations. There is the conversational goal of correcting misconceptions that arise in the student's talk. When the student articulates a misconception, AutoTutor acknowledges the error and corrects it. AutoTutor also gives short feedback on the quality of student contributions: positive, neutral, or negative. AutoTutor accommodates a mixed-initiative dialogue by attempting to answer the student's questions. The answers to the questions are retrieved from glossaries or from paragraphs in textbooks through intelligent information retrieval. AutoTutor asks counterclarification questions when it does not understand the students' questions.

Five-Step Dialogue Frame

This dialogue frame is prevalent in human tutoring (Graesser & Person, 1994; VanLehn et al., 2007) and is implemented in AutoTutor. The five steps of the dialogue frame are:

1. Tutor asks main question.
2. Student gives initial answer.
3. Tutor gives short feedback on the quality of the student's answer in #2.
4. Tutor and student collaboratively interact through expectation- and misconception-tailored dialogue.
5. Tutor verifies that the student understands.

Students often answer that they understand in step 5, when most do not. Human tutors would ideally press the student further by asking more questions to verify the student's understanding, but most tutors rarely do this. Most tutors end up giving a summary answer to the main question and then select another main question. Ideally, human tutors would ask the student to provide the summary (as in the example dialogue in Table 11.1) rather than providing it themselves, but again most human tutors rarely do that. Thus, AutoTutor has the potential to be an improvement over human tutors.

Managing One Conversational Turn

Each turn of AutoTutor in the conversational dialogue has three information slots (i.e., units, constituents). The first slot of most turns is short feedback on the quality of the student's last turn. The second slot advances the coverage of the ideal answer with prompts for specific words, hints, assertions with correct information, corrections of misconceptions, or answers to student questions. The third slot is a cue to the student for the shift from AutoTutor as the speaker to the student. For example, AutoTutor ends each turn with a question or a gesture to cue the learner to do the talking. Discourse markers (*and, also, okay, well*) connect the utterances of these three slots of information within a turn.

The three levels of AutoTutor go a long way in simulating a human tutor. AutoTutor can keep the dialogue on track because it is always comparing what the student says with the anticipated input (i.e., the expectations and misconceptions in the curriculum script). Pattern-matching operations and pattern completion mechanisms drive the comparison. These matching and completion operations are based on latent semantic analysis (Landauer, McNamara, Dennis, & Kintsch, 2007) and symbolic interpretation algorithms (Rus, McCarthy, McNamara, & Graesser, 2008) that are beyond the scope of this chapter to address.

AutoTutor cannot interpret student contributions that have no matches to content in the curriculum script. This of course limits true mixed-initiative dialogue. Thus, AutoTutor cannot explore the topic changes and tangents of students as the students introduce them. However, available studies of naturalistic tutoring (Chi et al., 2001; Graesser et al., 1995) reveal that (a) human tutors rarely nurture true mixed-initiative dialogue when students change topics that steer the conversation off course and (b) most students rarely change topics, rarely ask questions, and rarely grab the initiative to take the conversational floor. Instead, it is the tutor that takes the lead and drives the dialogue. AutoTutor and human tutors are very similar in these respects.

Current Version of AutoTutor

We have created many versions of AutoTutor that were designed to incorporate particular pedagogical goals, to cover different topics, and even to respond to student emotions. Figure 11.1 shows the interface of a version of AutoTutor that has an interactive simulation. This *AutoTutor-3D* version guides learners on using interactive simulations of physics microworlds (Graesser, Chipman, et al., 2005; Jackson et al., 2006). The student manipulates parameters of the situation (e.g., mass of objects, speed of objects, distance between objects) and then asks the system to simulate what will happen. Students are also prompted to describe what they see. Their actions and descriptions are evaluated with respect to covering the expectations or matching misconceptions. AutoTutor manages the dialogue with hints and suggestions that scaffold the learning process with dialogue.

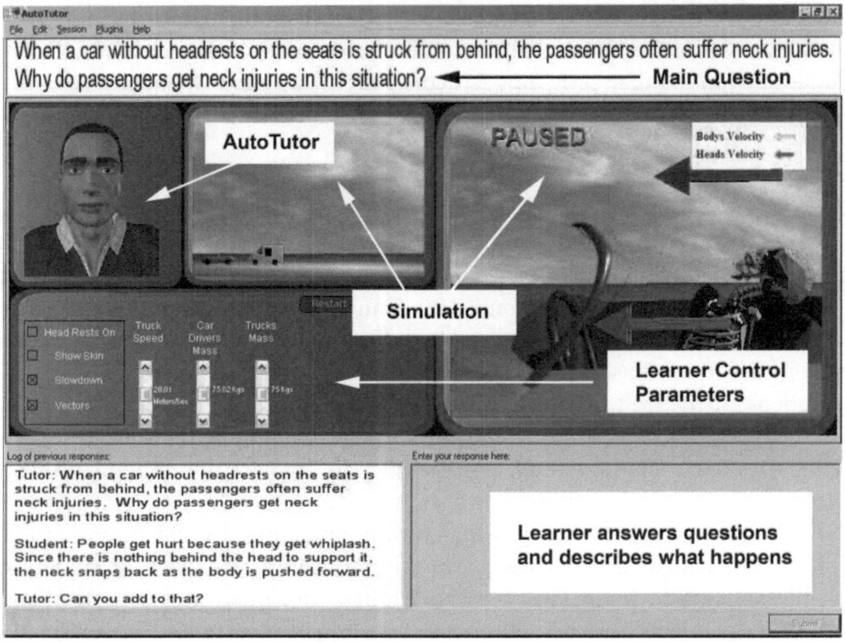

Figure 11.1 A Computer Screen of AutoTutor on Conceptual Physics with Interactive 3D Simulation.

Learning Gains with AutoTutor

The learning gains of AutoTutor have been evaluated in 15 experiments conducted during the last nine years. Assessments of AutoTutor on learning gains have shown effect sizes of approximately .8 standard deviation units in Newtonian physics (VanLehn et al., 2007), which is on par with or even superior to human tutors (Cohen et al., 1982; Graesser et al., 2009). The two primary measures used in assessing learning are (1) multiple-choice questions on deep knowledge (see Force Concept Inventory of Hestenes, Wells, & Swackhamer, 1992) and (2) the quality of answers to essay questions that involve near or far transfer from the training problems. A variety of comparison conditions to AutoTutor have uncovered the following findings.

1. *Versus reading a textbook.* Learning gains with AutoTutor are superior to reading from a textbook on the same topics for an equivalent amount of time.
2. *Reading a textbook versus doing nothing.* Learning gains are zero in both of these conditions when the tests tap deeper levels of comprehension.
3. *Versus expert human tutors.* Comparisons were made between AutoTutor and accomplished human tutors through computer-mediated communication. The learning gains were equivalent for students with a moderate degree of physics knowledge, but expert human tutors prevailed when the students had low physics knowledge and the dialogue was spoken.
4. *Zone of proximate development.* AutoTutor is most effective when there is an intermediate gap between the learner's prior knowledge and the ideal answers

of AutoTutor. AutoTutor is not particularly effective with students with high domain knowledge or when students with low knowledge receive problems that do not push them to new levels of understanding.

5 *Versus carefully prepared texts.* AutoTutor shows few if any advantages when compared with texts that succinctly answer the physics questions.

iDRIVE (Instruction with Deep-level Reasoning questions In Vicarious Environments)

Learning environments can also have pairs of agents (dyads) and larger ensembles of agents that exhibit ideal learning strategies and social interactions (McNamara et al., 2004; Millis et al., 2006). iDRIVE has agent dyads train students to learn science content by modeling deep-reasoning questions in question–answer dialogues. A student agent asks a series of deep questions about the science content and the teacher agent immediately answers each question. There is evidence that learning improves when learners have the mindset of asking deep questions that tap causal structures, complex systems, and logical justifications (Craig, Gholson, Ventura, Graesser, & the Tutoring Research Group, 2000; Rosenshine, Meister, & Chapman, 1996). However, the asking of deep questions and inquiry does not come naturally (Graesser, McNamara, & VanLehn, 2005) so the process needs to be trained or modeled by agents or humans (Azevedo & Cromley, 2004; Goldman et al., 2003). The iDRIVE system models the asking of deep questions with dialogues between agents. Virtually any content can be augmented by preceding each chunk of content with a deep question that motivates the content. When the content is laced with a large number of deep questions, the learner acquires the mindset of thinking deeply and this improves learning, as discussed below.

Most of the iDRIVE studies have been conducted on college students in the area of computer literacy (Craig, Sullins, Witherspoon, & Gholson, 2006). Learning gains on the effectiveness of iDRIVE on question asking, recall of text, and multiple-choice questions have shown effect sizes that range from .56 to 1.76 compared with a condition in which students listen to a monologue on the same content without questions. We have recently conducted a study on ninth and eleventh graders who learned Newtonian physics. The students were randomly assigned to one of three conditions: iDRIVE, AutoTutor, or monologues (Gholson et al., 2009). Attempts were made to achieve information equivalence by covering the same set of expectations in all conditions. The results of the study showed pre- to post-test effect sizes of .95 (iDRIVE), .43 (AutoTutor), and .58 (monologue) sigma. The fact that the iDRIVE dialogues showed the highest learning, even higher than AutoTutor, supports the claim that modeling deep question asking with agents can have a powerful impact on physics comprehension. Agent technologies can indeed improve science learning in young children.

Learning the Principles and Shedding the Misconceptions

One way of viewing Newtonian physics is that there is a set of physics principles that constrain the reasoning of students as they work on toy problems or real-world applications (Hunt & Minstrell, 1996; VanLehn et al., 2007). Students should follow these principles or articulate them when asked to explain their reasoning. Their actions and explanations should not exhibit the misconceptions that reflect everyday intuitions and faulty mental models.

We have identified approximately 40 principles and a similar number of misconceptions that are affiliated with Newtonian laws of motion and mechanics. These principles and misconceptions are frequently associated with fundamental physics concepts, problems, and experiments conducted in the laboratory. This content can be clustered in the following broad categories that arguably can be ordered on complexity and a prerequisite continuum. An example principle (P) and misconception (M) is provided for each category.

1 Meanings of displacement, velocity, acceleration, and kinematic relationships for single moving bodies.
 P: Constant velocity implies zero acceleration.
 M: Zero acceleration implies zero velocity.
2 Newton's first law (inertia)
 P: Air resistance can be ignored under certain circumstances.
 M: Air resistance can never be ignored.
3 Newton's second law (net $F = ma$, net force equals mass times acceleration)
 P: The same force will accelerate a less massive object more than a more massive object.
 M: Accelerations of both objects are equal during interaction.
4 Gravity and contact forces
 P: All objects in freefall have the same acceleration.
 M: Heavier objects fall faster.
5 Newton's third law (action and reaction forces)
 P: When A and B exert a force on each other, the magnitudes of the two forces are equal.
 M: A smaller object exerts no force on a larger object.

Many physics books order the chapters or lessons in the above order. It is unclear whether this is mere convention, a historical accident, or a custom motivated by pedagogy. It is intuitively obvious that there should be some ordering on prerequisites. It is difficult to imagine how Newton's laws could be understood without at least a preliminary understanding of the definitions of displacement, velocity, and acceleration. However, it is not intuitively obvious that Newton's first law must precede the second and third laws. We are convinced that the ordering of these lesson categories needs to be justified by a pedagogical theory that is strongly rooted in a development science. This volume, of course, was inspired by this proclamation.

Aside from the ordering of the principles in the sequence of lessons, it is important to assess how reliably the students apply these principles when they solve problems. In an ideal world, the students would consistently apply the principles correctly as they receive hundreds of problems (toy or real-world) over time. They would also discontinue being seduced by misconceptions and faulty mental models. However, the ideal world is a very long distance from reality. The best we can hope for is a statistical progression over time and the learning history that eventually increases the probability of correct application of principles and decreases the probability of misconceptions. Occasionally the students manifest all-or-none learning, but that is more of a special case than the normal state of affairs.

In either case, one of the challenges for a developmentally inspired pedagogy is to select example problems in an intelligent manner that optimizes learning. If a student has already mastered a principle consistently, it is presumably a waste of time to select problems that address that principle. Instead, the selected problems should address the deficits of the individual student. Computer technologies will undoubtedly be required to precisely track the learning history of individual students and to generate problems that optimize learning. It would be much too tedious for a human to keep track of such details.

We remain optimistic about the pedagogical value of learning environments with conversational agents, interactive simulations, and serious games. These should be the important routes to accelerated learning of physics in children. The agents are firmly grounded in social interaction, can model excellent learning, can intelligently respond to individual learners, and are captivating to young children. The interactive simulations make difficult concepts perceptually visible and allow the learner to actively manipulate the environment. The games keep the students motivated for hours on end as they simultaneously acquire difficult conceptualizations with academic and cultural value. All we need to do now, at this magic moment in the history of technology, is to build the systems, test them, and disseminate the successful ones to relevant educational communities. The learning, cognitive, discourse, and developmental sciences seem ready to hunker down and take up the challenge. As we take up the challenge, we will continue to remind ourselves of the long parade of Don Quixotes in education.

Acknowledgments

The research on AutoTutor was supported by the National Science Foundation (SBR 9720314, REC 0106965, REC 0126265, ITR 0325428, REESE 0633918), the Institute of Education Sciences (R305H050169, R305B070349, R305A080589), and the DoD Multidisciplinary University Research Initiative (MURI) administered by ONR under Grant N00014-00-1-0600. Any opinions, findings, and conclusions or recommendations expressed in this material are those of the authors and do not necessarily reflect the views of NSF, IES, DoD, or ONR.

References

Ainsworth, S. (2008). How do animations influence learning? In D. H. Robinson and G. Schraw (Eds.), *Recent innovations on educational technologies that facilitate student learning.* Charlotte, NC: Information Age Publishing. pp. 37–67.

Atkinson, R. K. (2002). Optimizing learning from examples using animated pedagogical agents. *Journal of Educational Psychology*, 94, 416–427.

Azevedo, R., & Cromley, J. G. (2004). Does training on self-regulated learning facilitate students' learning with hypermedia? *Journal of Educational Psychology*, 96, 523–535.

Baylor, A. L., & Kim, Y. (2005). Simulating instructional roles through pedagogical agents. *International Journal of Artificial Intelligence in Education*, 15, 95–115.

Chi, M. T. H. (2005). Commonsense conceptions of emergent processes: Why some misconceptions are robust. *Journal of the Learning Sciences*, 14, 161–199.

Chi, M. T. H., Siler, S. A., Jeong, H., Yamauchi, T., & Hausmann, R. G. (2001). Learning from human tutoring. *Cognitive Science*, 25, 471–533.

Cohen, P. A., Kulik, J. A., and Kulik, C. C. (1982). Educational outcomes of tutoring: A meta-analysis of findings. *American Educational Research Journal*, 19, 237–248.

Craig, S. D., Gholson, B., Ventura, M., Graesser, A. C., & the Tutoring Research Group (2000). Overhearing dialogues and monologues in virtual tutoring sessions: Effects on questioning and vicarious learning. *International Journal of Artificial Intelligence in Education*, 11, 242–253.

Craig, S. D., Sullins, J., Witherspoon, A., & Gholson, B. (2006). Deep-level reasoning questions effect: The role of dialog and deep-level reasoning questions during vicarious learning. *Cognition and Instruction*, 24, 565–591.

Dodds, P., & Fletcher, J. D. (2004). Opportunities for new "smart" learning environments enabled by next-generation web capabilities. *Journal of Educational Multimedia and Hypermedia*, 13(4), 391–404.

Gee, J. (2003). *What video games have to teach us about learning and literacy*. New York: Palgrave Macmillan.

Gentner, D., & Stevens, A. L. (Eds.) (1983). *Mental models*. Hillsdale, NJ: LEA.

Gholson, B, Witherspoon, A., Morgan, B., Brittingham, J., Coles, R., Graesser, A. C., Sullins, J., & Craig, S. D. (2009). Exploring the deep-level reasoning questions effect during vicarious learning among eighth to eleventh graders in the domains of computer literacy and Newtonian physics. *Instructional Science*, 37(5), 487–493.

Goldman, S. R., Duschl, R. A., Ellenbogen, K., Williams, S., & Tzou, C. T. (2003). Science inquiry in a digital age: Possibilities for making thinking visible. In H. van Oostendorp (Ed.), *Cognition in a digital world*. Mahwah, NJ: LEA. pp. 253–284.

Graesser, A. C., Chipman, P., Haynes, B. C., & Olney, A. (2005). AutoTutor: An intelligent tutoring system with mixed-initiative dialogue. *IEEE Transactions in Education*, 48, 612–618.

Graesser, A. C., D'Mello, S. K., & Person, N. K. (2009). Meta-knowledge in tutoring. In D. Hacker, J. Dunlosky, & A. C. Graesser (Eds.), *Handbook of metacognition in education*. Mahwah, NJ: Taylor & Francis/LEA.

Graesser, A. C., Jackson, G. T., & McDaniel, B. (2007). AutoTutor holds conversations with learners that are responsive to their cognitive and emotional states. *Educational Technology*, 47, 19–22.

Graesser, A. C., Lu, S., Jackson, G. T., Mitchell, H., Ventura, M., Olney, A., & Louwerse, M. M. (2004). AutoTutor: A tutor with dialogue in natural language. *Behavioral Research Methods, Instruments, and Computers*, 36, 180–193.

Graesser, A. C., McNamara, D. S., & VanLehn, K. (2005). Scaffolding deep comprehension strategies through Point & Query, AutoTutor, and iSTART. *Educational Psychologist*, 40, 225–234.

Graesser, A. C., & Person, N. K. (1994). Question asking during tutoring. *American Educational Research Journal*, 31, 104–137.

Graesser, A. C., Person, N. K., & Magliano, J. P. (1995). Collaborative dialogue patterns in naturalistic one-to-one tutoring. *Applied Cognitive Psychology*, 9, 359.1–28.

Hegarty, M. (2004). Dynamic visualizations and learning: Getting to the difficult questions. *Learning and Instruction*, 14, 343–351.

Hestenes, D., Wells, M., & Swackhamer, G. (1992). Force Concept Inventory. *The Physics Teacher*, 30, 141–158.

Hewitt, P. G. (1998). *Conceptual physics*. Menlo Park, CA: Addison-Wesley.

Hunt, E., & Minstrell, J. (1996). A collaborative classroom for teaching conceptual physics. In K. McGilly (Ed.), *Classroom lessons: Integrating cognitive theory and the classroom.* Cambridge, MA: MIT Press.

Jackson, G. T., Olney, A., Graesser, A. C., & Kim, H. J. (2006). AutoTutor 3-D simulations: Analyzing user's actions and learning trends. In R. Son (Ed.), *Proceedings of the 28th Annual Meetings of the Cognitive Science Society.* Mahwah, NJ: LEA. pp. 1557–1562.

Johnson, W. L., & Beal, C. (2005). Iterative evaluation of a large-scale intelligent game for language learning. In C. Looi, G. McCalla, B. Bredeweg, and J. Breuker (Eds.), *Artificial intelligence in education: Supporting learning through intelligent and socially informed technology.* Amsterdam: IOS Press. pp. 290–297.

Kalyuga, S., Chandler, P., & Sweller, J. (1999). Managing split-attention and redundancy in multimedia instruction. *Applied Cognitive Psychology*, 13, 351–371.

Klahr, D. (2002). *Exploring science: The cognition and development of discovery processes.* Cambridge, MA: MIT Press.

Kozma, R. (2000). Reflections on the state of educational technology research and development. *Educational Technology Research and Development*, 48, 5–15.

Landauer, T., McNamara, D. S., Dennis, S., & Kintsch, W. (Eds.) (2007). *Handbook on latent semantic analysis.* Mahwah, NJ: LEA.

Malone, T. W., & Lepper, M. R. (1987). Making learning fun: A taxonomy of intrinsic motivations for learning. In R. E. Snow & M. J. Farr (Eds.), *Aptitude, learning and instruction. Vol. 3: Conative and affective process analyses.* Hillsdale, NJ: LEA. pp. 223–253.

Mayer, R. E. (2005). *Multimedia learning.* Cambridge: Cambridge University Press.

McNamara, D. S., Levinstein, I. B., & Boonthum, C. (2004). iSTART: Interactive strategy trainer for active reading and thinking. *Behavioral Research Methods, Instruments, and Computers*, 36, 222–233.

van der Meij, J., & de Jong, T. (2006). Supporting students' learning with multiple representations in a dynamic simulation-based learning environment. *Learning and Instruction*, 16(3), 199–212.

Millis, K. K., Magliano, J., Britt, A., Wiemer-Hastings, K., Halpern, D., & Graesser, A. C. (2006). Acquiring research evaluative and investigative skills (ARIES) for scientific inquiry. Unpublished manuscript, Northern Illinois University.

Moreno, R., & Mayer, R. E. (2005). Role of guidance, reflection, and interactivity in an agent-based multimedia game. *Journal of Educational Psychology*, 97(1), 117–128.

O'Neil, H. F., Wainess, R., & Baker, E. L. (2005). Classification of learning outcomes: Evidence from the computer games literature. *The Curriculum Journal*, 16, 455–474.

Otero, J., Leon, J. A., & Graesser, A. C., (Eds.) (2002). *The psychology of science text comprehension.* Mahwah, NJ: LEA.

Ploetzner, R., & VanLehn, K. (1997). The acquisition of informal physics knowledge during formal physics training. *Cognition and Instruction*, 15, 169–206.

Reeves, B., & Nass, C. (1996). *The media equation: How people treat computers, televisions, and new media like real people and places.* Cambridge: Cambridge University Press.

Ritterfeld, U., Cody, M., & Vorderer, P. (Eds.) (2009). *Serious games: Mechanisms and effects.* Mahwah, NJ: Routledge.

Rosenshine, B., Meister, C., & Chapman, S. (1996). Teaching students to generate questions: A review of the intervention studies. *Review of Educational Research*, 66, 181–221.

Rus, V., McCarthy, P. M., McNamara, D. D., & Graesser, A. C. (2008). A study of text entailment. *International Journal on Artificial Intelligence Tools*, 17, 659–685.

diSessa, A. A. (1993). Towards an epistemology of physics. *Cognition and Instruction, 10*(2 & 3), 105–225.

Stein, N. L., Hernandez, M., & Gamez, P. (2007). Making the invisible visible: The conditions for the early learning of physics. Symposium presentation at the Society for Text and Discourse, Glasgow, Scotland, 7–10 July.

Tversky, B., Morrison, J. B., & Betrancourt, M. (2002). Animation: Can it facilitate? *International Journal of Human–Computer Studies*, 57, 247–262.

VanLehn, K., Graesser, A. C., Jackson, G. T., Jordan, P., Olney, A., & Rose, C. P. (2007). When are tutorial dialogues more effective than reading? *Cognitive Science*, 31, 3–62.

VanLehn, K., Lynch, C., Taylor, L., Weinstein, A., Shelby, R. H., Schulze, K. G., et al. (2002). Minimally invasive tutoring of complex physics problem solving. In S. A. Cerri, G. Gouarderes, & F. Paraguacu (Eds.), *Intelligent Tutoring Systems, 2002, 6th International Conference*. Berlin: Springer. pp. 367–376.

Virvou, M., Katsionis, G., & Manos, K. (2005). Combining software games with education: Evaluation of its educational effectiveness. *Educational Technology and Society*, 8, 54–65.

Vitale, M. R., & Romance, N. R. (2007). A knowledge-based framework for unifying the content-area reading comprehension and reading comprehension strategies. In D. McNamara (Ed.), *Theories of text comprehension: The importance of reading strategies to theoretical foundations of reading comprehension*. Mahwah, NJ: LEA. pp. 73–104.

White, B. Y., & Frederiksen, J. R. (1998). Inquiry, modeling, and metacognition: Making science accessible to all students. *Cognition and Instruction*, 16(1), 3–118.

Woolf, B. P. (2009). *Building intelligent interactive tutors: Student-centered strategies for revolutionizing e-learning*. Burlington, MA: Morgan Kaufmann Publishers.

Part III

Mathematical Learning

12 Emerging Ability to Determine Size

Use of Measurement

Janellen Huttenlocher, Susan C. Levine, and Kristin R. Ratliff

This chapter concerns an important aspect of quantitative development, namely the ability to determine amount ("extent") with respect to a single dimension such as length. Our focus is on measurement ability and its growth with age. Humans have developed schemes for measuring various quantities such as length, overall object size (area and volume), and distance between objects. Quantities obtained using these measures can then be maintained in memory and shared with others. Such schemes specify a system of standards with which amounts can be compared, contributing to accuracy in establishing the values of these amounts. Further, measurement schemes generally allow for the breaking down of continuous dimensions into discrete units that can be used to preserve quantitative information about individual amounts, and to compare among different amounts.

To trace measurement skills across development, one must begin early in life. Contrary to the Piagetian view claiming that young children are unable to code distance and length until school-age years, we argue that even infants are sensitive to extent. Additionally, we posit that children's understanding of measurement and units is malleable during the preschool and early elementary ages, particularly in response to direct training and presenting problems within familiar contexts during instruction. However, this development continues well into the school-age years, since use of conventional measurement tools such as rulers is acquired only then. In the present chapter, we provide evidence for early acquisition and development of three aspects that are fundamental to success in measurement: (a) acquisition of the basic process involved in determining extent along one dimension (e.g., length or height); (b) acquisition of the notion of a unit of measure (that units must be specified and be equal in size in order to measure and compare, including the notions of standardization and iteration of units, as well as understanding the inverse relation between unit size and number); and (c) acquisition of conventional measurement skills from experience with measurement instruments such as rulers in the early elementary school years.

Determining Extent along One Dimension

What is Entailed in Judging Extent?

The starting point for measurement skills is the judgment of extent for a single object. Although the Piagetian view of development recognizes infants' ability to directly compare objects, it also holds that measurement abilities such as determining lengths and distances do not emerge until the school-age years (Piaget & Inhelder, 1967). However, as we argue here, there is recent evidence showing that even infants are sensitive to variations in amount. That is, they exhibit an ability to distinguish different extents, differentiating lengths of objects, or distances among them. Although extent may seem to be a perceptually available property, particularly when differences between targets are quite large, accuracy in determining extent when targets differ only modestly involves comparisons in which targets are judged relative to a standard. A standard can be a conventional measurement device such as a ruler, or may consist of an arbitrary object to which a target is compared and judged to be shorter, equal, or longer. This is compared with a mature system of measurement that allows for coding of object size and shape and the relations between objects (e.g., distance, direction) when direct comparisons with a perceptually present standard are unavailable.

The simplest measurement situation involves comparison of a single target with a standard. If the target and standard are presented together and are aligned, a direct comparison can be made. However, if no direct comparison is possible, as when a target is displaced in space or time from a potential standard, either the target or the standard must be moved, physically or mentally, in order to make an estimate of the target. If only one object is shown, the standard can be used to establish its length. Often, two objects must be compared to determine which is longer or shorter than the other, or whether they are equal. If the objects are in quite different locations, a standard can be used to mediate comparison of the two. These situations illustrate the logic of measurement, and we will examine how these skills play out in children's developing ability to estimate length in particular tasks.

Coding Extent in Relation to a Standard

A basic issue in examining quantitative development is to determine the origin of the use of standards. Several experiments demonstrate that, in some circumstances, even infants can discriminate length (in both the horizontal and vertical planes) (Baillargeon, 1991). In our studies of the emergence of measurement, we predicted that it would be easier to judge extent in a visually complex situation, where a standard is present, than in a visually simpler situation, where a target is presented alone. For example, we reasoned it would be easier to determine extent when a target is in a container, because the container serves as a standard. When the target is presented alone, no standard of comparison is present, so judging length requires imposing a standard. We hypothesized that the mental processes involved in imposing a standard are not available to infants and young children.

In a series of studies using habituation methods, we examined infants' ability to discriminate different amounts of liquid. Six-month-old infants were repeatedly shown a particular stimulus (e.g., red liquid in a container that was either one-quarter or three-quarters full) until looking time decreased. Infants were then shown a new amount alternating with the old amount to determine if they looked longer at the new amount. After being habituated to one of these quantities, they looked longer at the new amount (Gao, Levine, & Huttenlocher, 2000). In subsequent studies, Huttenlocher, Duffy, and Levine (2002) contrasted infants' ability to code the length of a target stick (either 6 cm or 12 cm) when a standard was present or absent. In two conditions, a standard was present; the target stick was placed inside a glass container (18 cm tall) or next to a gray stick (18 cm tall). In the third condition, the target was presented alone. Once habituated to the display, infants were alternately presented with the target stick and a novel stick that differed only with respect to height (6 cm or 12 cm). When the target stick was in a container or alongside a standard, looking times were longer for the new stick. However, when the target was presented in isolation, looking times were equal for the old and new stick. These findings suggest that, without a standard, infants are unable to tell the difference between the 6-cm and 12-cm sticks. Thus, infants coded information about the stick's length, but only when a standard was present.

The ability to discriminate extent appears at an early age for distance as well as for length. In another study, infants coded the location of a toy in relation to a standard (e.g., the frame of a sandbox) (Newcombe, Huttenlocher, & Learmonth, 1999). Six-month-olds were placed in an infant seat and shown a narrow three-foot-long sandbox. They watched a hand hiding a target object in a particular location and retrieving it there four times. Then the toy was hidden in the same location, but was retrieved eight inches away from the original hiding spot. Infants looked longer at this event because it violated their expectation of where the toy should be, suggesting that they used distance information to code the location of the object and to discriminate between locations. An additional study found that toddlers (16 to 24 months of age) were remarkably accurate in locating a hidden object in a sandbox five feet long (Huttenlocher, Newcombe, & Sandberg, 1994). Accuracy remained high even when the child's view of the scene changed from encoding to retrieval (e.g., children saw the object being hidden and were then moved laterally to one end of the box before they pointed to its location), suggesting that they coded location relative to the sandbox, not themselves.

These studies with length and distance show an early ability to code linear extent that is present by six months of age. Additionally, the hypothesis that length and distance coding in infants and toddlers is constrained to situations with a present, salient standard was supported by our findings. They succeeded only when a target was contained within the standard (e.g., a stick placed within a container) or when the target was placed next to a standard (e.g., a target stick next to a different comparison stick).

Development of Extent Coding from Relative (Present Standard) to Absolute (Absent Standard)

In order to explore the development of children's ability to perform measurement operations by imposing a mental standard on targets, we next turned to preschool children. We wanted to determine at what age they would provide evidence of understanding measurement by discriminating length without a standard. The procedure we used was somewhat different from that used with infants. In a series of studies by Huttenlocher, Duffy, and Levine (2002; Duffy, Huttenlocher, & Levine, 2005), children were told about a dog named Toby, who had a favorite stick. They were shown the target stick. They were also told that the stick was lost, and were asked to help Toby find it. Children were then shown two sticks that differed in length and were asked which one was Toby's stick. We used the same conditions as with infants and toddlers. That is, sticks were presented either (a) alone, (b) in containers, or (c) alongside a standard stick. The sticks used in these studies ranged in size from 2.25 cm to 15.75 cm and the standard (container or aligned stick) was 18.0 cm tall. When the sticks were presented with an accompanying standard, children successfully chose the target stick in the discrimination task. However, if the sticks were presented in isolation, two-year-olds were not able to discriminate between the target and foil.

Before concluding that two-year-olds needed a standard, it was important to determine if they would be able to discriminate size without a standard if the difference between the lengths of two sticks were larger. Even with a very large difference (9 cm) between the sizes of the target and foil sticks, two-year-olds were unable to discriminate between them without an aligned standard. For four-year-olds, a difference of 4.5 cm was sufficient for them to succeed at the task, both with and without a standard, but, when the size difference was reduced to 2.25 cm, the four-year-olds could succeed only in the presence of an aligned standard. Although four-year-olds demonstrated at least a coarse-grained discrimination of extent (by correctly choosing the target stick even without directly comparing it to a present standard when differences were large, i.e., 4.5 cm), we do not know whether they are mature reasoners about extent, which involves maintaining a constant mental standard that does not rely on the relative length of the target in relation to a given standard. To show that they judge the extent of a target by imposing a constant standard, four-year-olds would have to do so even in a perceptually misleading situation where the standard is changed between initial exposure and the discrimination task.

To address this issue, Duffy, Huttenlocher, and Levine (2005) varied the size of the standard between the initial presentation and the discrimination task. Thus, four-year-olds and eight-year-olds were shown a target stick in a glass container and asked to find the same item—Toby's stick. Then the target and foil sticks were presented in containers that were equal in size, but that differed in size (either smaller or larger) from the original container in which the target stick was shown. If children choose Toby's stick by comparing it with a mental standard, the task should pose no problem. However, if they use the present frame to choose the target, they should be misled.

There were two experimental conditions: the "relative foil" condition and the "unrelated foil" condition. In both of these conditions, the correct target choice was the same in *absolute* size (length from one end of the stick to the other end) but different in *relative* size from the original (proportion of the container filled by the stick). However, the nature of the foil choice differed across the two conditions. In the relative foil condition, the foil was different in absolute size but the same in relative size compared with its container as the target. In contrast, in the unrelated foil condition, the foil was different in both absolute and relative size than the original target. In the control condition, the containers in the choice task were the same size as in the original presentation (see Figure 12.1).

The results showed that four-year-olds chose correctly at well above the chance level of .50 in the control condition (.82; $p < .001$). However, in the experimental conditions, four-year-olds' performance did not exceed chance in the unrelated foil condition (.56; $p > .15$). Moreover, in the relative foil condition they were badly misled, choosing the stick that was the same relative size, not the same absolute size, significantly less than chance (.33; $p < .001$). In contrast, eight-year-olds were above chance in judging extent in all conditions, even the misleading relative foil condition ($p < .001$ in all cases).

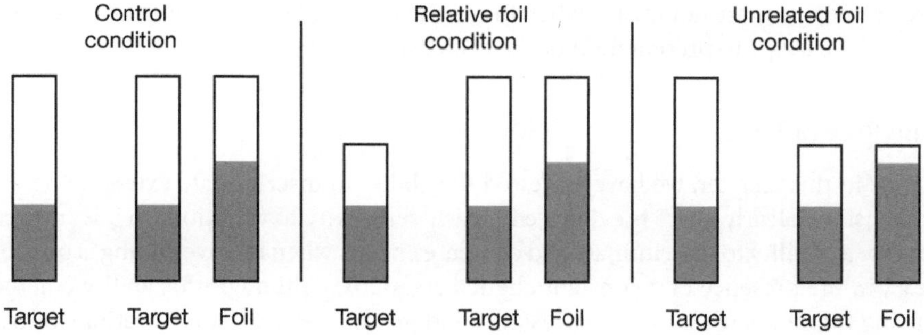

Figure 12.1 Three Conditions that Vary the Relation of a Target Length to a Standard Length in Order to Test whether Children (Four to Eight Years Old) Judge the Extent of a Target by Imposing a Constant Standard or They Use Relative Proportion. From "It's all relative: How young children encode extent" by S. Duffy, J. Huttenlocher, & S. Levine, 2005. *Journal of Cognition and Development*, 6(1), 51–63. Reprinted by permission of the publisher Taylor & Francis, Ltd. (http://www.tandf.co.uk/journals).

Thus, for four-year-olds, size relative to the container is a critical factor in judging length. Younger children depend more on the relation of the target to surroundings and less on the length of the target itself to determine quantity than do older children. Even though only the target itself was to be judged, the context in which a stimulus appears affects judgment well into the school years, some time between ages four and eight. Indeed, this sensitivity to context does not disappear altogether with age, as even adults are affected by the frame in judging extent (Rock & Ebenholtz, 1959). The way objectivity is achieved is through using measures that capture only the size of the target, excluding surrounding features. That achievement marks a mature form of measurement of extent. The operations involved in establishing the size of the target can prevent framing effects in judging target length.

Although we found that the ability to determine relative extent emerges earlier than the ability to determine absolute extent, the question arises as to whether relative judgments remain easier at later ages. Vasilyeva, Duffy, and Huttenlocher (2007) found that, even for seven- and nine-year-olds, it is easier to select a target on the basis of relative size than absolute size. This is consistent with studies of adults showing that the ability to maintain information about absolute size is more difficult than maintaining information about relative size when a salient standard is present (Rock & Ebenholtz, 1959).

To summarize, we have seen that coding and retention of information about continuous quantity is not based on simple perceptual responses. There is a critical intermediate step involved in assessing quantity, that is, the use of a standard to judge the target. As early as six months, infants can discriminate extent along one dimension (length or distance) when a standard is present (e.g., an aligned standard object or container) but not when a standard is lacking. The ability to distinguish two lengths in the absence of an aligned standard develops some time between two and four years of age. However, not even four-year-olds are successful in judging extent in the absence of a constant aligned standard unless the difference between two objects is at least 4.5 cm. By the age of eight, children are able to judge extent in the absence of an aligned standard, even when differences are smaller (i.e., 2.25 cm); when standards are present, they are not misled when the size of the standard changes (from presentation of the target to presentation of the choices).

The Role of Units

So far in this chapter, we have reviewed the ability to discriminate extent along one dimension. Clearly, there is evidence for early sensitivity to variations in linear extent and for an ability to discriminate and compare extents when relative coding is possible (e.g., in the presence of a constant aligned standard). Although this ability emerges during infancy, such length and distance judgments are not demonstrations of true measurement, and the relation of these skills to mature measurement abilities has yet to be identified. A mature system of measurement allows for coding of amount, such as objects that vary in size, shape, and orientation, and the relations between objects even when direct comparisons are unavailable. Such mature measurement requires conceptual understanding that young children lack. Key among the requirements is the concept of a *unit*. Children must understand several fundamental properties of units in order to measure objects based on quantity in the absence of a perceptually available standard.

Standardized Units and Inverse Relations

Children must develop the notion that objects of certain extents can be measured in relation to units of varying sizes. An understanding of units of measure entails the realization that the smaller the size of the unit, the larger the number of units a given object will encompass, and that changes in the units of measure change numerical answers (1 foot = 12 inches) but do not change the length of an object being measured. It also involves understanding the importance of maintaining a constant unit size when measuring an object or distance and when comparing the measures of different objects or distances.

Miller (1984) has shown that children between three and five years old lack a general understanding of the notion that the size of pieces (or units) must remain constant in a particular measurement situation. Preschool children were presented with a series of different materials—such as pieces of candy, strips of clay "spaghetti," clay squares of "fudge," and glasses of "kool-aid"—and were asked to evenly divide the materials among several puppets. The most common strategy was to cut the materials into pieces, regardless of size, and count them to ensure the same number of pieces. For example, if one puppet ended up with fewer pieces of "fudge" after dividing the given pieces, children considered it fair to break the last piece into several smaller parts until each puppet had an equal number of pieces. However, second- and fourth-grade children were more likely to cut the materials directly into portions of approximately equal size. This suggests that the preschool children were concerned not with a constant size of units (e.g., some pieces were bigger than others), but rather with the number of total units (e.g., that each puppet had the same number of pieces of "fudge").

Further evidence suggests that preschool children do not understand a crucial aspect of units, that the subdivision of a whole object into units can be variable, but that the size of the units does not change the overall amount that is being measured. For example, four- to five-year-old preschool children were shown an array of forks with some broken into two pieces and some whole, and were then asked to count the forks. Adults would count the two broken pieces as only one fork, whereas preschool children counted all the discrete parts separately, thus counting each part (e.g., the two pieces) as individual whole objects (e.g., two forks) (Shipley & Shepperson, 1990). Similarly, kindergarten children consider the total number of units when determining amount rather than the sum of all the units, suggesting a lack of conceptual understanding of relating the unit to a whole (Gal'perin & Georgiev, 1969). In this case, children were given two equal cups of rice and used either a tablespoon or a teaspoon to scoop out piles of rice until the cup was empty. When asked which piles contained more rice, one might think that children would consider the size of the piles themselves (e.g., the piles formed using the tablespoon are larger than the teaspoon piles) rather than by the number of piles, but this was not the case. Children instead solved this problem based on the number of individual piles and generally chose the teaspoon piles as containing more rice than the tablespoon piles simply because there were a greater number of teaspoon piles.

These studies demonstrate that children up to six years of age generally lack understanding that the total amount of a quantity, *X*, can be divided into different size units, but that the overall number of units has no effect on the amount of *X*. Further, at this age, children lack knowledge that unit size, not just number, is relevant to amount judgments. Thus, they are inclined to determine amount by simply counting the number of discrete units without taking unit size into account.

The Context of Learning Can Enhance Unit Understanding

Young children are capable of demonstrating some understanding of units when problems are presented using familiar contexts. Sophian and colleagues (1997) found that, although inverse relationships are a difficult concept for preschoolers, framing the problem in specific, familiar ways can improve performance. Specifically, five-year-olds were better able to understand inverse relations when asked to share a set of objects when the directions were framed as a subtractive problem (giving away a cupful to one versus two recipients) rather than in terms of a fractional problem (sharing equally among one versus two recipients). Similarly, allowing children to directly compare the results of dividing two equivalent quantities into different shares (e.g., divide an amount into two versus five shares) improved performance more than in a condition in which children were shown the divisions sequentially, not side by side. Thus it appears that five-year-olds can learn about the inverse relation between number of shares and share size but only when direct comparison is available and the problem is framed in a familiar context (Sophian, Garyantes, & Chang, 1997; Sophian, 2007).

Other training procedures have been shown to improve even younger children's understanding of important measurement principles. Sophian (2002) trained three- and four-year-olds to correctly estimate area and unit size by judging whether more small objects or more large ones would fit in a specific space. Like the five-year-olds in the study described above, three- and four-year-olds initially, and incorrectly, thought that more of the larger objects would fill a container. Following several training sessions in which the experimenter placed objects of the two sizes, one by one, into two identical containers, children dramatically improved in estimating area using different unit sizes. These results support the idea that young children have difficulty in dealing with units in relation to a whole, in particular with the inverse relation between unit size and number of units. However, they also show that young children benefit from interventions that highlight this relation (Sophian, 2004).

Similarly, Casey and colleagues (2008) found that using a storytelling context to present part–whole relations and basic geometric principles greatly improves performance on geometry-related tasks, particularly for children from lower socio-economic backgrounds. Kindergarten children received geometry instruction either with or without a story context that emphasized part–whole relation skills (e.g., "The tabletop fell off Tan's head and broke into two pieces . . . it had broken into two triangles that looked exactly the same."; Schiro, Casey, & Anderson, 2002). Children were then given geometry-based tasks designed to assess their knowledge of part–whole relations using puzzles with triangular pieces. The story-based intervention was more effective for lower-SES (socio-economic status) students, and benefited both girls' and boys' learning of early geometry principles. Thus, preschool and kindergarten children show some understanding of the concept of units and other fundamental notions relevant to measurement, when the concepts are explored in interactive and familiar contexts (e.g., through storytelling), or through direct training.

Conventional Measurement in Elementary School

As we have seen, the ability to code extent along a single dimension is present early on (as evidenced by infants' sensitivity to size), and develops throughout early childhood. During the preschool years, children generally lack fundamental notions about units, but, given certain kinds of instruction, they can acquire an understanding of part–whole inverse relations and unit standardization. However, the degree to which these early skills are related to a mature understanding of measurement is an empirical question that remains unanswered. With respect to conventional measurement abilities, children undergo tremendous development during the school-age years, and the activities engaged in during elementary education seem to be critical to acquiring mature measurement skills. As indicated previously, by eight years of age, children could distinguish two lengths without a direct comparison (i.e., they could impose a mental standard as a way to distinguish the lengths of different objects). It seems that this ability is more closely related to conventional measurement than the skill displayed from infancy to four years of age (Duffy et al., 2005). The increase in measurement skills from preschool to school age may be related to two kinds of change. The first is an increasing ability to form and maintain images with certain attributes in working memory (Bull, Espy, & Wiebe, 2008). The second is formal instruction in conventional measurement at school, including the role of standards in measurement, the use of units, and experience with proportional reasoning, which is important in understanding the inverse relation between unit size and number as well as scaling.

Experience Using Measurement Tools

During the early elementary school years, children develop an understanding of measurement tools and how to use them appropriately, often through direct instruction in school. Children between five and eight-and-a-half years of age show increases in understanding the process of measuring length (Boulton-Lewis, Wilss, & Mutch,

1996). These advances follow the principles seen in experimental studies and those encountered in traditional measurement instruction during early elementary school. Direct comparisons to determine if two objects seem equal in length emerge first (e.g., holding two pencils next to each other), followed by the use of measuring tools. In general, math curricula introduce the use of nonconventional measurement tools to determine length, prior to the use of standard measuring devices (e.g., traditional rulers instead of their fingers, hands, or pieces of string), although there is debate as to whether this ordering leads to optimal learning (Boulton-Lewis et al., 1996; Nunes, Light, & Mason, 1993).

When using rulers, children often make errors. In the first- to fourth-grade range, six- to ten-year-olds exhibit several patterns that primarily follow procedural strategies of using a ruler. Their lack of conceptual understanding of units is evident when they are asked to generalize the procedures to novel problems (Bragg & Outhred, 2001). In fact, a majority of students answer questions about length incorrectly (86 percent in third grade, 78 percent in fourth grade, 51 percent in seventh grade, and 37 percent in eighth grade) on the NAEP (Kamii, 2006; Martin & Strutchens, 2000). The difficulty is seen when targets are not aligned with the "0" point on a conventional ruler.

Two erroneous strategies can be seen when children measure misaligned items (Lehrer, Jenkins, & Osana, 1998). The first error is to read off the number on the ruler corresponding to the end of the item. This error suggests that the children do not understand that measurement involves a count of the units that overlap with the length of the object. The second error is to focus on the units and count the hash marks on the ruler, incorrectly starting with the mark where the misaligned target begins, a strategy that leads to a response that is one greater than the correct number. This error suggests that children do not represent the interval between the hash marks as units. The concept of an interval unit with a spatial extent is lacking in this case.

The suggestion that differences in input may be the source of differences in mastery is seen in the fact that SES is related to overall performance as well as to strategy differences. Preliminary results show that low-SES students are more likely to produce "read-off" errors whereas high-SES students are more prone to errors in counting hash marks, suggesting that, as children gain more knowledge about measurement, about conventional tools, and about units, their sophistication in solving measurement problems increases (Solomon, Vasilyeva, Levine, Huttenlocher, & Ratliff, under review). Older age groups and higher-SES groups perform better on problems in which the ruler and object are not aligned, shifting from a perceptual strategy (reading off the number corresponding with the end of an object) to a more conceptual understanding of measurement (a counting strategy but with incorrect counting of the units), and finally to a correct strategy of counting intervals.

A subsequent study with second-grade students implemented a training procedure aimed at improving children's ruler measurement and understanding of linear units. The training consisted of superimposing transparent "unit" pieces on a ruler to measure objects that were misaligned with the ruler. A control group received training that incorporated practices that are common in mathematics curricula. Specifically, children measured objects that were aligned with a ruler and they measured objects with "unit pieces" in two separate activities. The results showed significant improvement in the experimental group but not in the control group. Moreover, the improvement in the experimental group persisted after a week-long delay (Levine, Kwon, Huttenlocher, Ratliff, & Dietz, 2009).

Length, Area, and Volume

Measurement instruction during elementary school involves a typical progression of activities focusing first on length (along one dimension), then on area (two dimensions), and finally on volume (three dimensions) by grade 3 or 4. This trajectory relates to the developmental pattern described in the literature, where success in measuring length occurs prior to success in measuring area and then volume (Hart, 1984). Although there has been a large amount of research investigating measurement of length, area, and volume, most studies examine these emerging skills separately. A few studies have examined the relation between two types of measure, focusing on understanding of length and area (Outhred & Mitchelmoore, 2000), confusion between perimeter and area (Lehrer et al., 1998), and relating area measurement to volumetric measurement (Battista, 2003).

Recent research suggests that there are reasons, both theoretical and empirical, to expect that understanding of length, area, and volume measurement are related, and that perhaps these concepts should *not* be presented separately by grade during elementary instruction (Curry & Outhred, 2005). For example, all three forms of measurement involve a spatial component, the use and operation of units (standardized and iterated), and a unit structure (length along one dimension, area along two dimensions, and volume along three dimensions). The patterns of errors children make in measurement tasks are often the same for length (Bragg & Outhred, 2001) and area (Outhred & Mitchelmoore, 2000); these errors include overlapping the units of measure, leaving gaps between units, and using non-congruent units. Similarly, children often follow an identical path in learning to measure length, area, and volume—by first filling a space with multiple units, then measuring by iterating a single unit, and finally visualizing the unit without a present standard. This marks a difficult transition (e.g., going from using concrete units to fill a space to a more conceptual understanding of visualizing the unit structure) that takes place within all three forms of measurement: length (Bragg & Outhred, 2001), area (Battista, Clements, Arnoff, Battista, & Van Auken Borrow, 1998), and volume (Battista & Clements, 1996).

Perhaps the most striking evidence suggesting that the teaching of length, area, and volume should be integrated comes from a study that relates measurement in these three domains with equally demanding tasks presented to first- through fourth-grade students (Curry & Outhred, 2005). Children were asked to complete four measurement tasks (one each for length and area, and two for volume). For length, children were shown a ribbon of a certain length and asked how many ribbon lengths would be needed to measure the space alongside a line. The children then used the ribbon to check their answer. For area, children used tiles to cover a rectangular region and, for volume-by-packing, children used blocks to fill a rectangular box. An additional volume-by-filling task involved watching the experimenter fill a cup with rice and pour the cup into a jug. The child was then asked how many cupfuls of rice were needed to fill the jug completely. They were to demonstrate using the cup to check their answer.

The results suggest a distinct increase in understanding the unit structure of length, area, and volume across grades 1 through 4, with the following order from easiest to most difficult: (a) length and volume (as measured by filling), (b) area, (c) volume as measured by packing. Interestingly, measurement of length was *not* found to be a prerequisite for successfully measuring area in that some children scored higher on area tasks than they did on length tasks. Instead, measuring length and area both depend on precision in iterating units and both provide the foundation for later understanding of volume-by-packing. A critical discovery in this study was that, as early as first grade, children succeeded in measuring volume-by-filling, which was similar in its developmental trajectory to measuring length. Thus, an understanding of measuring volume is possible much earlier than the traditional schedule of measurement activities performed in school. Volume-by-filling provides the concept of measuring something that takes up space in three dimensions, but simplifies the measurement task itself by limiting the dimensions utilized (in this case, the height of the rice in the jug). In contrast, volume-by-packing requires children to utilize all three dimensions of space (length, width, and height) during the measurement task. By grade 4, children demonstrated equivalent understanding of length, area, and volume-by-filling but not an understanding of volume-by-packing, which was still well below the other three forms of measurement. This may reflect the difficult transition from using two-dimensional to using three-dimensional units, in that measuring length and even volume-by-filling can be accomplished along one dimension, whereas area requires two dimensions, and volume-by-packing requires three dimensions.

Conclusions

Until recently, it was widely believed that children do not encode continuous quantity (distance or amount) until well into middle childhood (e.g., Piaget & Inhelder, 1967). Doing so, it was said, requires measurement, and it was thought that infants and toddlers cannot measure. However, we argue that children have an immature form of measurement early in life. If infants and toddlers are like adults, they should be able to estimate extent *without* perceptually present standards. The fact that they cannot do this and that they fail to estimate extent when the frame of reference changes indicates that they do not possess the basic notions that underlie measurement. Thus, their inability to impose standards is consistent with Piagetian constructivist claims that coding of length does not emerge until later during the childhood period. Yet what we have shown here is that the ability to impose a standard may not arise as a totally new mental function. Rather, it may emerge from earlier forms of extent coding. However, the degree to which the abilities observed in infancy serve as precursors to mature measurement in the child is yet to be established.

Consistent with Bryant's (1974) views, relative coding of extent is available early in life. Moreover, this ability is not unique to humans, but rather shared with a variety of nonhuman animals that demonstrate an ability to discriminate relative length. Thus, for example, rats, pigeons, chicks, and even fish exhibit a remarkable sensitivity to geometric properties of enclosed spaces (Cheng, 1986; Gouteux, Thinus-Blanc, & Vauclair, 2001; Kelly, Spetch, & Heth,1998; Sovrano, Bisazza, & Vallortigara, 2002; Vallortigara, Zanforlin, & Pasti, 1990). They are able to discriminate among different corners on the basis of differences in relative lengths of sides. Thus the relative coding of extent reflects a competence that is broadly available across species, rather than one that is restricted to creatures who will later understand measurement in relation to units.

Whereas the early forms of extent coding may be necessary for true measurement, they are not sufficient. The emergence of mature measurement is guided by processes other than earlier skills, and, unlike the species-general sense of relative amount, is highly sensitive to input. Thus, young children's understanding of units can be dramatically improved through direct training or using familiar contexts during instruction. The role of experience using conventional measurement tools during elementary school also increases children's understanding of units and their general measurement skills. Further research is needed to address whether formal instruction of length, area, and volume would have greater impact if integrated rather than presented individually across the elementary grade levels. Similarly, research on this issue could inform educational preparation for teachers by highlighting the connections across the three areas of measurement, which may assist in linking the children's understanding of length, area, and volume.

References

Baillargeon, R. (1991). Reasoning about the height and location of a hidden object in 4.5- and 6.5-month-old infants. *Cognition*, 38, 13–42.

Battista, M. T. (2003). Understanding students' thinking about area and volume measurement. In D. Clements & G. Bright (Eds.), *Learning and teaching measurement*. Reston, VA: NCTM. pp. 122–142.

Battista, M. T., & Clements, D. H. (1996). Students' understanding of three-dimensional rectangular arrays of cubes. *Journal for Research in Mathematics Education*, 27, 258–292.

Battista, M. T., Clements, D. H., Arnoff, J., Battista, K., & Van Auken Borrow, C. (1998). Students' spatial structuring of 2D arrays of squares. *Journal for Research in Mathematics Education*, 29, 503–532.

Boulton-Lewis, G. M., Wilss, L. A., & Mutch, S. L. (1996). An analysis of young children's strategies and use of devices for length measurement. *Journal of Mathematical Behavior*, 15, 329–347.

Bragg, P., & Outhred, L. (2001). Students' knowledge of length units: Do they know more than rules about rulers? In M. van den Heuvel-Panhuizen (Ed.), *Proceedings of the 25th Annual Conference of the International Group for the Psychology of Mathematics Education* (Vol. 1). Utrecht: PME. pp. 377–384.

Bryant, P. (1974). *Perception and understanding in young children: An experimental approach.* New York: Basic Books.

Bull, R., Espy, K. A., & Wiebe, S. A. (2008). Short-term memory, working memory, and executive functioning in preschoolers: Longitudinal predictors of mathematical achievement at age 7 years. *Developmental Neuropsychology*, 33, 205–228.

Casey, B., Erkut, S., Ceder, I., & Young, J. M. (2008). Use of a storytelling context to improve girls' and boys' geometry skills in kindergarten. *Journal of Applied Developmental Psychology*, 29, 29–48.

Cheng, K. (1986). A purely geometric module in the rat's spatial representation. *Cognition*, 23, 149–178.

Curry, M., & Outhred, L. (2005). Conceptual understanding of spatial measurement. In P. Clarkson, A. Downton, D. Gronn, A. McDonough, R. Pierce, & A. Roche (Eds.), *Building connections: Theory, research, and practice, Proceedings of the 28th Annual Conference of the Mathematics Education Research Group of Australia, Melbourne*. Sydney: MERGA. pp. 265–272.

Duffy, S., Huttenlocher, J., & Levine, S. C. (2005) It's all relative: How young children encode extent. *Journal of Cognition and Development*, 6, 51–63.

Gal'perin, P. A., & Georgiev, L. S. (1969). The formation of elementary mathematical notions. In J. Kilpatrick & I. Wirszup (Eds.), *Soviet studies in the psychology of learning and teaching mathematics: Vol. 1. The learning of mathematical concepts*. Stanford, CA: School Mathematics Study Groups. pp. 189–216.

Gao, F., Levine, S. C., & Huttenlocher, J. (2000). What do infants know about continuous quantity? *Journal of Experimental Child Psychology*, 77, 20–29.

Gouteux, S., Thinus-Blanc, C., & Vauclair, J. (2001). Rhesus monkeys use geometric and nongeometric information during a reorientation task. *Experimental Psychology: General*, 130, 505–519.

Hart, K. (1984). Which comes first—length, area, or volume? *Arithmetic Teacher*, 31(9), 16–27.

Huttenlocher, J., Duffy, S., & Levine, S. C. (2002). Infants and toddlers discriminate amount: Are they measuring? *Psychological Science*, 13, 244–249.

Huttenlocher, J., Newcombe, N., & Sandberg, E. H. (1994). The coding of spatial location in young children. *Cognitive Psychology*, 27, 115–147.

Kamii, C. (2006). Measurement of length: How can we teach it better? *Teaching Children Mathematics*, 13, 154–158.

Kelly, D. M., Spetch, M. L., & Heth, C. D. (1998). Pigeons' (*Columba livia*) encoding of geometric and featural properties of a spatial environment. *Journal of Comparative Psychology*, 112, 259–269.

Lehrer, R., Jenkins, M., & Osana, H. (1998). Longitudinal study of children's reasoning about space and geometry. In R. Lehrer & D. Chazan (Eds.), *Designing learning environments for developing understanding of geometry and space*. Mahwah, NJ: LEA. pp. 137–167.

Levine, S. C., Kwon, M., Huttenlocher, J., Ratliff, K. R., & Dietz, K. (2009). Children's understanding of ruler measurement and units of measure: A training study. *Proceedings of the 31st Annual Cognitive Science Society, July 2009*. Amsterdam: Cognitive Science Society.

Martin, W. G., & Strutchens, M. E. (2000). Geometry and measurement. In E. A. Silver & P. A. Kenney (Eds.), *Results from the Seventh Mathematics Assessment of the National Assessment of Educational Progress*. Reston, VA: National Council of Teachers of Mathematics. pp. 193–234.

Miller, K. F. (1984). Child as the measurer of all things: Measurement procedures and the development of quantitative concepts. In C. Sophian (Ed.), *Origins of cognitive skills*. Hillsdale, NJ: LEA. pp. 193–228.

Newcombe, N. S., Huttenlocher, J., & Learmonth, A. (1999). Infants' coding of location in continuous space. *Infant Behavior and Development*, 22, 483–510.

Nunes, T., Light, P., & Mason, J. (1993). Tools for thought: The measurement of length and area. *Learning and Instruction*, 3, 39–54.

Outhred, L. N., & Mitchelmoore, M. C. (2000). Young children's intuitive understanding of rectangular area measurement. *Journal for Research in Mathematics Education*, 31, 144–167.

Piaget, J., & Inhelder, B. (1967). *The child's conception of space* (F. J. Langdon & J. L. Lunzer, Trans.). New York: Norton. (Original work published 1948.)

Rock, I., & Ebenholtz, S. (1959). The relational determination of perceived size. *Psychological Review*, 66, 387–401.

Schiro, M., Casey, B., & Anderson, K. (2002). *Tan and the shape changer*. Chicago: The Wright Group/McGraw-Hill.

Shipley, E. F., & Shepperson, B. (1990). Countable entities: Developmental changes. *Cognition*, 34, 109–136.

Solomon, T., Vasilyeva, M., Levine, S. C., Huttenlocher, J., & Ratliff, K. R. (under review). Sizing it up: What do elementary school children understand about linear measurement?

Sophian C. (2002). Learning about what fits: Preschool children's reasoning about effects of object size. *Journal of Research in Mathematics Education*, 33, 290–302.

Sophian, C. (2004). Mathematics for the future: Developing a Head Start curriculum to support mathematics learning. *Early Childhood Research Quarterly*, 19, 59–81.

Sophian, C. (2007). *The origins of mathematical knowledge in childhood.* New York: LEA.

Sophian, C., Garyantes, D., & Chang, C. (1997). When three is less than two: Early development in children's understanding fractional quantities. *Developmental Psychology*, 33, 731–744.

Sovrano, V. A., Bisazza, A., & Vallortigara, G. (2002). Modularity and spatial reorientation in a simple mind: Encoding of geometric and nongeometric properties of a spatial environment by fish. *Cognition*, 85, B51–B59.

Vallortigara, G., Zanforlin, M., & Pasti, G. (1990). Geometric modules in animal spatial representations: A test with chicks (*Gallus gallus domesticus*). *Journal of Comparative Psychology*, 104, 248–254.

Vasilyeva, M., Duffy, S., & Huttenlocher, J. (2007). Developmental changes in the use of absolute and relative information: The case of spatial extent. *Journal of Cognition and Development*, 8, 455–471.

13 Number Development in Context

Variations in Home and School Input During the Preschool Years

Susan C. Levine, Elizabeth A. Gunderson, and Janellen Huttenlocher

The ability to think mathematically is a central aspect of human cognition, and the development of this ability has been a topic of intense study. This literature, in large part, paints a general picture of the development of number knowledge. Much less attention has been paid to individual differences in number knowledge, to variations in number-relevant input, or to the relation between number knowledge and input. A confluence of research findings and societal goals is fueling an increased interest in understanding individual differences in early mathematics knowledge. First, it is now clear that individual differences in math knowledge emerge early, and that these early differences predict later achievement (e.g., Duncan et al., 2007). Second, early differences in mathematics achievement are associated with socio-economic status (SES) (e.g., Jordan, Huttenlocher, & Levine, 1992; Jordan, Levine, & Huttenlocher, 1994; Lee & Burkam, 2002; Saxe, Guberman, & Gearhart, 1987), which is an impediment to diversifying the workforce in the science, technology, engineering, and mathematics (STEM) disciplines (e.g., Arnold, Fisher, Doctoroff, & Dobbs, 2002). Finally, workforce demands for people with high levels of mathematical skill are increasing while American children lag children in other countries in math achievement (e.g., Gonzales et al., 2004; OECD, 2007). Research aimed at increasing our understanding of the factors that contribute to individual differences in mathematics achievement is central to addressing these issues.

In this chapter, we review research on early variations in numerical development, and the relation of these differences to variations in number-related input. These research efforts are grounded in socio-cultural views of development, which emphasize the impact of adult support in propelling children's development (e.g., Fluck, 1995; Rogoff, 1990; Vygotsky, 1978; Wertsch & Tulviste, 1992). We begin with a brief discussion of evidence supporting universal starting points. We then discuss the slow and effortful process through which children acquire an understanding of the verbal, symbolic number system. Second, we review evidence that children have already diverged in their level of mathematical knowledge by the start of school, and that this divergence predicts later academic achievement. Third, we review the evidence that parents and teachers vary widely in the number-related inputs they provide young children, and that these variations impact children's number knowledge. Finally, we consider directions for future research and implications for policy and practice.

Infants Show an Early Number Competency

Infants' sensitivity to variations in set size, for both large and small sets, has been demonstrated through habituation studies (e.g., Antell & Keating, 1983; Brannon, 2002; Starkey & Cooper, 1980; Strauss & Curtis, 1981; Wood & Spelke, 2005; Xu & Spelke, 2000). Various models have been proposed to account for the pattern of results obtained. Gallistel and Gelman (1991, 2000) proposed the accumulator model, based on Meck and Church (1983), which provides approximate, ratio-limited numerical representations for small and large sets. Thus, six-month-old infants can discriminate 8 vs. 16 but not 8 vs. 12 (Xu & Spelke, 2000). A different model, proposed by Feigenson, Dehaene, and Spelke (2004), posits that infants begin life with two core systems for representing numbers: a small number system and a large number system. The small number system is characterized by exact representation of set sizes up to three and offers an explanation for young infants' ability to discriminate between small set sizes such as 2 vs. 3 but not between larger set sizes with the same ratio such as 4 vs. 6 (e.g., Starkey & Cooper, 1980). The large number system, for numbers greater than or equal to 4, is characterized by ratio-dependent performance (2:1 ratio at six months), consistent with the accumulator model.

This early competency with numbers might suggest that young children need only map the verbal count words onto their underlying representation of numbers. However, this mapping develops in a slow and gradual manner. Wynn (1990) found that the majority of three-year-old children were able to count objects, but if the same children were asked "How many are there?" they seldom produced a response corresponding to the last number in their count. Instead, they typically re-counted the objects. Although it is possible that children simply misinterpreted the "how many" question (Zur & Gelman, 2004), young children also fail when asked to give a puppet a certain number of objects (known as a "give-a-number" task) (e.g., Wynn, 1990, 1992). In a longitudinal study, Wynn (1992) found that it takes about a year from succeeding in giving a set of "one" to being able to produce sets of four and above, at which time children become "cardinal principle knowers," as they can map all the numbers in their count list onto corresponding cardinal values.

Understanding the cardinal principle has important implications for understanding numerical relations (e.g., Carey, 2004; Le Corre, Van de Walle, Brannon, & Carey, 2006; Sarnecka & Carey, 2008). For example, it is related to attaining the concept of cardinality, knowledge that all sets with a given cardinal value (e.g., three candies, three jumps, three clouds) form an equivalence class. Children who understand the cardinal principle are able to make numerical matches for perceptually dissimilar sets, whereas those who have not yet acquired this understanding are able to make numerical matches only for perceptually similar sets (e.g., Mix, 2008). In addition, cardinal principle knowers have connected counting to determining the cardinal value of sets, as they are much more likely to count when asked to produce sets of four or more on the give-a-number task than children who have not reached this milestone (Le Corre et al., 2006). Further, only cardinal principle knowers understand the successor function as they know that adding exactly one item to a set means moving forward exactly one count word (Sarnecka & Carey, 2008).

These studies suggest that children's number concepts do not come "for free" based on their early numerical sensitivity (although see Gallistel & Gelman, 1992, for an argument that counting principles are present from birth). Rather, children gradually construct the natural numbers (Wynn 1990; Carey, 2004), raising the possibility that the number-related input they receive plays a substantial role in this accomplishment.

Children's Math Abilities Vary Widely by the Start of School

Preschool children show marked variation in their numerical knowledge, and these differences are especially apparent for children from different SES groups (e.g., Ginsburg & Russell, 1981; Lee & Burkam, 2002). At the start of preschool, low-income children score significantly lower than middle-income children on some mathematical tasks, but not others. For example, low-income preschoolers show less knowledge of the ordering of the numbers 1 through 10 in a number line task, correctly ordering only 61 percent of the numbers compared with 81 percent by middle-income children (Siegler & Ramani, 2008). However, Saxe et al. (1987) report SES differences among preschoolers only on tasks that require higher-level mathematical reasoning skills, such as knowledge of the cardinal principle and arithmetic calculations, but not on less complex tasks involving counting or the ability to read number symbols. Further, Ginsburg and Russell (1981) report SES-related differences among kindergarteners on complex but not simple addition problems.

The nature of the presentation format also impacts whether SES-related differences are found on calculation problems (Jordan et al., 1992, 1994). Thus, low- and middle-SES kindergarteners performed similarly on calculation problems administered nonverbally (children were asked to reproduce the resulting set after watching the experimenter lay out and cover a set of disks, and then remove or add disks without revealing the result). In contrast, low-SES kindergarteners performed worse than middle-SES kindergarteners when the same calculation problems were presented as story problems (e.g., "Beth has *m* balloons, Steve gives her *n* more balloons. How many balloons does Beth have altogether?") or number-fact problems (e.g., "How much is *m* and *n*?").

At the start of school, children from middle- and low-SES backgrounds also differ in their calculation strategies and error patterns on verbal calculation problems. Middle-income kindergarteners were more likely than lower-income children to use a finger-counting strategy, which at that age is associated with more accurate performance. Further, low-income children were more likely to make errors in the wrong direction than middle-income children (e.g., $4-2=5$), suggesting that they have a weaker conceptual understanding of numerical relations (Jordan et al., 1992).

These early differences in numerical skills are of course of concern only if these differences are related to long-term achievement patterns. In an important meta-analytic study of six longitudinal data sets, Duncan et al. (2007) have shown that that this is the case. In particular, children's math skills at the time of school entry predicted subsequent mathematics and reading achievement through the third grade.

What Do We Know about Numerical Input Provided by Parents During the Preschool Years?

The finding of early individual differences in numerical knowledge associated with SES raises the question of whether differential exposure to math input in the home environment may be an important factor in children's learning trajectories. This hypothesis is supported by evidence showing that cognitive and social factors in the home environment are more predictive of children's mathematics achievement than either SES or mothers' math test scores (although all three factors predict variance in children's mathematics achievement) (Crane, 1996). For reading and literacy, we have learned much about the kinds of parental inputs that support strong foundations for school achievement by examining early home environments (e.g., Sénéchal & LeFevre, 2002; Snow, Burns, & Griffin, 1998; Whitehurst & Lonigan, 1998). However, much less is known about the nature or frequency of early mathematically relevant parent–child interactions, or about the extent to which these interactions predict children's later math achievement. Nonetheless, we are beginning to build a knowledge base about variations in early parent–child interactions about number based on studies using questionnaires, checklists, structured observations, and naturalistic observations.

Questionnaires and Checklists

Several studies of the mathematical input that preschool children receive at home have relied on parental interviews or activity checklists. In a study that used a structured parental interview with mothers of two- and four-year-olds, Saxe et al. (1987) found that middle-SES mothers reported mathematics activities with more complex goal structures than lower-SES mothers. Complex activities included calculation and comparing the cardinal values of multiple arrays, whereas simpler activities involved labeling the cardinal values of single arrays, reciting numbers, and recognizing number symbols. The activities engaged in by middle-SES mother–child dyads also spanned a greater range of complexity than those engaged in by dyads from lower-SES backgrounds, suggesting that middle-SES families do not abandon simpler number activities in favor of more complex ones, but rather engage in both simple and more complex activities.

In another study, using an activity checklist, Blevins-Knabe and Musun-Miller (1996) asked parents to estimate the frequency of their kindergarten children's engagement in a large set of number-related activities during the previous week (e.g., using the concept "more" with the child, singing a number song, and encouraging the child to write numbers). They found that the frequency with which the parent or child used the words "one," "two," or "three" and the frequency with which the parent or child mentioned number facts (e.g., "1 + 1 = 2") were positively correlated with children's scores on the TEMA-2, a standardized test that focuses on number-related knowledge (Ginsburg & Baroody, 1990). Interestingly, the reported frequency of teaching children to recite the numbers 1 to 10 was *negatively* correlated with the child's performance on the TEMA-2, suggesting that rote counting activities may be less useful than activities that relate number words to set size.

Finally, Starkey et al. (1999) surveyed low- and middle-SES American parents about the types and frequency of math activities that their four-year-old children engaged in at home. Middle-SES parents reported that their children engaged in more types of math-related activities and received more parental math support than lower-SES parents. This was true both for math activities that required outlays of money (e.g., math books, math software, and purchased games) and for those that did not (e.g., math activities that were part of the home routine and made-up games). Starkey et al. hypothesized that these SES differences may be related to parents' education levels, exposure to math courses, and the value they place on their children's mathematical competence.

As noted by Starkey and Klein (2008), more research is needed to understand the causal factors underlying SES-related differences in mathematical input. The effects of poverty on parents' involvement in math learning are likely to be similar to those impacting their involvement in early literacy. These include financial strains, time constraints, and inadequate education. For mathematics, these issues are likely to be compounded by additional factors including parents' discomfort with their own math skills, lack of awareness about the importance of early math input, and inadequate knowledge about how to support children's math development (e.g., Barbarin et al., 2008; Clements & Sarama, 2007; McLoyd, 1990).

Structured Observations

Other studies have examined the kinds of number-related input parents provide during prescribed numerical activities. Saxe et al. (1987) observed mothers assisting their two- and four-year-old children in two tasks: one that involved counting an array of five or 13 dots, and one that involved producing a set of three or nine pennies to match the number of Cookie Monster cards in a given set. In both tasks, most mothers introduced the problem with a relatively high level of complexity, then adjusted their level of instruction to their child's ability level, with mothers of less knowledgeable children structuring less complex goals. Children in turn modified their behavior appropriately in response to their mothers' instruction, and many children who were unable to complete the tasks when unassisted were successful when scaffolded by their mothers' instruction. Furthermore, although the complexity of maternal instruction was highly related to the child's ability level, SES differences in mothers' instruction were found even after the child's age and ability were controlled. For example, on the task of producing a set of three pennies, working-class mothers were more likely to simplify the goal structure of the task in their initial instructions, even after controlling for child's age and ability. These findings suggest that mothers' instructional complexity is not completely dependent on children's ability levels, and therefore may play a causal role in shaping number development.

Fluck (1995) also observed the kinds of support mothers provide when asked to help their two- and three-year-old children count sets of objects. The results showed that the most common prompt mothers used included the word "count" (e.g., "Can you *count* them?"). Less commonly, mothers asked their children how many there were, which explicitly refers to the cardinal value of the set (e.g., "*How many* animals are there?") or used both "count" and "how many" (e.g., "Are you going to *count how many* there are?"). Regardless of the parents' prompt, children tended to count rather than respond with a one-word answer corresponding to the set size. The results also showed that mothers were more likely to state the cardinal value of a set (e.g., "There's five bricks there") after the child counted correctly than after the child counted incorrectly. Fluck hypothesized that this type of input may help the child connect counting and cardinality. In fact, maternal repetitions of the child's final count word were positively correlated with the child's counting competence. Although these findings suggest that this type of input is important, it is not possible to conclude that it is causally related to children's number understanding because of the correlational nature of the study.

Naturalistic Observations

Durkin, Shire, Riem, Crowther, and Rutter (1986) carried out a longitudinal, observational study of 10 mother–child dyads when children were 9 to 36 months of age that described the use of number words by mothers and children. The results showed that parents used number words throughout this developmental period, which overlaps extensively with the period during which preschoolers' counting skills are commonly studied (e.g., Gelman & Gallistel, 1978; Wynn, 1990). Mothers' number word uses were largely confined to the first four numbers, and frequency of use increased from 9 to 27 months and then leveled off. Common contexts in which number words were used included nursery rhymes, stories and songs, recitation of number strings, alternation between mother and child while counting, repetition and clarification of cardinality, and routines such as "one, two, three, go!"

Durkin et al. (1986) point out that many of these uses may be confusing to children. For example, parents sometimes ask their child to repeat each count word after them, resulting in a combined mother–child utterance of "one, one, two, two, three, three," and at other times ask their child to say the next number in the sequence, so that the combined utterance is the typical count string. Parents' number input also frequently contains elements such as "one, two, three, tickly" rather than "one, two, three, four." Findings also show that much of the child's use of number words occurs in joint dialogues with the mother. Given the noise in parents' number input, and children's difficulty learning the cardinal meaning of number words, it is likely that high amounts of exposure to number talk are helpful to the child in figuring out these meanings.

In a longitudinal study examining parent number talk in naturalistic parent–child interactions, we found such frequency effects (Levine, Suriyakham, Rowe, Huttenlocher, & Gunderson, 2010). Forty-four parent–child dyads were visited every four months at home beginning at child age 14 months. At each visit, parent–child interactions were videotaped for 90 minutes. Parent and child speech from the 14- to 30-month visits was transcribed in the laboratory, and all parent and child number talk was coded. At 46 months, children were given a test assessing their knowledge of the cardinal values of the numbers 1 through 6, commonly known as the Point-to-X task. On each item, the child was shown two cards, each with a set of dots on it, and was asked to, for example, "point to 3."

Our findings show dramatic variations in the amount of number talk that parents engage in with their children. For example, during the 7.5 hours of observation that occurred with children between ages 14 and 30 months, parental input varied from a low of four number words to a high of 257 number words. Further, we found that this variation in number talk was significantly related to children's talk about numbers and to their later knowledge of the cardinal values of the number words, even when we controlled for SES and for parents' overall amount of talk to their children. These findings indicate that it is specifically talk about numbers, not talk in general, that predicts children's later knowledge. Interestingly, the frequency of number word usage during this early period was not related to parents' SES; however, the nature of the input does vary with SES, as parents from low-SES backgrounds provided more number input involving counting and parents from high-SES backgrounds provided more number input about the cardinal value of sets.

What Do We Know about Numerical Input that Occurs in Preschool Classrooms?

As for research on numerical input at home during the preschool years, various approaches have been used to study the numerical input at school during this developmental period.

Survey and Questionnaire Studies

A survey of public and private preschool teachers in California about their beliefs and practices concerning children's mathematical development showed that they regarded the preschool learning environment as more important to children's mathematical development than the home environment (Starkey et al., 1999; Starkey & Klein, 2008). However, they also believed that general classroom enrichments rather than focused mathematical experiences were needed to support children's development in this domain. Despite their belief that the preschool learning environment was important for mathematical development, few preschool teachers had knowledge of the mathematics goals of the kindergarten curriculum used in their local schools. Consistent with these findings, in a survey of public preschool teachers in North Carolina, teachers generally expressed a lack of knowledge about how to support the development of children's numerical skills and concepts (Farran, Silveri, & Culp, 1991).

Naturalistic Studies

Observational studies of preschool classrooms report that activities supporting this development are rare (e.g., Graham, Nash, & Paul, 1997). Starkey and Klein (2008) report that time spent on math in preschool classrooms varies as a function of children's SES. On average, teachers in programs serving middle-income children provided 21 minutes of mathematically relevant support per day whereas teachers in programs serving low-income children provided less than 10 minutes per day. Most of this time was in the context of group activities such as calendar time. Moreover, based on survey data, Copley (2004) reports that prospective early childhood teachers feel uncomfortable with mathematics, find it difficult to teach, and generally ignore the subject except for counting and simple arithmetic operations. In contrast, these prospective teachers feel much more comfortable teaching reading and language skills.

In research in our laboratory, we have observed preschool classrooms to examine naturally occurring variations in teacher talk about number during circle time, and the relation of this variation to the growth of children's mathematical knowledge over the school year. Our studies included Head Start as well as tuition-based preschool classrooms. Our results provide an important piece of the puzzle concerning the relation of early input to the development of children's mathematical knowledge. That is, the studies involving parent input, although suggestive, leave open the possibility that the finding of a relation between parent input and child skill is attributable to the biological relation of the parent and child. The finding of a relation between teacher input and children's mathematical knowledge strengthens the argument that this type of input is causally related to the growth of children's knowledge, as teachers are not biologically related to their students. Further, by assessing children at the beginning and end of the school year, we were able to determine whether there was a relation of teacher input to children's start levels or only to the growth of their knowledge over the school year. If teacher input is related to the growth of children's mathematical knowledge but not to children's levels at the beginning of the school year, it eliminates many alternative explanations for the finding of teacher input effects on children's mathematical knowledge. Notably, this pattern of results suggests that input effects are not attributable to parents with higher levels of mathematical ability selecting preschools that emphasize mathematics.

The first school study carried out in our laboratory (Klibanoff, Levine, Huttenlocher, Vasilyeva, & Hedges, 2006) examined the growth of mathematical knowledge in 198 four-year-old children attending 26 different preschool and daycare classrooms in the Chicago area, which served children from a variety of SES backgrounds. Children were administered a short assessment of mathematical knowledge at the beginning and end of the school year. The items assessed knowledge of ordinality, cardinality, calculation, shape names, understanding "half," and recognizing conventional number symbols. We found main effects of Time of Test [F (1, 137) = 33.25, $p < .01$] and SES [F (2, 137) = 28.91, $p < .01$] on children's scores. High- and middle-SES children performed significantly better on the math assessment than those from low-SES families at both time points ($p < .01$).

This study went beyond the typical finding of SES differences by examining teachers' talk about numbers. One hour of teacher talk, which included circle time and the time surrounding circle time, was audiotaped and all speech was later transcribed and coded for number-related talk. Teacher number talk varied widely, from 1 to 104 instances. The most common type of teacher number talk referred to cardinal value of the number words (e.g., "What's the difference between these two beads?"; 48 percent of all instances). This was followed by labeling number symbols (e.g., "Can you get the blue 9?"; 17 percent of all instances), and counting (e.g., "Now we counted 1, 2, 3, 4, 5, 6, 7, 8, 9, 10 teeth to go in the top of your mouth"; 13 percent of all instances). Together, these three types of number input account for 78 percent of all the number instances teachers produced. Interestingly, there was no significant difference in the amount of mathematically relevant input provided by classrooms serving children from different SES backgrounds. Using hierarchical linear modeling (HLM) we showed that teacher number talk was related to the growth of children's mathematics knowledge over the school year, controlling for complexity of teacher syntax, classroom quality, and SES. Moreover, none of these other variables predicted math growth when math input was controlled. Effect size of teacher input was such that an additional 25 number words would lead to a gain of .21 standard deviations in children's scores and an additional 50 number words would lead to a gain of .42 standard deviations.

A second study, carried out by Ehrlich (2007) in her dissertation, examined teacher input in more detail. Ninety children attending eight tuition-based classrooms and nine Head Start classrooms were given the Test of Early Mathematics Achievement

(TEMA-3) (Ginsburg & Baroody, 2003) in the fall and spring. Ehrlich also looked in greater detail at the kinds of number concepts teachers talked about and examined whether variations in this input made a difference in students' growth patterns.

Replicating prior research findings, Ehrlich found that children attending Head Start classrooms performed significantly lower than children attending tuition-based preschool on the TEMA-3 (82 vs. 102 at pretest and 85 vs. 105 at post-test, $p < .001$ at both time points). As in the Klibanoff study, teachers varied dramatically in the amount of number talk they engaged in. Because Ehrlich observed each teacher on two occasions during the school year, she could examine whether this variation was stable. In fact, she found that the frequency of their number talk was correlated across time, such that teachers who engaged in a lot of number talk during the first visit were likely to engage in number talk during the second visit ($r = .60, p < .02$).

An HLM analysis showed that children's growth in TEMA-3 scores was related to the frequency of teachers' number words (number words per minute) but not to their overall verbosity (overall words per minute). In addition, teachers in classrooms in which students' TEMA-3 standard scores increased over the school year (12 classrooms) showed more number word utterances and more number elicitations than teachers in classrooms in which students' TEMA-3 standard scores decreased over the school year (five classrooms). Teachers in TEMA-increase classrooms were also more likely to elicit cardinal values (How many are there?), name number symbols (read the numeral "3"), match numbers (e.g., show me four fingers), and to elicit calculations from children (e.g., $2 + 1 = 3$) than teachers in TEMA-decrease classrooms. Thus, TEMA-increase and TEMA-decrease classrooms appear to differ not only in the frequency of number related input but also in the nature of this input.

Enrichment and Intervention Studies

Many studies of mathematical input in preschool classrooms have examined the impact of enrichment efforts on children's mathematical learning, and have generally found positive effects on children's mathematical knowledge. Some studies have provided children with high-quality educational programs without focusing on mathematical instruction per se. For example, Campbell, Pungello, Miller-Johnson, Burchinal, and Ramey (2001) found that children enrolled in high-quality child-care from infancy through five years of age had higher math achievement than control children who did not experience this intervention but who did receive the same nutritional supplements and social services as the intervention group. Moreover, these gains were long lasting, as they were still apparent during young adulthood. On the other hand, the Head Start Impact Study (U.S. Department of Health and Human Services, Administration for Children and Families, 2005) does not show differential gains in mathematics for children who attended Head Start programs compared with an SES-matched control group that did not. Thus, attending preschool does not appear to be enough to positively impact the mathematical development of low-SES children (Starkey & Klein, 2008).

Other studies have specifically focused on enriching mathematics instruction in preschool and kindergarten classrooms. The Rightstart program was developed to teach kindergarten children the "central conceptual structures" that underlie formal mathematics, such as the concepts of relative magnitude and continuous quantity

(Griffin, Case, & Siegler, 1994). In this program, children participated in 30 different number games for 20 minutes per day for a three- to four-month period. As an example, in the number line game a child rolls a die, determines the quantity represented, asks the banker for that number of chips, places the chips on each space on the board (which is labeled with the number symbols), and moves his or her playing piece until it rests on the last chip. Thus, the child's understanding of the concept "four" is reinforced by the visual array of chips, the distance moved along the number line, the amount of time it takes to move the piece, and the visual symbol "4." Children who participated in the Rightstart program showed gains in number concepts relative to control children. Impressively, these gains transferred to topics that were never mentioned in the Rightstart curriculum (e.g., counting money, telling time) and were present one year after the intervention had ended.

Another intervention that focused on integrating math activities into the daily routine of Head Start classrooms had similar success (Arnold et al., 2002). In this program, teachers were asked to implement many different math activities for six weeks. Relative to a control group that engaged in their typical activities, children in intervention classrooms showed significant gains in math scores (TEMA-2) and in math interest.

A final intervention example included school and home components for preschool children (Starkey, Klein, & Wakeley, 2004). The Pre-K mathematics curriculum (Klein, Starkey, & Ramirez, 2002) was implemented in preschool classrooms serving children from low- and middle-SES backgrounds. Control children attended the same classrooms a year before the intervention was implemented. The curriculum consisted of units linked to the Pre-K to second-grade standards specified by the National Council of Teachers of Mathematics (2000) as well as to research on early mathematical development. The school component included small-group activities, computer-based mathematics activities, and a math learning center. Teachers were provided with intensive professional development, including workshops and on-site training. The home component complemented the school mathematics units and included workshops for parents and children. The results showed typical SES differences, but both SES intervention groups showed significant gains in mathematics knowledge and performed better than their corresponding control groups. Moreover, children in the low-SES intervention group performed as well as children in the middle-SES control group in the spring of the preschool year.

Although these intervention programs are clearly effective, the inclusion of many different activities and the "business-as-usual" control groups makes it difficult to determine which aspects of the program are most effective. Intervention studies focusing on specific activities can shed light on this question. One such effort has found that a simplified version of the number line game, played for only four 15- to 20-minute sessions over the course of two weeks, can have a positive impact on low-income children's number concepts (Ramani & Siegler, 2008). The number line game consisted of a linear row of 10 colored squares, labeled with the number symbols above each square. Children were asked to use a spinner to determine whether to move one or two spaces, and to say out loud the space number as they moved (for example, if they were on "3" and spun a two, they would say "four, five"). In the control condition, the board was identical except that it was not labeled with number symbols, and children used colors instead of numbers to determine where to move their piece. Playing the number line game improved low-income preschoolers' ability to count to 10, recognize number symbols, compare magnitudes, and represent numbers linearly. Moreover, this improvement was present at a nine-week post-test.

Conclusions

Existing research supports several strong conclusions concerning the early development of numerical knowledge. First, it is clear that there are marked individual differences in number knowledge as early as the preschool years, and these differences are associated with children's SES backgrounds. Second, there are large differences in the number input children receive at home and preschool that are associated with early differences in math knowledge. These differences involve talk about numbers as well as number-related activities such as board games involving numbers.

Early differences in number knowledge are of concern because levels of math knowledge at school entry predict later mathematical achievement. Fortunately, early number development is highly malleable. Number-relevant input at home and at preschool is associated with children's number concepts and skills. Further, intervention studies show substantial improvement in children's developmental trajectories when supports for mathematical learning are put into place in preschools and home environments. Thus, children who fall behind in math do not have to stay behind if they receive effective instruction.

Key Research Questions and Policy Implications

Although existing research indicates that variation in input accounts for at least some of the individual differences in young children's numerical knowledge, several questions remain. A key question concerns the kinds of input that are most effective in teaching children important mathematical concepts. We know that mathematics curricula in preschool classrooms are effective in increasing children's mathematical knowledge and that amount of input is an important predictor of growth. However, we do not know which kinds of input are most effective. There are some hints in the literature that children who engage in more complex mathematical problem solving, such as calculation, may show greater growth than those who do not. There are also hints that number talk that explicitly connects counting to set size may promote development. However, these hints are provided by correlational data, and thus are only hypotheses. In order to identify the kinds of number-related input that are most helpful to children at certain ages and knowledge states, we need to carry out carefully designed experimental studies using pretest/post-test designs and then test these findings in classroom settings.

Consistent with the recommendation of the National Council of Teachers of Mathematics (2000), existing data support the importance of concerted efforts to augment the mathematical experiences of preschool children in their home and school environments. This will involve large changes in preschool curricula and a commitment to preservice and in-service teacher training. In addition, it will involve efforts to increase public awareness about the importance of early mathematical talk and activities, paralleling efforts that have been made to increase awareness about the importance of early language and literacy to children's academic achievement. Such efforts to increase both the frequency and the quality of children's early mathematical experiences have great potential to impact children's academic success in elementary school and beyond.

References

Antell, S. E., & Keating, D. P. (1983). Perception of numerical invariance in neonates. *Child Development*, 54, 695–701.

Arnold, D. H., Fisher, P. H., Doctoroff, G. L., & Dobbs, J. (2002). Accelerating math development in Head Start classrooms. *Journal of Educational Psychology*, 94(4), 762–770.

Barbarin, O., Early, D. M., Clifford, R. M., Bryant, D. M., Frome, P., Burchinal, M. Howes, C., & Pianta, R. (2008). Parental conceptions of school readiness: Relations to ethnicity, socioeconomic status, and children's skills. *Early Education and Development*, 19(5), 671–701.

Blevins-Knabe, B., & Musun-Miller, L. (1996). Number use at home by children and their parents and its relationship to early mathematical performance. *Early Development and Parenting*, 5(1), 35–45.

Brannon, E. M. (2002). The development of ordinal numerical knowledge in infancy. *Cognition*, 83(3), 223–240.

Campbell, F. A., Pungello, E. P., Miller-Johnson, S., Burchinal, M., & Ramey, C. T. (2001). The development of cognitive and academic abilities: Growth curves from an early childhood educational experiment *Developmental Psychology*, 37, 231–242.

Carey, S. (2004). Bootstrapping & the origin of concepts. *Daedalus*, 133(1), 59–68.

Clements, D. H., & Sarama, J. (2007). Early childhood mathematics learning. In F. K. Lester (Ed.), *Second handbook of research on mathematics teaching and learning*. New York: Information Age Publishing. pp. 461–555.

Copley, J. V. (2004). The early childhood collaborative: A professional development model to communicate and implement the Standards. In D. H. Clements & J. Sarama (Eds.), *Engaging young children in mathematics: Standards for early childhood education*. Mahwah, NJ: LEA. pp. 401–414.

Crane, J. (1996). Effects of home environment, SES, and maternal test scores on mathematics achievement. *Journal of Educational Research*, 89(5), 305–314.

Duncan, G. J., Dowsett, C. J., Claessens, A., Magnuson, K., Huston, A. C., Klebanov, P., et al. (2007). School readiness and later achievement. *Developmental Psychology*, 43(6), 1428–1446.

Durkin, K., Shire, B., Riem, R., Crowther, R. D., & Rutter, D. R. (1986). The social and linguistic context of early number word use. *British Journal of Developmental Psychology*, 4, 269–288.

Ehrlich, S. B. (2007). The preschool achievement gap: Are variations in teacher input associated with differences in number knowledge? Chicago: University of Chicago, unpublished doctoral dissertation.

Farran, D. C., Silveri, B., & Culp, A. (1991). Public preschools and the disadvantaged. In L. Rescorla, M. C. Hyson, & K. Hirsh-Pasek (Eds.), *Academic instruction in early childhood: Challenge or pressure? New directions for child development*. San Francisco: Jossey-Bass. pp. 65–73.

Feigenson, L., Dehaene, S., & Spelke, E. (2004). Core systems of number. *Trends in Cognitive Sciences*, 8(7), 307–314.

Fluck, M. J. (1995). Counting on the right number: Maternal support for the development of cardinality. *Irish Journal of Psychology*, 16(2), 133–149.

Gallistel, C. R., & Gelman, R. (1991). Subitizing: The preverbal counting process. In F. Craik, W. Kessen, & A. Ortony (Eds.), *Essays in honor of George Mandler*. Hillsdale, NJ: LEA Associates. pp. 65–81.

Gallistel, C. R., & Gelman, R. (1992). Preverbal and verbal counting and computation. *Cognition*, 44, 43–74.

Gallistel, C. R., & Gelman, R. (2000). Non-verbal numerical cognition: From reals to integers. *Trends in Cognitive Sciences*, 4, 59–65.

Gelman, R., & Gallistel, C. R. (1978). *The child's understanding of number*. Cambridge, MA: Harvard University Press.

Ginsburg, H. P., & Baroody, A. J. (1990). *Test of Early Mathematics Ability* (2nd edn.). Austin: Pro-Ed.

Ginsburg, H. P., & Baroody, A. J. (2003). *Test of Early Mathematics Ability* (3rd edn.). Austin: Pro-Ed.

Ginsburg, H. P., & Russell, R. L. (1981). Social class and racial influences on early mathematical thinking. *Monographs of the Society for Research in Child Development*, 46, 1–69.

Gonzales, P., Guzmán, J. C., Partelow, L., Pahlke, E., Jocelyn, L., Kastberg, D., & Williams, T. (2004). *Highlights from the Trends in International Mathematics and Science Study (TIMSS) 2003* (NCES 2005-005). U.S. Department of Education, National Center for Education Statistics. Washington, DC: U.S. Government Printing Office.

Graham, T. A., Nash, C., & Paul, K. (1997). Young children's exposure to mathematics: The child care context. *Early Childhood Education Journal*, 25, 31–38.

Griffin, S. A., Case, R., & Siegler, R. S. (1994). Rightstart: Providing the central conceptual prerequisites for first formal learning of arithmetic to students at risk for school failure. In K. McGilly (Ed.), *Classroom lessons: Integrating cognitive theory and classroom practice*. Cambridge: MIT Press. pp. 25–50.

Jordan, N. C., Huttenlocher, J., & Levine, S. C. (1992). Differential calculation abilities in young children from middle-income and low-income families. *Developmental Psychology*, 28(4), 644–653.

Jordan, N. C., Levine, S. C., & Huttenlocher, J. (1994). Assessing early arithmetic abilities: Effects of verbal and nonverbal response types on the calculation performance of middle- and low-income children. *Learning and Individual Differences*, 6, 413–432.

Klein, A., Starkey, P., & Ramirez, A. (2002). *Pre-K mathematics curriculum*. Glendale, IL: Scott Foresman.

Klibanoff, R. S., Levine, S. C., Huttenlocher, J., Vasilyeva, M., & Hedges, L. V. (2006). Preschool children's mathematical knowledge: The effect of teacher "math talk." *Developmental Psychology*, 41(1), 59–69.

Le Corre, M., Van de Walle, G., Brannon, E. M., & Carey, S. (2006). Re-visiting the competence/performance debate in the acquisition of the counting principles. *Cognitive Psychology*, 52(2), 130–169.

Lee, V. E., & Burkam, D. T. (2002). *Inequality at the starting gate: Social background differences in achievement as children begin school*. Washington, DC: Economic Policy Institute.

Levine, S. C., Suriyakham, L. W., Rowe, M. L., Huttenlocher, J., & Gunderson, E. A. (2010). What counts in the development of young children's number knowledge? *Developmental Psychology*, 46(5),1309–1319.

McLoyd, V. C. (1990). The impact of economic hardship on black families and children: Psychological distress, parenting, and socioemotional development. *Child Development*, 61(2), 311–346.

Meck, W. H., & Church, R. M. (1983). A mode control model of counting and timing processes. *Journal of Experimental Psychology: Animal Behavior Processes*, 9(3), 320–334.

Mix, K. S. (2008). Surface similarity and label knowledge impact early numerical comparisons. *British Journal of Developmental Psychology*, 26, 13–32.

National Council of Teachers of Mathematics (2000). *Principles and standards for school mathematics*. Reston: National Council of Teachers of Mathematics.

Organisation for Economic Co-operation and Development (OECD). (2007). *PISA 2006: Science competencies for tomorrow's world. Vol. 1: Analysis*. Paris: OECD.

Ramani, G. B., & Siegler, R. S. (2008). Promoting broad and stable improvements in low-income children's numerical knowledge through playing number board games. *Child Development*, 79(2), 375–394.

Rogoff, B. (1990). *Apprenticeship in thinking: Cognitive development in social context*. New York: Oxford University Press.

Sarnecka, B. W., & Carey, S. (2008). How counting represents number: What children must learn and when they learn it. *Cognition*, 108, 662–674.

Saxe, G. B., Guberman, S. R., & Gearhart, M. (1987). *Social processes in early number development*. Chicago: University of Chicago Press.

Sénéchal, M., & LeFevre, J.-A. (2002). Parental involvement in the development of children's reading skill: A five-year longitudinal study. *Child Development*, 73(2), 445–460.

Siegler, R. S., & Ramani, G. B. (2008). Playing linear numerical board games promotes low-income children's numerical development. *Developmental Science*, 11(5), 655–661.

Snow, C. E., Burns, S., & Griffin, P. (1998). *Preventing reading difficulties in young children*. Washington, DC: National Academies Press.

Starkey, P., & Cooper, R. G. J. (1980). Perception of numbers by human infants. *Science*, 210, 1033–1035.

Starkey, P., & Klein, A. (2008). Sociocultural influences on young children's mathematical knowledge. In O. N. Saracho & B. Spodek (Eds.), *Contemporary perspectives on mathematics in early childhood education*. Charlotte, NC: Information Age Publishing, Inc.

Starkey, P., Klein, A., Chang, I., Dong, Q., Pang, L., & Zhou, Y. (1999, April). Environmental supports for young children's mathematical development in China and the United States. Paper presented at the Society for Research in Child Development, Albuquerque, NM.

Starkey, P., Klein, A., & Wakeley, A. (2004). Enhancing young children's mathematical knowledge through a pre-kindergarten mathematics intervention. *Early Childhood Research Quarterly*, 19, 99–120.

Strauss, M. S., & Curtis, L. E. (1981). Infant perception of numerosity. *Child Development*, 52(4), 1146–1152.

U.S. Department of Health and Human Services, Administration for Children and Families (2005). *Head Start Impact Study: First year findings*. Washington, DC: U.S. Department of Health and Human Services, Administration for Children and Families.

Vygotsky, L. S. (1978). *Mind in society: The development of higher mental processes*. Cambridge, MA: Harvard University Press.

Wertsch, J. V., & Tulviste, P. (1992). L. S. Vygotsky and contemporary developmental psychology. *Developmental Psychology*, 28(4), 548–557.

Whitehurst, G. J., & Lonigan, C. J. (1998). Child development and emergent literacy. *Child Development*, 69(3), 848–872.

Wood, J. N., & Spelke, E. S. (2005). Infants' enumeration of actions: Numerical discrimination and its signature limits. *Developmental Science*, 8(2), 173–181.

Wynn, K. (1990). Children's understanding of counting. *Cognition*, 36, 155–193.

Wynn, K. (1992). Children's acquisition of the number words and the counting system. *Cognitive Psychology*, 24, 220–251.

Xu, F., & Spelke, E. S. (2000). Large number discrimination in 6-month-old infants. *Cognition*, 74, B1–B11.

Zur, O., & Gelman, R. (2004). Young children can add and subtract by predicting and checking. *Early Childhood Research Quarterly*, 19, 121–137.

14 Analogy and Classroom Mathematics Learning

Lindsey E. Richland

A young child sits down with blocks to solve a new problem the teacher has given her as a follow-up to earlier instruction on addition. The child exclaims: "Oh, I can do this one, this is sort of like that problem we did before."

This child's simple statement reflects a sophisticated recognition of analogical similarity between the mathematical structure of two instances, separated by time and context. Supporting the flexible, generative understanding reflected in this child's analogy lies at the heart of high-quality mathematics instruction. The domain structure of mathematics creates an epistemology of necessary classroom mathematical knowledge that is quite different from retention of verbatim details, as might be privileged in other academic domains such as geography or spelling. In fact, information taught in mathematics classrooms is rarely instructed with the intention that children retain the verbatim details (e.g., the context or numbers used in problem 4). Rather, mathematical proficiency is more directly related to learners' ability to draw inferences from prior knowledge and instruction to represent and solve previously unseen problems [National Research Council (NRC), 2001; National Mathematics Advisory Panel (NMAP), 2008].

Mathematics is a system for rule-based manipulation of numbers, or "anything that plays by the rules" (Gallistel & Gelman, 2005), that is accessible to even very young children (Gelman & Gallistel, 1978/1986). The rules themselves combine into structured systems that can be instantiated in widely varied representations. Once the structured systems have been instantiated into varied representations, however, recognizing their similarity is not a trivial cognitive act. Varied representations may include multiple mathematical problems, abstract concepts and a problem context, graphical or physical manipulatives as representations. Some of these representations appear quite similar at a surface level, using similar-sized numbers and mathematical form (e.g., "3 + 4 = ?" and "5 + 3 = ?"), whereas others appear different at a surface level (e.g., an equation and a word problem with the same mathematical composition). Many novice learners are misled by surface, or featural, characteristics of mathematical representations, and tend to either fail to notice commonalities between representations or draw false parallels between them (e.g., using the same procedure to solve two mathematically different problems about trains).

In spite of these difficulties, recognizing commonalities in mathematical structure across contexts is a critical skill, and is a key element of mathematical proficiency (NMAP, 2008; NRC 2001). The ability to notice commonalities between representations allows learners to build on prior instruction to solve new homework or test problems, as well as draw more sophisticated connections between concepts. Whereas much research on transfer and generalization points to the challenges of fostering this ability, either as a general reasoning skill or within particular content areas, the cognitive underpinnings of relational reasoning are less frequently discussed in the educational literature. Drawing from basic cognitive research on analogical reasoning and development allows for new insights into strategies for teaching analogical thinking in mathematics.

This chapter reviews a line of research on analogy that draws from basic studies of children's cognition and observations of classroom practices of analogy to generate classroom-feasible pedagogical practice recommendations. Analogy is first defined and its relations to classroom mathematics proficiency are discussed. Basic research on analogical reasoning and problem solving in adults and children is next reviewed. Third, an international study of mathematics teaching by analogy is described, in which teaching practices were examined in light of the basic research. The analysis led to practice recommendations that derive from everyday teaching in the U.S. and two higher-achieving countries, China (Hong Kong) and Japan. Finally, controlled experiments are reported in which these recommendations were tested and shown to positively impact learners' mathematical proficiency in instructed topics. Overall, this chapter argues that U.S. mathematics teachers' practices of analogy need strengthening, and that doing so by adding elaborative cues could have broad implications for improving children's mathematical proficiency.

Defining Analogical Reasoning

Analogical reasoning may be a uniquely human capacity that is central to complex reasoning and learning (Gentner, 2003). Although many people associate the term "analogy" with the form "a" is to "b" as "c" is to "d", the cognitive skill is widely recognized to be a much more integral part of the way humans process our environment. Infants attend to relations very early (e.g., Baillargeon & Hanko-Summers, 1990), and show problem solving by analogy in the first year of life (Chen, Sanchez, & Campbell, 1997). This skill seems to provide a bootstrapping function, enabling children to draw on their prior knowledge to comprehend and reason about novel and increasingly complex environments (Gentner, 2003).

Definitions of analogical reasoning have taken several forms. Gentner (1983) proposed the structure-mapping model of analogy as the process of matching system-wide correspondences between the structured relations that comprise two or more entities. Thus the system of relations within one analog (e.g., a hen and a chick) is recognized as corresponding to the system of relations within another analog (e.g., a mare and a foal). Individual elements within the systems then can be aligned and mapped together (e.g., the hen is like the mare).

An important element of this definition is the distinction between similarity based upon relational correspondences (e.g., a maternal relationship) and object correspondences (e.g., hens do not look like horses). Analogies may be formed between two structures that share no surface features, or those that share both surface and structural similarity (Gentner, 2003). Analogies are therefore partial similarities between different situations that support further inferences. These may be asymmetrical systems, such that the base is better known than the target, or they may be equally well known. An analogy may result in novel inferences about the target, or about the commonalities or differences between the representations.

Holyoak and colleagues have taken a related position, though they have focused on the role of pragmatics. Specifically, they consider the ways in which context and reasoners' goals impact source analog retrieval and structure mapping (see Holyoak & Thagard, 1995; Spellman & Holyoak, 1996). Holyoak and Thagard (1995) proposed the multiconstraint theory of analogy, positing that reasoners settle upon particular correspondences based on their goals for the analogy. In the math domain, for example, a mathematics teacher might develop a different relational mapping between two problems when attempting to show students how to find a solution than when seeking to help her students to better understand the common conceptual structure.

Defining analogy for consideration in the mathematics classroom context is best accomplished through a combination of these approaches. For the remainder of the chapter, analogy is treated as a goal-directed cognitive act of aligning and mapping relational correspondences between structural systems. We turn next to the relationship between analogical reasoning and classroom mathematical knowledge.

Relational Thinking and Mathematical Proficiency

Mathematical proficiency, as defined by the NRC (2001, p. 16) and recently endorsed by the NMAP (2008), involves five strands. These are (1) conceptual understanding, (2) procedural fluency, (3) strategic competence, (4) adaptive reasoning, and (5) productive disposition. Applying the analytical lens of analogical reasoning reveals that at least the first, second, and fourth of these strands, as articulated by the NRC, clearly engage relational thinking. The implications of relational thinking for these three aspects of mathematical proficiency are discussed briefly.

In the conceptual understanding strand, students must understand relations between concepts and operations. The ability to integrate new rules into learners' larger, stored relational structures relies on drawing structural correspondences between previous and new instruction. Deeply integrated knowledge provides a foundation for conceptual understanding.

In the procedural fluency strand, students must demonstrate the ability to use procedures appropriately, which often requires identifying structural relations between novel problems and previously solved (or instructed) problems. The authors also note that procedural fluency involves comparatively analyzing the similarities and differences between problem features.

The strategic competence strand includes the ability to represent mathematical problems based on conceptual structure rather than on surface features. Relationally speaking, this can be considered as the importance of differentiating between object features and mathematical structure. Deep attention to structure should help students recognize that changes in surface feature do not alter solution strategies.

In a broader way, relational reasoning lies at the heart of the NRC's (2001) multistrand definition of mathematical proficiency, in which it is argued that deep understanding and productive problem solving require that learners connect mathematical knowledge across these multiple strands (p. 118). Although both teachers and educational researchers largely agree on the goal to lead students to develop richly connected knowledge, designing and implementing such instruction is challenging and often less than successful (Hiebert et al., 2003). This chapter next reviews a cognitive perspective on factors that facilitate or constrain comparative thinking.

Analogical Reasoning in Problem Solving and Learning

Much basic research indicates that analogy is a fundamental part of the way children and adults reason about their world. Despite famous cases of analogy use in scientific discoveries, most problem solving by analogy happens within mundane everyday reasoning, and involves smaller leaps of inference. Children learn to solve problems by analogy within the first year of life (Chen et al., 1997), and analogies are a regular part of classroom mathematical discourse (see English, 1997). Learners are also quite good at structure mapping between source and target representations when they are aware that they should be doing so (e.g., Brenner et al., 1997; Gick & Holyoak, 1980; Novick & Bassok, 2005).

Learning from Structure Mapping

Analogical reasoning can facilitate problem solving, inferential thinking, and learning new strategies as long as participants are provided with key support (e.g., see Brenner et al., 1997; Chen & Klahr, 2008; Novick & Bassok, 2005; Rittle-Johnson & Star, 2007). In a mathematics study that illustrates this potential, Novick and Holyoak (1991) provided participants with a problem and solution, and then evaluated their later performance on an analogous test problem when given one of three types of hints with varying levels of specificity, or no hint.

All hints led to initially more analogical transfer than no hints, and there was a direct correlation between the specificity of the hint and participants' likelihood of noticing and effectively using the source analog as a base for the analogy. The more specific the hint, the better the likelihood that participants performed analogical transfer. Importantly, those who were successful later showed enhanced transfer rates on delayed final problems when solving them without any cues or hints. These data suggested that learning by doing analogical reasoning, even with high support by an instructor, such as a very explicit hint, may lead to increasingly schematized, generalizable knowledge representations. The data also indicate that the nature of cues supporting instructional analogies may crucially impact learning.

Rittle-Johnson and Star (2007) have recently shown similar success with facilitating middle school students' comparisons between two accurate solution strategies to a single problem. Such comparisons led to higher performance on measures of retention as well as on measures of conceptual, schematized understanding. Providing learners with the same information in serial order, on different pages of a packet, did not produce the same benefits.

Instructional Comparisons Are Risky

Despite the evidence that analogies can facilitate problem solving both directly and through schema induction, providing an analogical reasoning opportunity to reasoners is not enough to guarantee learning or transfer. Rates of spontaneous usage of analogies are remarkably low in experimental contexts (e.g., Gick & Holyoak, 1980; Reed, 1989). Although this may under-represent the reasoning that is performed in everyday contexts in which reasoners have more expertise, classroom learning contexts are akin to laboratory contexts in which reasoners are relative novices. Retrieval searches for relevant source analogs are closely tied to one's knowledge base. Novices are more likely to conduct a search of stored potential analogs on the basis of surface features of the test problem, whereas experts are more likely to search on the basis of relational structure (see Chi & Ohlsson, 2005). As a consequence, novices who have not received sufficient training to view problems more like experts and notice the key structural elements may fail to notice the relevance of a stored problem (see Novick & Bassok, 2005).

Further, instructional analogies that are not well defined can lead to overextensions or misconceptions (Zook & Di Vesta, 1991). Because analogies are not isomorphs, there are always both similarities and differences between the representations. Thus, learners must receive strong scaffolding to ensure that they are making valid inferences based on the structure mapping, rather than being misled by surface or irrelevant source characteristics.

Processing Demands on Analogy

Some of the difficulty and potential for missteps from analogical reasoning may be attributable to the high processing demands of representing and manipulating complex relational structures. These demands are enhanced for novices whose grasp of the relevant representations is weaker. Dual task and cognitive neuropsychological methodologies have produced evidence that working memory and executive function are critically involved in two aspects of analogical reasoning: representing and integrating relevant relations (relational integration), and controlling attention to competitive, irrelevant information (interference resolution).

Relational integration refers to the number of relations that must be held active simultaneously in order to process a complex analogy, and Halford and colleagues have hypothesized that processing demand increases as the number of relations to be integrated increases (Halford, 1993; Halford, Wilson, & Phillips, 1998). Interference resolution refers to the ability to control attention and inhibit activated but irrelevant, or misleading, features of source and target analogs (e.g., attempting to map between two mathematically dissimilar word problems about trains). Experimental tasks requiring both interference resolution and relational integration showed that these demands share competitive cognitive resources. Increasing either kind of demand when both were required raised undergraduates' reaction times (Cho, Holyoak, & Cannon, 2007).

Learning from analogy in instructional contexts may present even more of a cognitive challenge since resources for controlling attention and manipulating information in working memory are already taxed by lack of background knowledge. Further, children are well known to have more limited working memory and executive function resources than adults. The relations between such processing considerations and children's development of analogical reasoning are next discussed.

Development of Analogical Reasoning

Whereas early Piagetian work on analogy suggested that analogical, higher-order reasoning was not available to children until at least early adolescence, the past two decades have revealed substantial evidence that children's analogical reasoning emerges in early childhood (see Goswami, 2001). Thus, capitalizing on children's relational reasoning capacity provides a powerful resource for aiding children in building well-structured, generalizable knowledge. In the mathematics domain, early analogical reasoning ability lays the foundation for acquiring deeply conceptual knowledge and high mathematical proficiency.

In the earliest empirical documentation of analogical transfer and problem solving, Chen and colleagues (1997) designed four experiments in which 10- and 13-month-old infants solved three isomorphic problems with varying levels of object similarity. Despite this early ability to reason analogically, children's relational thinking does not approximate adults' until adolescence (Halford, 1993; Richland, Morrison, & Holyoak, 2006). Children's reasoning appears to differ from adults' along two dimensions. First, the rates of attending preferentially to object similarity versus relational similarity differ, and have been charted developmentally (Gentner, 1988; Gentner & Rattermann, 1991; Richland et al., 2006). Second, children's ability to process increasingly complex relations improves with time (Halford, 1993). Thus a more nuanced awareness of children's skills is necessary to best design learning environments without overtaxing children's ability.

Theories of Analogy Development

Relational Knowledge

Understanding the mechanisms underlying children's growth in analogical reasoning over time lends insight into optimal strategies for facilitating this development. Several explanatory theories have been proposed, centering either on the explanatory role of relational knowledge or on processing capacity. The relational primacy theory (see Goswami, 2001) posited that children's ability to reason relationally is available very early, but that effectiveness improves with children's experience. In particular, knowledge of the relations and objects present in a particular reasoning context are hypothesized to increase the likelihood that a child notices relational correspondences (see Goswami, 2001). For example, understanding the relation "cut" is necessary before a child can solve the analogy: "*bread* is to *a bread slice* as *apple* is to ?"

In a hypothesis also related to children's knowledge, Gentner (1988) and colleagues posited that, although general structure mapping skills are available to young children, their reasoning in a novel context proceeds from relying upon object similarity to reasoning on the basis of relational similarity (Gentner & Rattermann, 1991). In what is termed the "relational shift" hypothesis, children with less knowledge are expected to notice and draw comparisons based on object features rather than on relational features, whereas children with greater knowledge would preferentially attend to relations. Evidence comes from an array of stimuli including formal analogies (e.g., " *bread* is to *a bread slice* as *apple* is to ?"). Children, before the relational shift, would be expected to select an object similarity match, "ball" to replace the question mark because an apple and ball are round and red. Children, after the relational shift, would be expected to select a "cut apple slice" because this shared the same relationship as in the source.

Background knowledge is thus clearly an important part of analogical reasoning. At the same time, although knowledge improves the likelihood that children will be able to reason about and learn from analogies, children who demonstrate the pertinent domain knowledge still fail on analogical reasoning tasks (Richland et al., 2006). Particularly in a learning context, where domain knowledge is incomplete by definition, other mechanisms must contribute to development.

Processing Constraints

Research with adults has demonstrated the high processing loads on working memory and executive function for relational integration and interference resolution. Studies with children show that these processing constraints may also impact the developmental trajectory. Processing capacity has been proposed to constrain children's development of analogical reasoning in two ways. Halford and colleagues have focused on the role of working memory (WM) capacity, arguing that growth in WM capacity enables children to process increasingly complex analogies with age (Halford, 1993). Richland and colleagues (2006) additionally posited the role of executive function—particularly inhibitory control of attention (see Diamond, 2002).

Data from U.S. children solving scene analogy problems indicate that these cognitive capacities both have distinct roles in children's analogical reasoning development that function above and beyond the role of prerequisite domain knowledge (Richland et al., 2006). The scene analogy task separately tests the developmental effects of relational similarity and ability to control distraction from object-based similarity, and uses counterbalancing to hold domain-specific knowledge largely constant. Pairs of meaningful visual scenes were used as stimuli in which common relations were depicted using different objects (e.g., chase, drop, kiss, pull). As shown in Figure 14.1, one object was highlighted in a top source picture (big monkey), and children were asked to find the corresponding object in the bottom, target picture (little girl).

Four counterbalanced versions were constructed for each of the 20 picture sets by varying two dimensions. Figure 14.1 shows the four versions constructed for the relation "hang." The relational shift was tested by varying the presence of a distractor—an object that appeared very similar to the highlighted source object within the target picture (Distractor condition; monkey in the bottom picture of Figure 14.1B and D). Second, children's ability to handle relational complexity was tested by varying the number of instances of the relevant relations within a scene that needed to be mapped [One Relation (1-R) or Two Relations (2-R)]. In Figure 14.1, the 1-R problems contained the single relation *hang from* (baby monkey, adult monkey) with the elephant as an independent entity (Figure 14.1A and B). In the 2-R problems the elephant was engaged to depict the two-part relational structure: *hang from* (baby monkey, adult monkey, elephant) (Figure 14.1C and D).

Richland et al. (2006) tested the scene analogy problems with children aged 3–14. In a knowledge check of the materials, children in the youngest age group (three and four years) showed over 90 percent accuracy in identifying the relevant relations. This meant that any developmental differences could not be attributed to a lack of prerequisite knowledge.

Figure 14.1 Sample Stimuli for Four Versions of the "Hang" Relation Problems.

Across varied instructions, the youngest children (three to four years) always showed above-chance performance, demonstrating basic structure-mapping skills and requisite knowledge of the relations. Importantly, however, their performance was significantly impacted by moving from a binary to a ternary level of relational complexity, and by adding a featural similarity distractor. Similar but less strong effects were demonstrated for six- to seven-year-olds, with both effects lessening with age. By 9–11 years of age both effects were minimal, though 13- to 14-year-olds in one sample showed a significant effect of relational complexity with these materials.

Thus, in spite of prerequisite knowledge of the tested relations, children's analogical processing varied along the same dimensions identified in more complex tasks as constraining adult and aging populations' relational reasoning. These data suggest that, although children have the capacity to identify and map structure across analogs, their ability to do so is limited by available resources to integrate complex relations and control responses to irrelevant object properties.

Implications for Classroom Mathematics Teaching by Analogy

Consideration of developmental constraints is therefore crucial to harnessing the potential of instructional analogies for improving children's mathematical proficiency. Instructors must ensure that analogical learning opportunities do not overtax background knowledge, adequate working memory resources, or ability to avoid distraction from surface similarity. Precisely what this means to classroom teachers, however, is not immediately evident. To make practice recommendations that were more directly relevant to the complexities of classroom teaching, subsequent studies used a cognitive lens to examine teachers' typical mathematical instructional use of comparisons and analogies with respect to the learning constraints noted above. Middle-school teaching was analyzed in the U.S. (Richland, Holyoak, & Stigler, 2004) and an international sample of typical U.S., Hong Kong, and Japanese lessons (Richland, Zur, & Holyoak, 2007).

These data were sampled from the Third International Mathematics and Science Study (TIMSS: Stigler, Gonzales, Kawanaka, Knoll, & Serrano, 1999) and the subsequent Trends in International Mathematics and Science Study (Hiebert et al., 2003). The TIMSS studies are a unique video 'survey' of typical classroom teaching both in the United States and internationally with approximately 100 teachers videotaped in each country. Each teacher and lesson was selected as a random probability sample of all lessons taught in a given school year across the country, rather than as a more typical convenience sample. Both of the original TIMSS studies showed that countries have normative pedagogical patterns. Despite some inevitable individual differences across teachers the variance between teachers within a country was much less than the variance across countries.

One pattern identified in the original TIMSS 1999 is important to the current discussion of analogy. In an analysis of problems that drew connections between mathematics concepts, procedures, or representations, U.S. teachers were less likely than their international peers to capitalize on these learning opportunities. Teachers in all countries, including the U.S., regularly administered such problems. However, close analyses of the ways in which these problems were solved and discussed revealed that the highest-achieving countries all drew out these connections and engaged the students in making connections more frequently than U.S. teachers (Hiebert et al., 2003). In fact, this was the only systematic difference between teaching in the U.S. and all higher-achieving countries.

This revealing divergence was further illuminated in a secondary analysis of the TIMSS data specifically focusing on those cases in which teachers made instructional analogies (Richland et al., 2007). Analogy use was examined in the U.S. and two high-achieving regions that did not share many commonalities in normative teaching patterns: Japan and China (Hong Kong). Ten lessons were randomly selected from the dataset for each country, each taught by a different teacher.

Analogies were identified using an integration of the structure-mapping and pragmatic definitions of analogy from the basic research literature and observations of the classroom practices. The result was a situated definition of analogy within a mathematics classroom context. Mainly, a comparison was identified if there were readily identifiable source and target representations that shared relational structure, and there was some evidence of drawing a comparison between these representations. Connections between representations based on surface features ("this solution looks like a mess") were not coded because they do not tax analogical reasoning. Additionally, source and target representations each were required to function as a whole within the pragmatic goal structure of the analogy. For this reason, if the learners' goal was to graph an equation or use a solution strategy to solve a problem, neither of these situations would be coded. Although a graph and an equation are different representations, they function as two parts of a single problem goal.

After identification, each analogy was then coded in many different ways. Codes were developed to reflect teachers' common practices that aligned with the cognitive factors outlined above. Codes sought to capture frequency of instructional decisions that could be expected to reduce processing load, facilitate attention to relational structure of target problems, draw learners' attention to relations versus object features, reduce competitive interference, and encourage learners to draw on prior knowledge. As codes, these translated to (yes/no): (1) produced a visual representation of a source analog versus only a verbal one, (2) made a visual representation of the source analog visible during comparison with the target, (3) spatially aligned written representations of the source and target analogs to highlight structural commonalities, (4) used gestures that moved comparatively between the source and target analogs, (5) constructed visual imagery, and (6) used likely well-known source analogs.

Achievement was clearly correlated with classroom analogy practices. Whereas teachers in all three countries used approximately the same number of analogies per lesson (there were no significant differences), *how* they organized the instructional context differed significantly. Japanese and Chinese teachers used practices of analogy that were closely aligned with the practice recommendations outlined above. As shown in Figure 14.2, U.S. teachers were reliably less likely to use the coded principles than either Japanese or Hong Kong teachers.

These data thus reveal that U.S. teachers regularly invoke analogies in their mathematics instruction, which could serve as potent opportunities for improving mathematical proficiency. However, there are many reasons to believe that students are not benefiting from these opportunities for relational reasoning. As reviewed in this chapter, analogies do not automatically benefit learners. In particular, analogies frequently fail learners who do not notice the relational correspondences or draw misconceptions or overextensions. Rather, certain elaborative conditions of the environment must be present. U.S. teachers' infrequent use of such supportive cues during instructional analogies is likely to reduce their efficacy.

So far, these data are suggestive and theoretically grounded, but correlational. No learning data were directly tied to the videotaped classroom lessons. The following section describes experiments that directly tested the prediction that adding instructional, elaborative cues to episodes of mathematical instructional analogy would improve relational reasoning, resulting in greater mathematical proficiency with the instructed topic.

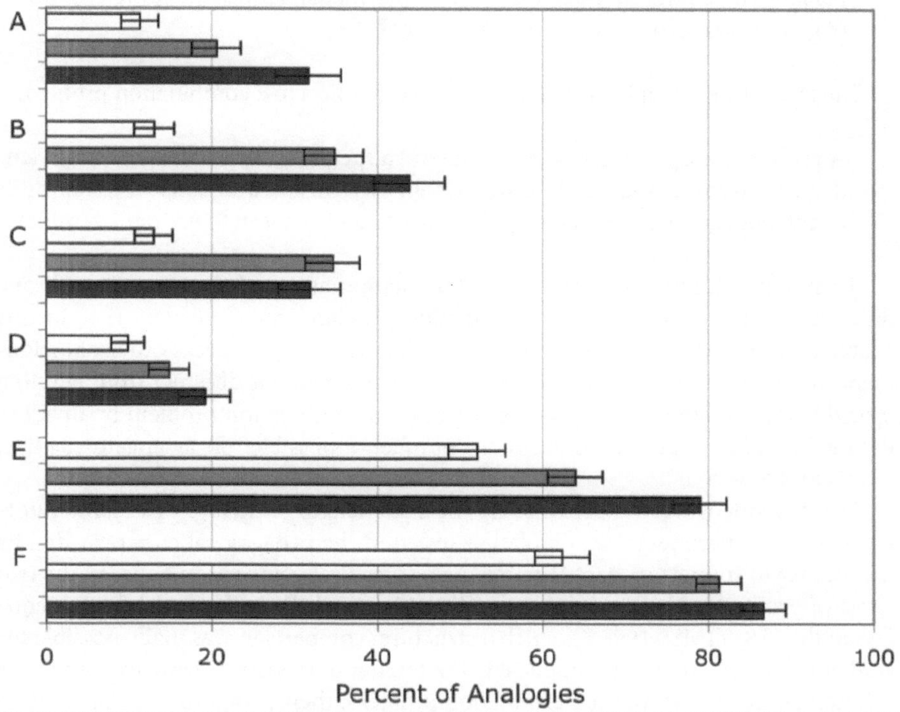

Figure 14.2 Percent of Analogies by Region Containing Cognitive Supports: (A) Visual and Mental Imagery, (B) Comparative Gesture, (C) Visual Alignment, (D) Use of a Familiar Source, (E) Source Visible Concurrently with Target, (F) Source Presented Visually. White denotes U.S. teachers, gray denotes Chinese teachers, black denotes Japanese teachers. From Richland, Zur, and Holyoak (2007).

Experimental Tests of Pedagogical Support for Instructional Analogies

Three experiments in separate mathematical content areas showed benefits for teaching by analogy, and all studies further revealed that adding instructional cues to support the analogy led to more flexible, generalizable knowledge representations. Two studies were conducted with undergraduates learning Graduate Record Exam concepts (Richland & McDonough, 2010), and the general pattern of results was replicated in a sample of children in the fifth grade learning fraction operations (Richland, under review).

Videotaped instruction was used in all three studies to provide control over the instructional manipulations. In the first experiment of the series (Richland & McDonough, 2010, Experiment 1), undergraduates were randomly assigned to one of two conditions: analogy with high support cues or analogy with low support cues. In both conditions a videotaped teacher first taught and demonstrated a solution to a permutation problem:

> Suppose there are five people running in a race. The winner of the race will get a gold medal, the person who comes in second will get a silver medal, and the

> person who comes in third will get a bronze medal. How many different orders of gold–silver–bronze winners can there be?

The teacher next taught and demonstrated a solution to a combination problem:

> A professor is choosing students to attend a special seminar. She has eleven students to choose from, but she only has four extra tickets available. How many different ways are there to make up the four students chosen to go to the seminar?

Permutation and combination problems share mathematical structure with one difference. All assigned roles in combination problems are equivalent (i.e., in this problem, it doesn't matter which ticket a student receives), whereas order of assignment to roles in permutations is critical (i.e., winning gold is different from winning bronze). Thus, mathematically, one must finish a combination problem by dividing the total number of permutations by the number of possible role arrangements (i.e., in this problem, four).

The two videos were approximately the same length and taught the same information. The experimental manipulation rested in the pedagogical cues provided by the teacher to support students in drawing a structural comparison between the two types of problems. The low-cueing condition invoked the pedagogical form identified in the U.S. TIMSS 1999 in which a structural comparison was made possible for students but was not highly supported. The teacher demonstrated and explained the solution strategy to solve the permutation problem, then erased the board. He then stated that he would next show a related but different kind of problem, and demonstrated and explained the solution strategy to solve the combination problem. The serial sequence and immediate proximity of the problems would lead some students to compare their structure. However, the student would have to retrieve the source representation (permutation problem) while considering the target combination problem, and recognize the structural similarities and differences by aligning them in mental imagery.

In contrast, the teacher in the high-cueing condition left the source problem on the board while teaching the target problem, and used explicit cues to help students align the two representations. Both problems were written on the board in a parallel way such that the structure was aligned visually. The teacher also used broad gestures to move between the two representations to draw students' attention to the paired analogs.

Two types of problems were included on the post-test. High-similarity problems matched both the mathematics of the instructed problems (permutation, combination) and the surface context (winning a race and tickets to a lecture). Misleading similarity problems cross-mapped mathematics and surface contexts, such that the permutation problem was set in the context of tickets to a lecture, and the combination problem was set in the context of winning a race. The high-similarity problems assessed participants' learning of and ability to implement instructed strategies. The misleading similarity problems were a more nuanced assessment of flexible, conceptual understanding. These captured learners' ability to represent the target problem based on mathematical structure versus surface features, and their ability

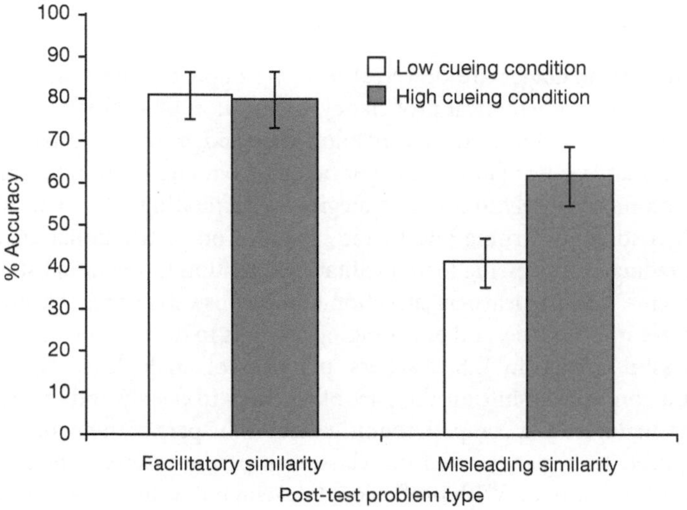

Figure 14.3 The Effects of High Versus Low Cueing of an Instructional Analogy on Post-test Problems with Varying Similarity to Instructed Problems. Adapted from Richland, L. E., & McDonough, I. (2010). Learning by analogy: Discriminating between potential analogs. *Contemporary Educational Psychology*, 35(1), 28–43.

to distinguish between source and target correspondences based on structural versus surface similarities.

As evident in Figure 14.3, the data revealed an interaction between instructional condition and problem type. Participants in both instructional conditions benefited from the instructional analogy, showing approximately 80 percent accuracy on the facilitory similarity problems (baseline performance with the same population was 10 percent). In contrast, the high-cueing condition significantly outperformed the low-cueing condition on the cross-mapped, misleading similarity problems (baseline level 7 percent). This pattern indicates that any instructional analogy was beneficial, but that adding pedagogical cues to support learners' analogical thinking led to more flexible, conceptual knowledge representations.

The same interaction between cueing and post-test problem similarity was identified in two additional studies. The second study revealed a very similar result with undergraduates learning to solve proportion word problems through an analogy between a correct solution and a common but invalid solution: use of the linearity assumption (Richland & McDonough, 2010, Experiment 2). This third study replicated the result in a classroom context with school-age children learning division of rational numbers by analogy with division of natural numbers (Richland, under review).

Overall, these data indicate a reliable finding that high-quality analogies can be effective learning tools, but that including additional pedagogical support strategies maximizes their impact. When given additional cues, learners seem to have developed more conceptual, schematized representations of the instructed concepts and/or more adaptive proficiency in representing new problems.

Conclusions

In conclusion, analogies are powerful learning opportunities that can deepen and shape students' mathematical proficiency. Instruction by analogy is not straightforward, however, since limits in relevant knowledge and processing capacity increase the likelihood that learners fail to notice or benefit from analogies in teaching. Aligning instruction more closely to tested strategies for facilitating relational thinking could strengthen student learning and better capitalize on instructional analogies. These include reducing processing load, facilitating attention to relational structure of target problems, drawing learners' attention to relations versus object features, reducing competitive interference, and encouraging learners to draw on prior knowledge.

Successful change in U.S. teachers' practices of analogies is unlikely to come without a conceptual shift on the part of teachers to deeply and explicitly consider everyday analogies as a complex cognitive act on the part of their students. However, the proposed strategies derive from classroom practices and involve minimal time or resource investment. With professional development, such practices could greatly impact teachers' already common use of analogy, in turn profoundly affecting students' mathematical proficiency.

Acknowledgments

The Office of Naval Research Grant N000140810186 partially supported the experiments reported herein. This material is also based upon work supported by the National Science Foundation under Grant No. 0757646. Any opinions, findings, and conclusions or recommendations expressed in this material are those of the author(s) and do not necessarily reflect the views of the granting agencies.

References

Baillargeon, R., & Hanko-Summers, S. (1990). Is the top object adequately supported by the bottom object? Young infants' understanding of support relations. *Cognitive Development*, 5, 29–53.

Brenner, M. E., Mayer, R. E., Moseley, B., Brar, T., Durán, R. Reed, B. S., et al. (1997). Learning by understanding: The role of multiple representations in learning algebra. *American Educational Research Journal*, 34(4), 663–689.

Chen, Z., & Klahr, D. (2008). Remote transfer of scientific reasoning and problem-solving strategies in children. In R. V. Kail (Ed.), *Advances in child development and behavior*. Amsterdam: Elsevier. pp. 419–470.

Chen, Z., Sanchez, R., & Campbell, T. (1997). From beyond to within their grasp: Analogical problem solving in 10- and 13-month-olds. *Developmental Psychology*, 33, 790–801.

Chi, M. T. H., & Ohlsson, S. (2005). Complex declarative learning. In K. J. Holyoak, & R. G. Morrison (Eds.), *Cambridge handbook of thinking and reasoning*. New York: Cambridge University Press. pp. 371–399.

Cho, S., Holyoak, K. J., & Cannon, T. D. (2007). Analogical reasoning in working memory: Resources shared among relational integration, interference resolution, and maintenance. *Memory & Cognition*, 35(6), 1445–1455.

Diamond, A. (2002). Normal development of prefrontal cortex from birth to young adulthood: Cognitive functions, anatomy, and biochemistry. In D. T. Stuss & R. T. Knight (Eds.), *Principles of frontal lobe function*. London: Oxford University Press. pp. 466–503.

English, L. (Ed.) (1997). *Mathematical reasoning: Analogies, metaphors, and images*. Mahwah, NJ: LEA.

Gallistel, C. R., and Gelman, R. (2005). Mathematical cognition. In K. Holyoak & R. Morrison (Eds.), *Cambridge handbook of thinking and reasoning*. Cambridge: Cambridge University Press. pp. 559–588.

Gelman, R., and Gallistel, C. R. (1978/1986). *The child's understanding of number*. Cambridge, MA: Harvard University Press.

Gentner, D. (1983). Structure-mapping: A theoretical framework for analogy. *Cognitive Science*, 7, 155–170.

Gentner, D. (1988). Metaphor as structure mapping: The relational shift. *Child Development*, 59, 47–59.

Gentner, D. (2003). Why we're so smart. In D. Gentner & S. Goldin-Meadow (Eds.), *Language in mind: Advances in the study of language and thought*. Cambridge, MA: MIT Press. pp. 195–235.

Gentner, D., & Rattermann, M. J. (1991). Language and the career of similarity. In S. A. Gelman & J. P. Byrnes (Eds.), *Perspectives on thought and language: Interrelations in development*. London: Cambridge University Press. pp. 225–277.

Gick, M. L., & Holyoak, K. J. (1980). Analogical problem solving. *Cognitive Psychology*, 12, 306–355.

Goswami, U. (2001). Analogical reasoning in children. In D. Gentner, K. J. Holyoak, and B. N. Kokinov (Eds.), *The analogical mind: Perspectives from cognitive science*. Cambridge: MIT Press. pp. 437–470.

Halford G. (1993). *Children's understanding: The development of mental models*. Hillsdale, NJ: LEA.

Halford, G. S., Wilson, W. H., & Phillips, S. (1998). Processing capacity defined by relational complexity: Implications for comparative, developmental, and cognitive psychology. *Behavioral and Brain Sciences*, 21(6), 803–831.

Hiebert, J., Gallimore, R., Garnier, H., Givvin, K. B., Hollingsworth, H., Jacobs, J., et al. (2003). *Teaching mathematics in seven countries: Results from the TIMSS 1999 video study* (NCES 2003-013). Washington, DC: U.S. Department of Education.

Holyoak, K. J., & Thagard, P. (1995). *Mental leaps: Analogy in creative thought*. Cambridge, MA: MIT Press.

National Mathematics Advisory Panel (2008). *Foundations for success: The final report of the national mathematics advisory panel*. Washington, DC: U.S. Department of Education.

National Research Council (2001). Science, evidence, and inference in education: Report of a workshop. In L. Towne, R. J. Shavelson, & M. J. Feuer (Eds.), *Committee on Scientific Principles in Education Research*. Washington, DC: National Academies Press.

Novick, L. R., & Bassok, M. (2005). Problem solving. In K. J. Holyoak & R. G. Morrison (Eds.), *Cambridge handbook of thinking and reasoning*. New York: Cambridge University Press. pp. 321–349.

Novick, L. R., & Holyoak, K. J. (1991). Mathematical problem solving by analogy. *Journal of Experimental Psychology: Learning, Memory, and Cognition*, 17, 398–415.

Reed, S. K. (1989). Constraints on the abstraction of solutions. *Journal of Educational Psychology*, 81, 532–540.

Richland, L. E. (under review). Teaching by analogy: Instructional strategies improve reasoning about fraction operations.

Richland, L. E., Holyoak, K. J., & Stigler, J. W. (2004). Analogy generation in eighth grade mathematics classrooms. *Cognition and Instruction*, 22(1), 37–60.

Richland, L. E., & McDonough, I. (2010). Learning by analogy: Discriminating between potential analogs. *Contemporary Educational Psychology*, 35(1), 28–43.

Richland, L. E., Morrison, R. G., & Holyoak, K. J. (2006). Children's development of analogical reasoning: Insights from scene analogy problems. *Journal of Experimental Child Psychology*, 94, 249–271.

Richland, L. E., Zur, O., & Holyoak, K. J. (2007). Cognitive supports for analogies in the mathematics classroom. *Science*, 316, 1128–1129.

Rittle-Johnson, B., & Star, J. R. (2007). Does comparing solution methods facilitate conceptual and procedural knowledge? An experimental study on learning to solve equations. *Journal of Educational Psychology*, 99, 561–574.

Spellman, B. A., & Holyoak, K. J. (1996). Pragmatics in analogical mapping. *Cognitive Psychology*, 31, 307–346.

Stigler, J. W., Gonzales, P., Kawanaka, T., Knoll, S., & Serrano, A. (1999). *The TIMSS videotape classroom study: Methods and findings from an exploratory research project on eighth-grade mathematics instruction in Germany, Japan, and the United States* (NCES1999-074). Washington, DC: U.S. Department of Education, NCES.

Zook, K. B., & Di Vesta, F. J. (1991). Instructional analogies and conceptual misrepresentations. *Journal of Educational Psychology*, 83, 246–252.

15 Gestures in the Mathematics Classroom

What's the Point?

Martha W. Alibali, Mitchell J. Nathan, and Yuka Fujimori

Communication is an integral part of teaching. Many factors influence whether students comprehend and learn from instructional communication, including whether students have a shared understanding of the referents used by the teacher (Mortimer & Wertsch, 2003), and whether the ideas addressed in a lesson connect to students' prior knowledge (Schwartz & Bransford, 1998). Another potentially important factor that has received limited research attention is the nonverbal support for comprehension provided by teachers' gestures. Gestures are movements of the hands and body that are produced in the act of speaking and that are closely synchronized with speech (McNeill, 1992). Gestures include pointing movements that indicate objects or locations, depictive movements that illustrate the content of speakers' thoughts, and rhythmic movements that mirror the cadence of speech.

Previous studies in noneducational settings have shown that speakers' gestures facilitate listeners' comprehension of speech. However, surprisingly little is known about how teachers use gestures in instructional settings, or about whether teachers' gestures influence students' learning. As Roth (2002) stated in *Review of Educational Research*:

> It is curious . . . that there exists very little educational research concerned with the role of gesture in learning and teaching, particularly in subject areas that have been characterized as dealing with abstract matters such as science and mathematics. The few existing studies that focus on gesture in an education context . . . suggest that such research might be of tremendous importance. (p. 365)

We agree with Roth's assessment, and, in this chapter, we report on a line of research that begins to address this gap.

The chapter proceeds in three parts. First, we review existing research on gesture in instructional settings and whether it matters for students' learning. Second, we present findings from a study of how teachers gesture in mathematics lessons. Third, we argue that teachers' gestures serve to connect mathematical ideas in their instruction, and we present illustrative examples drawn from classroom mathematics lessons. Our broad aim in this chapter is to document how practicing teachers actually use gestures in mathematics instruction.

Do Teachers' Gestures Matter for Students' Learning?

Gesture Affects Comprehension of Speech

Although some investigators have downplayed the communicative importance of gestures (e.g., Krauss, Morrel-Samuels, & Colasante, 1991), there is abundant evidence that gestures affect listeners' comprehension of speech (see Kendon, 1994). When gestures convey the same information as the accompanying speech, comprehension is facilitated (e.g., Goldin-Meadow & Sandhofer, 1999). For example, when asked to "find the block that has an arrow pointing up", preschool children chose the correct block more often when the speaker used a gesture that reinforced speech (i.e., an index finger pointing up) than when she used speech alone (McNeil, Alibali, & Evans, 2000). Gestures make a greater contribution to comprehension for complex or ambiguous verbal messages than for simpler ones (Graham & Heywood, 1976; McNeil et al., 2000). Thus, it seems likely that gestures are particularly important in instructional discourse that presents complex concepts and uses unfamiliar terms. In addition, classrooms are often noisy, with multiple individuals speaking at once. Under such challenging circumstances, gestures that reinforce speech may be crucial to aid comprehension (see Rogers, 1978).

Not all gestures reinforce the content of the accompanying speech, however. Speakers sometimes express information in gestures that is not expressed in the accompanying speech (McNeill, 1992). For example, in explaining her solution to a liquid conservation task, a child might say, "This cup is taller," while indicating the *width* of the container in gesture. Such "mismatching" gestures also influence listeners' comprehension of the speech they accompany. Listeners comprehend speech less well when it is accompanied by mismatching gestures than when it is accompanied by no gesture or by matching gestures (e.g., McNeil et al., 2000).

Furthermore, both adults and children often detect information that is expressed uniquely in mismatching gestures (e.g., Kelly & Church, 1997). In one study of this issue, Alibali, Flevares, and Goldin-Meadow (1997) asked adults to view video clips of children explaining mathematical equivalence problems (e.g., $3+4+5=3+\_\_$). In some clips, children conveyed information uniquely in gestures (e.g., pointed to addends they did not mention in speech). Adults often detected the information that children expressed uniquely in gestures, sometimes reiterating it in their own gestures, and sometimes translating it into speech. These findings suggest a likely explanation for the finding that mismatching gestures hinder speech comprehension: when gesture mismatches speech, people sometimes detect the message expressed in gesture, rather than the one expressed in speech.

Thus, a substantial body of evidence indicates that gestures play an important role in communication. Based on this evidence, it seems likely that gestures are important in instruction, when effective communication is crucial.

Gesture Affects Learning from Lessons

Only a handful of studies have directly examined the effects of teachers' gestures on students' learning. Most have focused on whether children learn more from lessons that include gestures than from lessons that do not. Two such studies investigated third- and fourth-grade students learning to solve mathematical equivalence problems (e.g., 3 + 4 + 5 = 3 + __). In one, the lessons were delivered by an experimenter (Perry, Berch, & Singleton, 1995), and in the other by video (Church, Ayman-Nolley, & Alibali, 2001). In both, students showed deeper learning (i.e., generalization to new problem types, retention over a one-month interval) from lessons with gestures. In fact, Church et al. (2001) found that nearly twice as many students displayed deep learning after the speech-plus-gesture lesson as after the speech-only lesson (71 percent vs. 37 percent).

Another study compared third- and fourth-grade students learning about equivalence problems from videotaped lessons with no gesture, matching gestures, and mismatching gestures (Singer & Goldin-Meadow, 2005). In the lessons with mismatching gestures, the instructor described one strategy in speech (e.g., make both sides sum to the same total), and another strategy in gesture (e.g., add the numbers on the left side and subtract the number on the right, expressed in gesture with pointing gestures to the numbers on the left side, then a flick-away gesture to the number on the right). Children learned more from the lessons with mismatching gestures than from the lessons with matching gestures or no gesture, which did not differ from one another. These data suggest that gestures that serve to link ideas (such as different strategies for solving problems) may be particularly beneficial for students' learning.

Studies of other age groups and concepts have also documented beneficial effects of instructional gesture on learning. Church, Ayman-Nolley, and Mahootian (2004) examined first-grade students learning about Piagetian conservation from videotaped lessons. For native English speakers, 91 percent learned (i.e., added new *same* judgments) from a speech-plus-gesture lesson, compared with 53 percent from a speech-only lesson. For native Spanish speakers with little English proficiency, 50 percent learned from the (English) speech-plus-gesture lesson, compared with 20 percent from the (English) speech-only lesson.

Valenzeno, Alibali, and Klatzky (2003) studied preschoolers learning about symmetry from videotaped lessons. Children viewed either a speech-only lesson or a speech-plus-gesture lesson. The lessons used the same audio track, and differed only in the teachers' use of gesture. The speech-plus-gesture lesson included pointing and tracing gestures that indicated the example shapes, delineated the center of each shape, and compared the contours of the two sides of each shape. At post-test, children judged illustrations of real-world objects as symmetrical or asymmetrical, and explained their judgments. Children in the speech-plus-gesture lesson group outperformed children in the speech-only lesson group at post-test (mean = 2.08 vs. mean = .85).

Taken together, these studies provide compelling evidence that gesture matters for students' learning. However, these studies also lack ecological validity. Most utilize videotaped lessons or lessons delivered by an experimenter, rather than lessons delivered by real teachers in realistic instructional settings. Further, these studies hinge on a comparison that is not realistic. In most experimental studies, the "control" lesson—typically a *speech-only* lesson—is not like any lesson that might actually occur in a real classroom, because real teachers *do* produce gestures when they teach. These controlled experiments have established that gesture matters for learning, but they do not provide guidance for teachers about how best to use gestures to promote student learning.

Our ultimate goal is to understand how teachers' gestural behavior relates to student learning, so that we can make empirically validated recommendations about instructionally effective gestures. However, before we can test whether variations in teachers' behavior matter, we need to understand how teachers actually use gestures. Unfortunately, little is known about how much teachers actually gesture, about what kinds of gestures they produce, and about the functions these gestures serve. To formulate hypotheses about how gesture matters for students' learning, we need more knowledge about how teachers actually gesture during instruction. Thus, we turn next to research that investigates teachers' gestures in naturalistic, classroom settings, with a specific focus on mathematics classrooms.

How Do Teachers Use Gestures in Naturalistic Mathematics Instruction?

Teachers routinely use gestures as part of their instructional communication. Many descriptions of teachers' behavior mention gestures or include gestures in transcripts of lessons (e.g., Núñez, 2005; Roth & Bowen, 1999; Yackel & Cobb, 1996); however, systematic analyses of gestures in instructional communication are scarce (for exceptions, see Roth & Lawless, 2002, on ecology lectures and Corts & Pollio, 1999, on psychology lectures). Few studies of teachers' gestures have focused on mathematics, and fewer still have focused specifically on the role of gestures in fostering students' mathematics understanding. Gestures may be particularly important in mathematics instruction, because mathematics involves spatial representations (e.g., graphs, number lines), relations between ideas (e.g., links between different representations of mathematical information, such as graphs and equations), and embodied concepts (e.g., arithmetic is motion along a path) (Lakoff & Núñez, 2001). Gestures are adept at communicating spatial, relational, and embodied concepts (Alibali, 2005; Hostetter & Alibali, 2008).

The few existing studies of gestures in naturalistic mathematics instruction document that gestures are pervasive. For example, Flevares and Perry (2001) found that first-grade teachers used five to seven "nonspoken representations" per *minute* in lessons about place value, and most of these involved gestures. Alibali and Nathan (2007) examined a middle-school early algebra lesson, and found that 74 percent of the teacher's utterances about the instructional task included gesture. Richland, Zur, and Holyoak (2007) examined American, Japanese, and Hong Kong mathematics teachers' use of gesture when they made analogies during their instruction. Teachers' use of gesture in analogies varied across cultures, with roughly 15 percent of analogies receiving gestural support in the United States, and 45 percent in Japan. Thus, gestures appear to be an integral part of teachers' instructional communication. However, further research characterizing the role of gesture in mathematics instruction is needed. To address this need, we undertook an examination of teachers' gestures in elementary mathematics lessons.

Source of Data

We analyzed videotapes of five fifth-grade geometry lessons that were collected by James Stigler and Giyoo Hatano for a cross-national study of mathematics education (Stigler, Fernandez, & Yoshida, 1996). We coded lessons from three American teachers and two Japanese teachers. Given the substantial differences in lesson planning, lesson organization, and teaching methods between the United States and Japan (e.g., Stevenson & Stigler, 1992; Stigler & Hiebert, 1999), we expected that this cross-cultural sample would represent a wide range of instructional styles, and would therefore be informative about the range of variation in teachers' gestures. All five lessons focused on finding the area of a triangle, and each lasted 40–45 minutes.

We made a full verbal transcript of each lesson, and in this transcript we identified the beginning of the main body of the lesson, defined as the moment when the teacher explicitly introduced the main topic of the lesson. For example, one teacher said, "And now, I'd like us to try to figure out how to get the area of a triangle." Teachers varied greatly in how much they spoke during the main body of the lessons. To insure an adequate behavioral sample from each teacher, we identified the first 100 utterances (complete statements or speaking turns) produced by each teacher in the main body of the lesson, excluding student utterances and off-camera utterances. All gestures accompanying these 100 utterances were transcribed and coded.

Coding Gesture Form

We classified each gesture based on its form, using a system based on that described by McNeill (1992). McNeill's system has been widely used in past research, and the primary coder (YF) received extensive training before performing the coding. Gestures that were difficult to classify were reviewed and discussed by two coders (YF and MWA). Examples are presented in Figure 15.1a–e.

1 Deictic gestures indicate their referents by pointing, typically with the index finger but sometimes with other fingers or the whole hand. Teachers used deictic gestures to indicate a variety of referents, including inscriptions, objects, and students. For example, a teacher might point to an angle to refer to that angle (Figure 15.1a).
2 Hold-up gestures display concrete objects or diagrams by holding them up. These gestures are functionally similar to deictic gestures in that they indicate a specific referent. For example, a teacher might hold up a paper triangle (Figure 15.1b).
3 Representational gestures depict semantic content through handshape or motion. For example, a teacher might depict the action of cordoning off an area (Figure 15.1c).
4 Hold-up + action gestures involve holding up and manipulating concrete objects or diagrams. For example, a teacher might hold up two identical triangles and move them together to show that two triangles form a rectangle (Figure 15.1d). These gestures are functionally similar to representational gestures, in that they depict meaning through action.
5 Beat gestures are motorically simple, rhythmic gestures that do not convey semantic content, but instead mirror the rhythm or cadence of speech (Figure 15.1e). For example, when saying "The area is base times height," a teacher might produce beat gestures on the words "base" and "height."
6 Emblems are gestures that have a conventional, culturally specified form and meaning. For example, a teacher might hold up her palm to ask students to "stop."

Figure 15.1 (a) Deictic gesture. (b) Hold-up gesture. (c) Representational gesture. (d) Hold-up-plus-action gesture. (e) Beat gesture.

Coding Gesture Function

We inferred the communicative function of each gesture based on the gesture form, the accompanying speech, and the instructional context. We identified four primary functions: managing interaction, expressing emphasis, conveying information, and guiding attention.

First, teachers often used gestures to *manage interaction* in the classroom. Two types of gestures were coded as managing interaction: emblems that sought to regulate students' behavior (e.g., conventional gestures meaning "shhh", "stop"), and deictic gestures used to call on students or regulate turn taking. For example, one teacher pointed at a student while asking, "What is perimeter, Jason?" Another teacher produced the "stop" gesture while saying, "Let's stop for a second."

Second, teachers used gestures to *express emphasis* or to "underscore" important parts of their speech. Beat gestures were coded as serving this function. For example, one teacher said, "Area is measured in *square* units" and (along with verbal emphasis) produced a beat gesture on the word *square*.

Third, teachers used gestures to *convey substantive information* relevant to the lessons. Two types of gestures were coded as serving this function: representational gestures and hold-up-plus-action gestures. These gestures depicted mathematical concepts visually or invoked real-world applications of mathematical ideas. For example, one teacher depicted a line in gesture while saying, "That would be a line measurement." Another teacher used a hold-up-plus action gesture to demonstrate that two identical paper triangles could make a rectangle, saying, "You put these together, like so."

Fourth, teachers used gestures to *guide students' attention* to portions of the instructional context. Two types of gestures were coded as serving this function: hold-up gestures and deictic gestures. Hold-up gestures were used to guide students' attention to objects. For example, one teacher held up a paper triangle while saying, "Here's a triangle." Deictic gestures frequently served to guide students' attention to objects or inscriptions; for example, one teacher indicated two sides of a right-angled triangle in gesture while saying, "Two sides are straight."

The Functions of Gesture in Instruction

For each of the five teachers, the predominant function of gestures was to guide students' attention to features of the instructional context. Across teachers, an average of 57 percent (SE = 7.0) of all gestures were used to guide attention. Gestures for managing interaction (mean = 8 percent of all gestures, SE = 1.4), expressing emphasis (mean = 16 percent, SE = 5.7) and conveying information (mean = 19 percent, SE = 1.8) were used much less frequently. However, all of the teachers used some gestures from each of the four functions, with the exception of one teacher who used gestures solely to express emphasis.

Japanese and American Teachers' Use of Instructional Gestures

We also compared instructional gestures in the two cultures. Of course, given the small sample size, the findings should be interpreted cautiously. Figure 15.2 presents the average number of gestures produced for each function by Japanese and American teachers. In both cultures, gestures were most often used to guide attention, and least often used to manage interaction. However, the rate of gesture production was lower among Japanese teachers for each function. American teachers produced an average of 83.3 gestures (range 71–98) over the 100 utterances, whereas Japanese teachers produced an average of 48.5 gestures (range 33–66). Thus, on average, American teachers produced 1.72 times as many gestures as Japanese teachers. American teachers

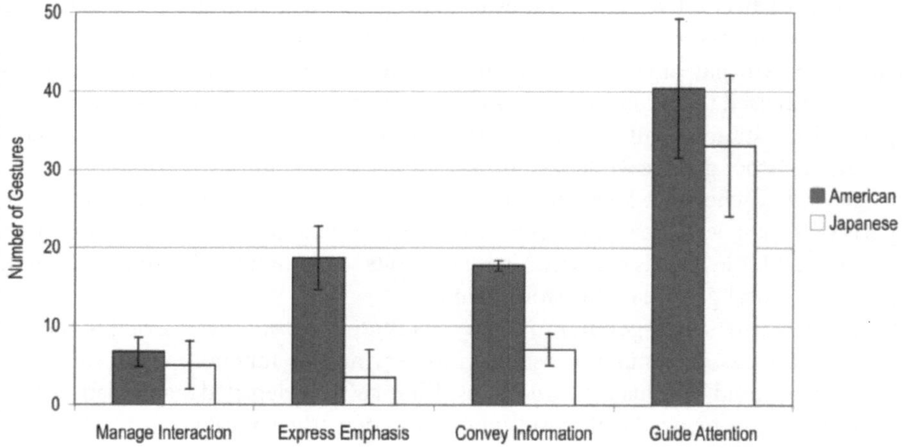

Figure 15.2 Mean Number of Gestures for Each Function Produced over 100 Utterances by American and Japanese Teachers. The error bars represent standard errors.

used many more gestures to convey substantive information [American mean = 17.7, SE = .7, vs. Japanese mean = 7.0, SE = 2.0; t(3) = 6.20, p = .008], and they also used more gestures to express emphasis [American mean = 18.7, SE = 4.1 vs. Japanese mean = 3.5, SE = 3.5; t(3) = 2.59, p = .08]. American and Japanese teachers used similar numbers of gestures to guide attention and manage interaction.

Despite these differences in gesture rates, it bears emphasizing that teachers in both cultures used gestures in largely similar ways. In both cultures, the most common function of gesture was to guide attention, and the least common function was to manage interaction.

Links between Representations

The focus of the lessons in this corpus (as in many math lessons) was on links between different representations of mathematical information. Specifically, the lessons focused on links between diagrams of geometric shapes (primarily triangles and rectangles) and formulae for calculating the areas of those shapes. Although the focus of our functional analysis was on individual gestures, we occasionally observed teachers using sets of gestures, along with speech, to link representations and to highlight correspondences among them. Such "linking episodes" often captured the mathematical goals of the lessons, and as such, they seem particularly significant.

One of the linking episodes we observed involved links between a diagram of a rectangle with length 12 and width 4, the general formula for the area of a rectangle ($A = l \times w$), and the " instantiated" formula $A = 12 \times 4$ (Figure 15.3). These links were made using speech and gesture to connect the diagram and the formula, and writing to generate the instantiated formula. In the excerpt below, brackets indicate speech that co-occurs with the gesture indicated in the lines beneath it.

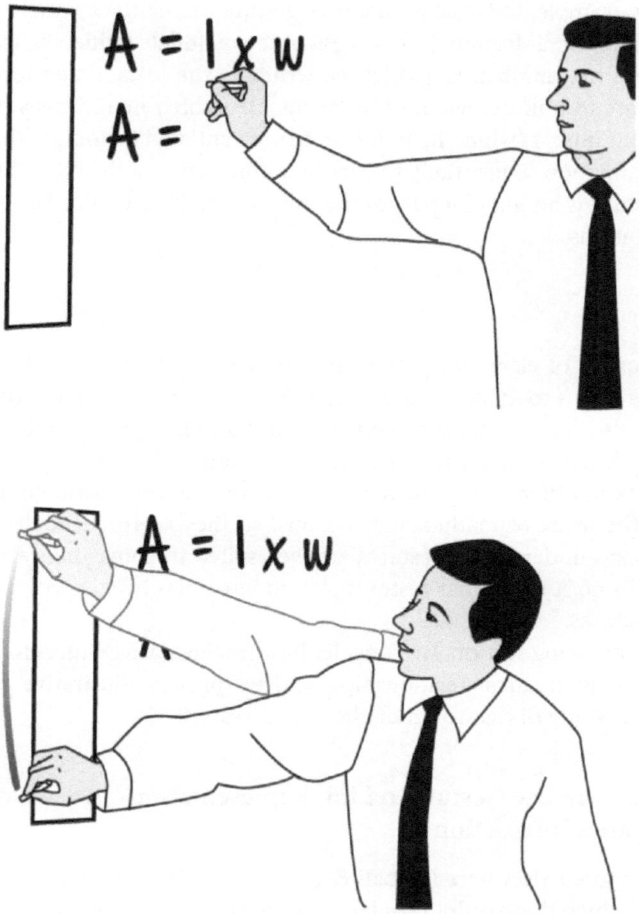

Figure 15.3 Teacher Uses Gesture to Link Area Formula and Diagram of Rectangle.

Speech:	Now we substitute,	[area equals]	[length] (pause)
Writing:		A =	
Gesture:			point to *l* in formula

Speech:	[12]	[(pause)]
Writing:	12	
Gesture:		trace length of long side of rectangle

Speech:	[times] the	[width, four]	[(pause)]
Writing:	×	4	
Gesture:			indicate short side of rectangle

In this example, the teacher first uses gesture to link the symbol *l* (which he indicates in the general formula), the *length* of the rectangle (which he traces on the diagram), and the number *12* (which he writes in the instantiated formula). He then uses gesture to link the *width* of the rectangle (which he indicates on the diagram) with the number *4* (which he writes in the instantiated formula). The gestures serve to guide attention sequentially to corresponding parts of the related representations. Thus, gesture is an integral part of the links the teacher establishes among the three representations.

Summary

In this corpus of elementary mathematics lessons, the primary function of teachers' gestures was to guide students' attention to features of the instructional context. Teachers also used gestures to convey information, express emphasis, and manage classroom interaction. Further, teachers sometimes used sets of gestures to highlight links between different representations of mathematical information. Such links are often at the heart of mathematics lessons, so they seem particularly important to examine and understand. Gestures are well suited to conveying relational information, so it is no surprise that gestures play an integral role in expressing links between representations.

In the following section, we consider how teachers use gestures to effectively communicate mathematical relationships, and we present illustrative examples drawn from a new study of classroom mathematics lessons.

How Teachers Use Gesture to Link Representations in Mathematics Instruction

We have argued elsewhere (Alibali & Nathan, 2007) that teachers' gestures are one means by which they scaffold student understanding of complex mathematical ideas. We based this claim on an analysis of an early algebra lesson. The teacher used gesture (a) more frequently for new material than for review material, (b) more frequently in response to students' questions than before such questions, and (c) more frequently for abstract referents than for concrete referents. Because links between representations are usually abstract, and because they often involve information that is new and potentially difficult for students, we hypothesize that teachers use gesture frequently when they communicate about such links. In an effort to better understand how teachers link representations, we have collected a corpus of 24 middle-school classroom mathematics lessons, and we are analyzing linking episodes within these lessons.

Our analysis thus far has revealed two primary ways in which teachers use gestures to establish links between different representations of mathematical information (e.g., equations, graphs, manipulatives): (1) teachers utilize sets of deictic gestures to highlight corresponding aspects of related representations and (2) teachers produced gestural catchments (i.e., repeated features in sets of representational gestures; see McNeill & Duncan, 2000) in order to show relatedness. We illustrate each of these gestural devices in turn. The examples presented here were drawn from two different lessons focusing on beginning algebra from the same sixth-grade mathematics teacher. However, it is important to note that we have observed linking episodes that utilize these techniques in all of the teachers analyzed to date. The examples we offer here are representative of those in the corpus as a whole.

Example 1: Sets of Deictic Gestures

Teachers frequently use sets of deictic gestures to highlight corresponding aspects of related representations. One representative example occurred in a lesson in which the teacher introduced a new way of using equations to model a story problem situation. The students were familiar with generating an equation that could be used to derive a solution, such as $(42 - 18) \div 4 = n$ (termed the *solution equation*). The lesson sought to build on this prior knowledge to help students generate a related equation that could be used to model the problem situation, namely, $4 \times n + 18 = 42$ (termed the *situation equation*).

In the lesson, the two equations were written side by side on the whiteboard at the front of the classroom. The teacher asked the students what was similar about the equations, and she revoiced the students' responses and produced deictic gestures to guide attention to the relevant, corresponding parts of the two equations. The following excerpt illustrates the teacher's use of deictic gestures to link "$4\times$" and "$\div 4$" and to link "$+ 18$" and "$- 18$". Gestures 6 and 7 are illustrated in Figure 15.4.

Student:	Timesing was there and dividing's there . . .			
Teacher:	Okay, so	[times],	[and] . . . so	[times four]
		1	2	3
Teacher:	and then	[divide by]	[four], cool	
		4	5	
Student:	and then plus, and then the minus over there.			
Teacher:	[Plus 18]	[and minus 18].		
	6	7		

1 Right-hand point to times sign in situation equation.
2 Right-hand point to division sign in solution equation.
3 Right-hand point toward situation equation.
4 Right-hand point to division sign in solution equation.
5 Right-hand point to 4 in solution equation.
6 Right-hand flat palm under + 18 in situation equation.
7 Right-hand flat palm under – 18 in solution equation.

In this example, the teacher used deictic gestures to establish mappings between the familiar solution equation and the less familiar situation equation, by guiding attention sequentially to corresponding aspects of the two representations. Specifically, she used gestures to delineate the correspondences between values and inverted operations across the two equations.

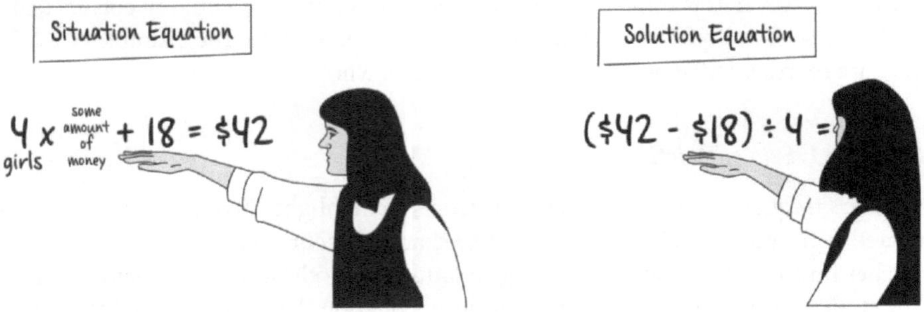

Figure 15.4 Teacher Uses Deictic Gestures to Link Corresponding Parts of Related Equations.

Example 2: Catchment of Representational Gestures

The teacher also sometimes highlighted relationships among entities with representational gestures, by repeating gesture features, such as handshape or motion, over a series of gestures. This is called a gestural *catchment*, and has been described in past research by McNeill and Duncan (2000) as a way of making connections in discourse.

The teacher produced a gestural catchment to link representations as she guided students to use equations to model a physical system, namely, a pan balance with objects on each side. The teacher gives meaning to the idea of subtracting unknown but equal quantities from both sides of an equation using a repeated gesture. She first mimes removing the same object from both sides of the pan balance, and then repeats the gesture by "removing" (i.e., pretending to pick up) the corresponding variables from both sides of the equation. These gestures are depicted in Figure 15.5.

I am gonna take away [a sphere from each side . . .]
1
[Instead of taking it off the pans]
2
[I am going to take it away from this] equation
3

1 Both hands cupped over pan balance picture, as if holding spheres on each side.
2 Both hands cupped over pan balance picture; hands move up and out to mime removing a sphere from each side.
3 Both hands cupped over equation, as if holding *s* on each side.

In this example, the teacher repeated the grasping gesture over the pan balance and then over the equation to highlight the conceptual connections between the two representations. Thus, the teacher established the correspondence between removing similar objects from both sides of the pan balance, and subtracting the same value from both sides of the equation.

It is also worth noting that, in this example, the teacher uses a *simulated action*

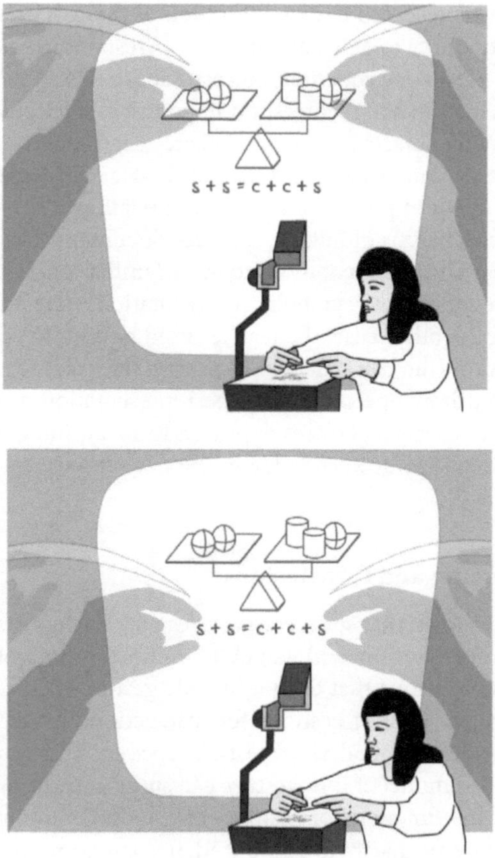

Figure 15.5 Teacher Uses a Catchment of Repeated Representational Gestures to Show that Removing the Same Object from Both Sides of the Pan Balance Corresponds to Removing the Same Value from Both Sides of the Equation.

expressed in gesture (see Hostetter & Alibali, 2008) to ground an abstract mathematical concept. Specifically, the teacher uses the familiar action of picking up two objects, represented here in gesture, to give meaning to the notion of subtracting the same value from both sides of the equation. The simulated action depicts the meaning of a potentially unfamiliar idea in terms of a familiar action.

Summary

These examples illustrate two gestural techniques that teachers use to link representations. First, teachers use sets of attention-guiding deictic gestures to delineate correspondences, as we saw both in the episode in which a teacher linked the diagram of the triangle to the formula for finding the area of the triangle and in the episode in which a teacher linked two different equations that represented the same mathematical situation. Second, teachers use repeated gesture features to highlight conceptual links, as we saw in the episode in which a teacher linked a picture of a pan balance and an equation that represented the configuration of the pan balance. Overall, gestures during instruction serve to guide attention to corresponding aspects of related representations (e.g., + 18 in one equations corresponds to – 18 in the other), and to convey substantive information about corresponding ways to manipulate related representations (e.g., removing objects from a pan balance corresponds to subtracting from an equation). These techniques are thus part of the teacher's repertoire of methods for communicating about important mathematical relations.

Our qualitative analysis of linking episodes documents teachers' use of gesture to connect new ideas and representations to more familiar ones. In one case, the teacher used gesture to connect two symbolic representations: one a familiar form and one a novel form. In the other case, the teacher used gesture to connect a familiar, concrete representation (the pan balance) to a target, abstract representation (the equation), and showed how operations on one representation are comparable to those on the other. The teacher's gestures appeared to be an integral part of her effort to bridge from students' prior knowledge to the new knowledge that was the target of her instruction.

General Discussion and Conclusions

Past research has shown that speakers' gestures play an important role in their listeners' language comprehension, and that children learn more from lessons that include gestures than from lessons that do not include gestures. Experimental work to date highlights the importance of gestures for instruction, but lacks ecological validity, and also lacks the detailed analysis of gesture types and functions that will allow us to construct a detailed model of how gesture can support learning.

Our study of the functions of gestures in classroom communication shows that teachers use gesture primarily to guide students' attention to aspects of the instructional context—often aspects of the context that are mathematically important. Teachers also use gesture to convey substantive information, express emphasis, and manage classroom interaction.

Mathematics lessons often focus on links between different representations of mathematical information, and teachers frequently use gesture to highlight such links. In many cases, this involves linking an abstract, mathematical representation (e.g., an equation) to a more concrete, grounded representation (e.g., a physical system or story). In other cases, this involves linking a less familiar mathematical representation to a more familiar one. Teachers use a variety of gestural devices to establish these links.

As yet, we do not know for certain whether variations in teachers' gestures make a difference in students' comprehension and understanding of those links. However, based on past research about the role of gesture in comprehension and learning, it seems highly probable. In ongoing research, we are testing this possibility, and attempting to ascertain what types of gesture help to convey links between mathematical ideas most effectively. By experimentally manipulating the ways in which relations between mathematical ideas are conveyed, and exploring the consequences for learning, we will gain a deeper understanding of the cognitive processes involved in acquiring mathematical understanding. Eventually, we hope to be in a position to make research-based recommendations about how teachers can use gestures most effectively. Knowledge about instructionally effective gestures is an inexpensive and valuable tool that teachers can add to their "toolkit" of methods for effective communication.

Acknowledgments

We are grateful to many people for their contributions to this line of research, including Breckie Church, Matthew Wolfgram, Eric Knuth, Autumn Hostetter, Kristen Bieda, Chelsea Johnson, Amy French, Laura Valenzeno Scheinholz, Elisabeth Sylvan, Jim Stigler, and Giyoo Hatano. We thank Beth Atkinson for preparing the illustrations. We are also grateful to the agencies and programs that have funded this work: the Institute for Education Sciences, the James S. McDonnell Foundation, the Wisconsin Alumni Research Foundation, and the Carnegie Mellon University Small Undergraduate Research Grant program. Finally, we are most grateful to the teachers and students who made this work possible.

References

Alibali, M. W. (2005). Gesture in spatial cognition: Expressing, communicating and thinking about spatial information. *Spatial Cognition & Computation*, 5, 307–331.

Alibali, M. W., Flevares, L., & Goldin-Meadow, S. (1997). Assessing knowledge conveyed in gesture: Do teachers have the upper hand? *Journal of Educational Psychology*, 89, 183–193.

Alibali, M. W., & Nathan, M. J. (2007). Teachers' gestures as a means of scaffolding students' understanding: Evidence from an early algebra lesson. In R. Goldman, R. Pea, B. Barron,& S. J. Derry (Eds.), *Video research in the learning sciences*. Mahwah, NJ: LEA. pp. 349–365.

Church, R. B., Ayman-Nolley, S., & Alibali, M. W. (2001). Multimodal representation and deep learning. Poster presented at the biennial meeting of the Cognitive Development Society, Virginia Beach, Virginia, October 26, 2001.

Church, R. B., Ayman-Nolley, S., & Mahootian, S. (2004). The role of gesture in bilingual education: Does gesture enhance learning? *International Journal of Bilingual Education and Bilingualism*, 7, 303–319.

Corts, D. P., & Pollio, H. R. (1999). Spontaneous production of figurative language and gesture in college lectures. *Metaphor and Symbol*, 14, 81–100.

Flevares, L. M., & Perry, M. (2001). How many do you see? The use of nonspoken representations in first-grade mathematics lessons. *Journal of Educational Psychology*, 93, 330–345.

Goldin-Meadow, S., & Sandhofer, C. M. (1999). Gesture conveys substantive information to ordinary listeners. *Developmental Science*, 2, 67–74.

Graham, J. A., & Heywood, S. (1976). The effects of elimination of hand gesture and of verbal codability on speech performance. *European Journal of Social Psychology*, 5, 189–195.

Hostetter, A. B., & Alibali, M. W. (2008). Visible embodiment: Gestures as simulated action. *Psychonomic Bulletin and Review*, 15, 495–514.

Kelly, S. D., & Church, R. B. (1997). Can children detect conceptual information conveyed through other children's nonverbal behaviors? *Cognition and Instruction*, 15, 107–134.

Kendon, A. (1994). Do gestures communicate? A review. *Research on Language and Social Interaction*, 27, 175–200.

Krauss, R. M., Morrel-Samuels, P., & Colasante, C. (1991). Do conversational hand gestures communicate? *Journal of Personality and Social Psychology*, 61, 743–754.

Lakoff, G., & Núñez, R. (2001). *Where mathematics comes from: How the embodied mind brings mathematics into being*. New York: Basic Books.

McNeil, N. M., Alibali, M. W., & Evans, J. L. (2000). The role of gesture in children's comprehension of spoken language: Now they need it, now they don't. *Journal of Nonverbal Behavior*, 24, 131–150.

McNeill, D. (1992). *Hand and mind: What gestures reveal about thought*. Chicago: University of Chicago Press.

McNeill, D., & Duncan, S. (2000). Growth points in thinking-for-speaking. In D. McNeill (Ed.), *Language and gesture*. Cambridge: Cambridge University Press. pp. 141–161.

Mortimer, E. F., & Wertsch, J. V. (2003). The architecture and dynamics of intersubjectivity in science classrooms. *Mind, Culture, and Activity*, 10, 230–244.

Núñez, R. (2005). Do real numbers really move? Language, thought, and gesture: The embodied cognitive foundations of mathematics. In F. Iida, R. Pfeifer, L. Steels, & Y. Kuniyoshi (Eds.), *Embodied artificial intelligence*. Berlin: Springer-Verlag. pp. 54–73.

Perry, M., Berch, D., & Singleton, J. (1995). Constructing shared understanding: The role of nonverbal input in learning contexts. *Journal of Contemporary Legal Issues*, 6, 213–235.

Richland, L. E., Zur, O., & Holyoak, K. J. (2007). Cognitive supports for analogies in the mathematics classroom. *Science*, 316, 1128–1129.

Rogers, W. T. (1978). The contribution of kinesic illustrators toward the comprehension of verbal behavior within utterances. *Human Communications Research*, 5, 54–62.

Roth, W.-M. (2002). Gestures: Their role in teaching and learning. *Review of Educational Research*, 71, 365–392.

Roth, W.-M., & Bowen, G. M. (1999). Complexities of graphical representations during ecology lectures: An analysis rooted in semiotics and hermeneutic phenomenology. *Learning and Instruction*, 9, 235–355.

Roth, W.-M., & Lawless, D. V. (2002). When up is down and down is up: Body orientation, proximity, and gestures as resources. *Language in Society*, 31, 1–28.

Schwartz, D. L., & Bransford, J. D. (1998). A time for telling. *Cognition & Instruction*, 16, 475–522.

Singer, M. A., & Goldin-Meadow, S. (2005). Children learn when their teacher's gestures and speech differ. *Psychological Science*, 16, 85–89.

Stevenson, H. W., & Stigler, J. W. (1992). *The learning gap: Why our schools are failing and what we can learn from Japanese and Chinese education*. New York: Simon & Schuster.

Stigler, J. W., Fernandez, C., & Yoshida, M. (1996). Cultures of mathematics instruction in Japanese and American elementary classrooms. In T. Rohlen & G. LeTendre (Eds.), *Teaching and learning in Japan*. Cambridge: Cambridge University Press. pp. 213–247.

Stigler, J. W., & Hiebert, J. (1999). *The teaching gap: Best ideas from the world's teacher for improving education in the classroom*. New York: Simon & Schuster.

Valenzeno, L., Alibali, M. W., & Klatzky, R. L. (2003). Teachers' gestures facilitate students' learning: A lesson in symmetry. *Contemporary Educational Psychology*, 28, 187–204.

Yackel, E., & Cobb, P. (1996). Sociomathematical norms, argumentation, and autonomy in mathematics. *Journal for Research in Mathematics Education*, 27, 458–477.

16 Perceptual Learning and Adaptive Learning Technology

Developing New Approaches to Mathematics Learning in the Classroom

Christine M. Massey, Philip J. Kellman, Zipora Roth, and Timothy Burke

Most humans, both young and old, are capable of remarkable feats of learning in their everyday lives, and yet, all too often, the news from classrooms in the United States is about perennial difficulty and persistent failure for large numbers of students in achieving the learning goals set out in local, state, and national standards. Although the causes and potential cures are many and varied, in this chapter we consider an approach that addresses dimensions of learning that have been studied for decades in the learning sciences but have received little to no attention in K–12 classrooms. Specifically, we examine *perceptual learning* as a form of learning that contributes to the insight and fluency that characterize expertise across many settings and domains. We introduce what perceptual learning (PL) is, findings about PL that have emerged from several different lines of research, and how PL might be brought into K–12 classrooms as a significant complement to other forms of instruction. Two empirical studies of PL interventions using specially designed learning software, known as Perceptual Learning Modules (PLMs), illustrate some of its key characteristics and effects on students' learning in mathematics.

What Is Perceptual Learning?

The classic definition of perceptual learning, offered by Eleanor Gibson (1969), is "an increase in the ability to extract information from the environment, as a result of experience and practice with stimulation coming from it" (p. 3). With practice, in virtually all domains of human experience, people become significantly better and faster at extracting relevant information, ignoring irrelevant information, making fine discriminations, and perceiving higher-order structure and relationships. It is particularly noteworthy that PL emphasizes the pick-up of information that is demonstrably present in the external environment (though typically unnoticed or inefficiently processed by novices). The changes that result from such learning are changes in the ability to recognize or distinguish significant features, structures, or relationships. (For a more detailed discussion of PL in relation to other taxonomies of learning see Gibson & Gibson, 1955, and Kellman & Garrigan, 2008. See Kellman, 2002, and Goldstone, 1998, for general reviews of contemporary research on PL.)

PL involves the optimization of attention, so that the learner becomes increasingly selective in what information is attended to and what is disregarded. It is also characterized by increasing specificity of discrimination, such that experience allows the learner to make fine distinctions among features or structures that initially appeared to be the same. Although the term "perceptual" may evoke a sense that the information being extracted involves only low-level sensory inputs, in concert both with contemporary theories of perception (J. Gibson, 1979; Marr, 1982) and with classic treatments of PL (E. Gibson, 1969) we see perception and perceptual learning as being crucially involved with the extraction of higher-level invariants and abstract structures. Indeed, an aim of this chapter is to extend applications of PL to the abstract symbolic domain of mathematics.

Perceptual Learning and Expertise

Studies of expertise across a variety of domains have provided a particularly rich set of phenomena illustrating the power of PL. Studies in diverse areas, including aviation (Kellman & Kaiser, 1994), radiology (Kundel & Nodine, 1975), and chess (Chase & Simon, 1973; DeGroot, 1965), have documented the development of the ability to detect and discriminate relevant features, patterns, and relationships, typically after extensive practice involving many examples in the course of carrying out a meaningful task. Experts differ from novices not just in their declarative knowledge or their procedural competence; one of the most distinctive features of their expertise is that they seem to "just see" (or hear, taste, etc.) differences, patterns, structures, and relationships in the environmental input that either are unavailable to novices or require effortful and error-prone analysis.

Whereas the novice typically attends to superficial or irrelevant information as much as relevant information, the expert's attention is selective. Experts tend to process both very fine-grained levels of detail when relevant, as well as abstract higher-order "chunks" and relationships (Chi, Feltovich, & Glaser, 1981). Experts are also more fluent and may show automaticity (Schneider & Shiffrin, 1977), processing remarkably large amounts of task-relevant information quickly and effortlessly in a way that is relatively insensitive to cognitive load.

Perceptual Learning in Everyday Contexts

Perceptual learning does not arise only in esoteric domains involving advanced training and education; it underlies many capacities that we take for granted in our everyday lives. Our ability to learn to recognize the faces of students in a class or to immediately identify the voices of friends and family on the phone are instances of PL.

It is likely that PL underlies our acquisition of many everyday concepts, as well. Consider, for instance, the case of a young child learning the extension of the word "dog." The child's task is to learn to pick out all and only the dogs in the world, despite (1) the very great variety among different types of dogs (for instance, a Saint Bernard vs. a hairless Chinese Crested); and (2) the strong resemblance between some dogs and animals of other species (e.g., a German Shepherd and a wolf, or a furry Pomeranian and a fluffy Persian cat). Restated as a more general principle of PL, the child must learn to extract the invariances that underlie the different instances of the target category or concepts, and must also learn to discriminate the key structures from negative cases that may appear similar, but which do not belong to the category. Very young children spontaneously engage in this kind of learning and become remarkably proficient without much in the way of explicit or systematic instruction.

What Promotes Perceptual Learning?

Interestingly, even an accomplished adult learner may or may not be able to articulate the grounds on which he or she recognizes and discriminates key structures and relationships. Experimental studies of PL using unfamiliar laboratory stimuli indicate that learners may approach perfect performance in being able to detect or discriminate target structures without being able to explain how they are doing so. Moreover, didactic instruction is generally not an effective way to promote PL. Instead, what seems to promote PL most effectively is processing many examples of instances that possess the features, structures, or relationships of interest and seeing them across a widely varying background of contexts and contrasting cases. This kind of experience may accrue naturally as an individual spends much time immersed in the practice of a hobby or profession, or it can be deliberately structured in a training or educational program.

Perceptual Learning in Middle-School Mathematics Instruction

Accelerating Expertise: Perceptual Learning Modules

In the intervention studies that follow, we use custom-designed learning software known as Perceptual Learning Modules (PLMs) as a way to operationalize a learning environment based on PL principles. PLMs rest on the assumption that human perceptual learning abilities advance not from the passage of time, the acquisition of declarative verbal knowledge, or the rehearsal of fixed procedures, but from making discriminations and classifications relevant to some task. The critical learning activity in the designed PLMs involves asking the learner to recognize or discriminate a target relational structure or to map related structures across different representations (e.g., graphs and equations) or across transformations (e.g., algebraic transformations performed on an equation) (Kellman et al., 2008). Several principles, derived from the research literature on PL, guide the way we structure materials and activities to make learning trials effective.

First, the PLM software is designed to engage the learner in *large numbers of brief classification episodes*—not just a small number of examples. This approach departs from common practice in math classrooms in two notable ways. First, learners see many more instances of the target structures and relationships and in more contexts than would normally be presented in a typical scenario (where a teacher works one or two problems with the whole class and students then go on to work problems that are similar to the model).

Second, when PL is the instructional goal, a much greater percentage of students' time and effort is devoted to *problem recognition and classification*, rather than completing calculations and procedures to solve problems. Learning episodes go quickly; a student might complete a dozen or more trials in the time it would take to work a problem by completing all of the calculations to arrive at a solution. Because specific instances seldom or never repeat in PLMs, structural invariance is learned, and learning generalizes to new instances.

It is also critical that the learning trials present multiple instances that incorporate *several kinds of systematic variation* across classification episodes. First, invariant relations characterizing key structure must appear in a variety of contexts so that the learner will come to recognize these structures independent of local contexts or surface variations. Second, irrelevant aspects of problems need to vary, so as not to be mistakenly correlated with the target structure. Contrasting classifications from which the learner must differentiate the target structure(s) should also appear. This is especially important when the learning involves differentiating similar structures, since learning to discriminate among a set of items that at first seem all to look alike is a frustrating learning problem commonly faced by novices. What is more, this learning problem is commonly underestimated by experts who have already automatized the discriminations.

What Perceptual Learning Is Not

It is important to consider approaches to learning that contrast with PL. Because of the large number of learning trials and the emphasis on having the learner interact with them, PL might be confused with some other well-known approaches. Although learners using the PLMs see many individual trials, the goal of the learning is not to memorize individual instances or a fixed set of items. The learning sets are typically very large, and the goal is to enable the learner to recognize or map the targeted structures in new contexts and novel instances.

PLM trials are also not intended to promote stimulus–response associations, as would be the goal in behaviorist approaches. Although having the learner actively engage with the items is often useful for maintaining attention and motivation, the specific behavioral responses employed are not the goal of the intervention. Instead, PL is aimed at the discovery and efficient extraction of important structures in novel, varying inputs. The fundamental question the learner is addressing in PL interventions is "What do I *see* in this problem?", not "What should I *do*?" This ability to recognize and discriminate between relevant structures and relationships may be incorporated into any number of behavioral repertoires, including complex problem solving. Indeed, we argue that the gains in being able to quickly and accurately select the most useful information for a meaningful task can reduce processing demands in a way that enables the learner to build up more sophisticated and flexible problem-solving strategies, as opposed to stereotyped behavioral routines.

The two studies that follow test the application of PL interventions, in the middle grades, to mathematics learning of fractions and measurement. These mathematical topics were selected for two reasons. First, they are areas in mathematics that many students fail to master despite their prominent representation in many curricula. Second, we believe that successful learning in these areas involves structure extraction, pattern recognition, and the achievement of fluency—thus making them good test cases for the kinds of learning for which PL interventions may provide a useful complement to other forms of instruction.

Study 1: Finding Structure in Fraction Problems

Many students struggle with word problems, and their difficulties often seem to increase when the problems involve fractions. Of course, successfully solving written problems involving rational numbers is a complex process requiring coordination of different kinds of knowledge and strategies, and there can be many sources of learning difficulties. This study investigates one aspect: whether instruction that emphasizes extracting underlying problem structures can help students learn to reliably discriminate between contrasting problem types and thus improve their ability to solve such problems.

Consider the following problems:

1. A class went on a 72-mile road trip. They stopped for lunch after driving three-quarters of the way. How many miles did they drive before stopping for lunch?
2. A class went on a road trip. After driving 72 miles, the driver announced that they had driven three-quarters of the way. How many miles are in their entire trip?

Both problems include the same quantities, involving a whole number (72 miles) and a fraction (3/4), which appear in the same order. Although the problems look alike, they have contrasting underlying structures. The first, which could be restated as "How many miles is three-quarters of 72 miles?," is what we term a "find the part" problem. The total quantity is known, and the question involves finding a fractional part of that total. The second, which could be stated in simplified form as "72 miles is three-quarters of how many miles?," is a "find the whole" problem, in which the amount of the fractional quantity is known and the task is to use that to figure out the total quantity. Many students fail to see the structural distinction between these two problems. They may see them as identical and attempt to solve them in the same way, or they may simply pull out the numbers and attempt to perform an operation on them, seemingly at random or based on the convenience of the calculation (Bell, Fischbein, & Greer, 1984). Our overarching hypothesis in this study is that actively recognizing the underlying problem structure and mapping that structure across word problems, graphic representations, and numerical representations (i.e., number sentences used to calculate a solution) are critical components in problem solving; furthermore, learning interventions that focus on structural relations and mapping will transfer to significant improvements in open-ended problem solving.

To examine whether PL approaches can help students improve at this kind of structural analysis and mapping, we developed a series of classroom lessons and two versions of a computer-based PL training module. Different versions of the PLM software were designed to address an additional experimental question related to the structure and sequence of how the content was introduced. It is common in many K–12 classrooms and curricula to break material down into component pieces and to build from simpler constituents to more complicated ones. Following this approach, we created one instructional sequence that first introduced unit fractions (fractions with a numerator of 1) and then introduced nonunit fractions. In a contrasting condition, students were introduced to both unit and nonunit fractions in classroom lessons and then used a version of the software that mixed unit and nonunit trials throughout. This approach to organizing instruction comes from the research literature on memory, motor learning, and training in industrial settings, where the issue of blocked versus randomized learning trials has been the subject of considerable study. In a review of a number of training studies, Schmidt and Bjork (1992) argue that mixing item types ultimately produces more durable learning as well as better transfer of learning, although it might slow the course of learning during training.

One group of students participated only in the classroom lessons, without using the PLM, while two other groups participated in both classroom lessons and individual PLM training. This design allowed us to ask whether teacher-led instruction that deliberately developed fraction concepts and problem-solving strategies from a structural point of view is effective at all in promoting students' learning and problem solving, and, if so, whether it is sufficient in itself or whether additional computer-based PL training could lead to further learning gains. We predicted that all groups participating in the study would improve, but that the groups in the PLM conditions would show larger or more durable learning gains. Motivating these predictions was the hypothesis that the PLM training would further consolidate students' ability to extract and map the target structures and would allow the development of fluency.

Method

PARTICIPANTS

Participants were 76 seventh-grade students (44 female, 32 male) enrolled in an urban public school serving a predominantly minority low-income population.

DESIGN AND PROCEDURE

All students completed a custom-designed pretest and were then randomly assigned to one of three conditions with the constraint that the means and ranges of pretest scores were approximately equal for each group. Students in all three conditions participated in a series of 16 classroom lessons during their regular math classes designed and led by Zipora Roth, an experienced middle-school math teacher and curriculum specialist. For students in the No-PLM Control group, there was no further learning intervention after the classroom lessons. Students in the Unit First PLM Condition completed the first sequence of seven classroom lessons, which involved unit fractions, then worked for several individual sessions with a version of the PLM software that contained only unit fraction problems. They then completed the last sequence of nine classroom lessons, which introduced nonunit fractions, before returning to individual sessions with PLM software that included both unit and nonunit problems. Students in the Mixed PLM Condition completed the entire sequence of 16 classroom lessons and then began training in individual sessions using a version of the PLM software that included both unit and nonunit fractions mixed from the beginning.

Students in the two PLM conditions worked individually for sessions of about 30–40 minutes approximately once per day until they either retired all categories by meeting predetermined speed and accuracy criteria or exhausted the time allowed for the study. Because the software adapts to each learner by allowing the student to retire problem categories as they are mastered, the number of sessions completed was highly variable, ranging from a minimum of two sessions (for students who demonstrated mastery quickly) to as many as 15 sessions. At the end of each group's learning intervention, students were given an immediate post-test. A delayed post-test was administered approximately nine weeks later, with no intervening study-related instruction.

MATERIALS

Classroom lessons Classroom instruction consisted of 16 lessons specially designed to provide a foundational introduction to fractions with a particular focus on structural relations underlying fraction concepts. Although the students in this school had completed curriculum units on fractions from current standards-based curricula, their performance on standardized assessments indicated that their math foundation was very weak and that a comprehensive review was in order. The lessons in this sequence introduced fractions involving both discrete and continuous quantities with activities emphasizing partitioning into units and iterating units to construct fractions to represent different quantities. Instruction also introduced four representations—word problems, simplified questions, fraction strips, and number sentences—that were also used in the PLM software.

PLM software The PLM software presented learners with many short trials wherein their task was to select which of several choices in one representational format matched a target presented in a different representational format. Any of the representational formats could appear in either target or choice positions. Choices were designed to include distractor items that represented common errors.

Designing trials so that the learner must map across representations challenges the learner to extract an abstract relational structure for which there are very few similarity cues. The choices, which were always of the same representational type, resembled each other much more than any one of them resembled the target. Trials were drawn from a large base of unique items, so that the user could not solve the task by memorizing answers to items that were repeated. Instead, on each trial, the student had to detect a common structure across stimuli with very different appearances (the target and its corresponding choice) and to discriminate among stimuli with similar appearances (the choices). Users received feedback on each trial as to whether they were correct or incorrect; if they were incorrect, the correct response was illustrated with a short interactive feedback sequence.

The PLM recorded each student's responses and reaction time on each trial and tracked their performance on various subcategories of problems against predetermined speed and accuracy criteria. Subcategories of items were based on bidirectional cross-mappings of each combination of representational types. When a student met the criteria for a given category of items (e.g., mapping word problems to corresponding fraction strips), that category was retired from the learning set.

PRETEST/POST-TEST FRACTION ASSESSMENT

Three equivalent forms of a 27-item pencil and paper test were created and administered in counterbalanced order as the pretest, immediate post-test, and delayed post-test. The assessment emphasized transfer items that did not directly resemble the kinds of tasks and problems used in either the classroom instruction or the PLM intervention. A few items were more similar to problems the students saw during training, but none was identical. The assessment included open-ended word problems requiring the students to calculate complete solutions, as well as multiple-choice items probing various aspects of their understanding of fractions, such as comparing and ordering fractions.

Results

The main results for this study are presented in Figure 16.1, which shows that students in all three groups demonstrated significant gains from pretest to immediate post-test and from pretest to delayed post-test. Students in the two PLM groups performed similarly to each other on the immediate post-test and outperformed the No PLM control group. By the time of the delayed post-test, the Mixed PLM group showed the best performance, fully maintaining learning gains over the nine-week delay. Mean scores on the delayed test for the Unit First PLM condition showed some decline from their immediate post-test scores. Students in the control group maintained their learning gains from immediate to delayed post-test, but did not reach levels as high as the Mixed PLM group.

These observations were tested statistically with a two-way repeated measures ANOVA performed on students' proportion of correct responses on the fraction assessment, with Test Phase (Pretest, Immediate Post-test, Delayed Post-test) as a within-subjects factor and Condition (Unit First PLM, Mixed PLM, and No PLM Control) as a between-subjects factor. There was a main effect of Test Phase [$F(2, 138) = 89.66$, $p < .001$], confirming that students across all groups showed significant learning gains. There was no reliable main effect of Condition, but there was a significant Condition by Test Phase interaction [$F(4, 138) = 5.396$, $p < .001$], indicating reliably different patterns of learning effects across conditions.

A series of planned comparisons was carried out to look at the condition differences in more detail. Both of the PLM groups showed significantly greater improvement

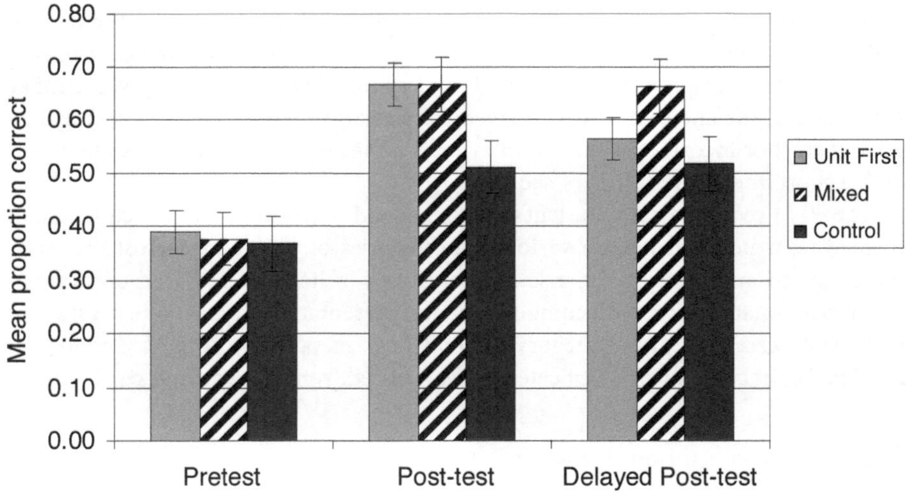

Figure 16.1 Mean Accuracy by Condition and Time of Test on the Fraction Assessment in Study 1. Error bars indicate ± one standard error of the mean. Adapted from Kellman, P. J., Massey, C., Roth, Z., Burke, T., Zucker, J., Saw, A., Aguero, K. E., & Wise, J. A. (2008). Perceptual learning and the technology of expertise: Studies in fraction learning and algebra. *Pragmatics & Cognition*, 16(2), 356-405.Copyright 2008 John Benjamins Publishing Company.

from Pretest to Immediate Post-test compared with the No PLM Control group [$t(51) = 2.60$, $p < .02$ for the Unit First PLM vs. Control, and $t(47) = 3.07$, $p < .01$, for the Mixed PLM vs. Control]. The two PLM groups did not differ from each other in improvement from Pretest to Immediate Post-test [$t(48) = .34$, n.s.], but the Mixed PLM group significantly outperformed both the Unit First PLM and the No PLM Control groups from Pretest to Delayed Post-test [$t(47) = 2.15$, $p < .04$, for the Mixed PLM vs. Unit First PLM groups and $t(43) = 2.86$, $p < .01$, for the Mixed PLM vs. No PLM Control groups]. The Unit First PLM group and the No PLM Control group did not differ significantly in learning gains from the pretest to the delayed post-test [$t(47) = -.528$, n.s.].

Discussion

The pattern of results obtained in this study generally supports our hypotheses that (a) instruction focused on structural patterns and relationships does lead to significant gains in students' ability to solve a variety of fraction problems in all conditions tested; and (b) supplementing classroom lessons with PLM training adds additional value to more typical classroom instruction methods. The classroom lessons included all of the same essential material, presented in a similar way—the PLM did not introduce any new content—yet students in the PLM conditions (particularly the Mixed PLM group) performed better than students in the No PLM Control group. Although the classroom lessons "covered" the material very systematically, most students had not learned it to the point that they demonstrated high levels of either accuracy or fluency when they first encountered the same ideas in the PLM. Both the classroom lessons and the software shared a focus on structural relationships and patterns as expressed in different representational formats. Critical differences between them were that students using the software saw a much larger and more varied set of examples. The PLM was also designed to help students extract target patterns and relationships on their own by interacting with them in a structured way, rather than having them explained by the teacher.

We submit that declarative and procedural components in instruction can be usefully supplemented by learning activities that focus specifically on practicing the extraction of important structural patterns and fluency. The number of PLM sessions completed by individual students was highly variable, indicating that students have very different needs for time and practice to consolidate this learning. PLM software holds significant promise as an efficient and effective way to customize instructional time to students' individual learning needs, and to track and certify to some standard the learning that is being accomplished.

The finding that students in the Mixed PLM group showed the strongest and most durable learning gains is also noteworthy. We suggest that mixing unit and nonunit fraction problems from the beginning of the PLM training may encourage learners to compare and contrast them and their relationship to a whole quantity from the beginning. That is, learners may be challenged to construct a more comprehensive and relational understanding at the outset rather than build what may be an overly simple approach to unit fractions that then has to be reworked and expanded when problems with nonunit fractions are introduced.

Study 2: Constructing Units of Linear Measurement

The second study we discuss extends our investigations using PL to another area known to be a problem in middle-school math classrooms in the United States: linear measurement. Many children come to measurement with mathematical conceptions tied to counting of discrete entities (Hartnett & Gelman, 1998). When they look at a ruler, they are looking for something to count and tend to focus on either the integers or the hash marks. This is related to the common observation in classrooms that children are extremely puzzled as to why one lines up the extent to be measured starting from the edge rather than from the number 1. Many children learn fixed procedures for measuring with rulers but continue to be confused about units of linear measurement and are unable to explain why the procedures are the way they are.

For example, when children are asked to measure with nonstandard rulers, they make characteristic errors. A compelling illustration comes from a released item from the 2003 National Assessment of Education Progress (National Center for Education Statistics, 2008) in which children were asked to measure a 2½-inch toothpick with a broken ruler that started at the 8-inch mark. Some students err by saying the toothpick is 10½ inches long. These students have probably learned a routine procedure for lining an item up with the end of the ruler and reading off the rightmost point. This kind of response indicates that their procedural knowledge is shallow, inflexible, and not supported by a mathematically correct understanding of units of linear measurement.

A more common error among both fourth and eighth graders is choosing 3½ inches as the correct answer. These students are most likely counting hash marks on the ruler, starting with the hash mark at the leftmost point of the toothpick as "1." Both of these incorrect responses suggest that students do not conceive of units of linear measurement as having *extent*. In other words, they do not recognize that an inch (or centimeter) on a ruler is the *entire length* between the hash marks that demarcate the unit, not just the point where the numbered hash mark is located.

A second learning challenge in elementary and middle-school mathematics is mastery of concepts related to fractions and operations with fractions. The National Mathematics Advisory Panel's (2008) recently released report singled out mastery of fractions as the most critical need in improving elementary mathematics education in the United States. Fractions and measurement are deeply connected—they involve the foundational mathematical concepts required for working with continuous quantities. It is through measurement, rather than discrete counting, that continuous quantities of all kinds (e.g., distance, area, volume, temperature, etc.) are numerically quantified (Schwartz, 1988); and accurate measurement frequently involves working with fractional parts of units (Lehrer, Jaslow, & Curtis, 2003). Indeed, measurement contexts can motivate meaningful understanding of fractions as numbers that represent quantities.

In the present study, we bring together linear measurement and fractions. The study involves a classroom intervention in which students participated in an introductory lesson and then worked individually with specially designed PLM software. The specific learning goals of this intervention were (1) understanding that any start point can be seen as zero on a scale; (2) distinguishing between position and distance on a ruler (e.g., starting at "3" as a position vs. moving three units as a distance);

(3) measuring distances accurately and flexibly on a ruler, including fractional distances; (4) improving students' understanding of units of linear measurement, unit structure, and fractional partitions of units, as represented on rulers; and (5) strengthening students' intuitions about the relationship between ruler problems and addition and subtraction operations, especially with fractions. For example, the number sentence "12½ – 10½ = ?" can be modeled on the ruler as starting at 12½ and moving a distance of 10½ units to the left.

Method

PARTICIPANTS

Participants were 63 sixth graders who participated in a PLM instructional intervention, plus 49 seventh graders and 29 eighth graders who served as uninstructed control participants. All participants were enrolled in an urban public middle school serving a predominantly low-income neighborhood.

MATERIALS

The PLM software presented learners with a graphic display showing a ball on top of a ruler and a billiard cue poised to strike it. Learners were presented with four types of trial formats, which varied what information was given and what information was to be found (e.g., given the start point and endpoint, find the distance traveled; or given the start point and distance traveled, find the endpoint). On some trials the user entered responses by keying them in using an onscreen interface. On other trials the user entered a response by dragging a marker on the ruler to the desired point. Once the learner had entered his or her response and pressed a button labeled "strike," the billiard cue would carry out the event on the screen. Animated feedback was provided on each trial. On incorrect trials, the feedback illustrated the correct answer in a way that allowed the student to compare the response entered with a correct response.

The learning items in the database varied the types of values involved, whether the rulers were fully versus partially labeled, and whether they were partitioned in the most economical way to solve the problem or were overpartitioned (e.g., a ruler marked in units of one-sixteenth for a problem involving eighths). Movement on the ruler could be in either the rightward or leftward direction, though the endpoint was never a negative number. The quantities involved varied from single digits into the hundreds and included both fractions and integers. For larger values, the ruler did not start at zero but represented a segment of a ruler marked for numbers in the relevant range. Items in the learning set were classified into eight categories, including both fraction and integer problems.

DESIGN AND PROCEDURE

At the start of the study, sixth-grade students completed a 44-point pencil and paper assessment with a variety of items related to linear measurement with integers and fractions, and adding and subtracting fractions. A control group of seventh and eighth graders, who did not participate in any study-related instruction, were administered the same test just once, providing a baseline comparison for the sixth graders' scores.

After completing the pretest, sixth graders participated in a single introductory classroom lesson conducted by Zipora Roth. They then used the PLM software for a series of individual computer-based sessions. Students continued sessions with the software until they had either met the mastery criteria for all eight categories of problems in the PLM or completed six sessions. Within one to two days of completing their last computer session, students completed a post-test assessment. Four months later, the sixth graders completed a delayed post-test, with no study-related activities occurring in the interim.

PRETEST/POST-TEST ASSESSMENT

Participants were administered a 44-point pencil and paper assessment, with items distributed across a number of subscales. The subscales were designed to assess children's ability to use a partitioned number line to express the length of a line in generic units; to use both conventional and broken rulers to measure lengths in inches and centimeters; to use conventional and broken rulers to construct extents of varying lengths; to solve addition and subtraction problems with fractions; and to solve open-ended word problems involving linear measurements. Virtually all of the items were transfer items that did not directly resemble the trials presented to students during the PLM training. Three parallel forms of the assessment were created to allow equivalent tests to be administered to the same sixth graders in the intervention group as a pretest, an immediate post-test, and a delayed post-test. The forms were counterbalanced across children and times of administration, so that the child saw a different form each time the assessment was administered.

Results

As Figure 16.2 indicates, prior to instruction, the sixth graders and the seventh- and eighth-grade control groups scored similarly on the assessment, with no evidence of improvement through middle school. At the time of the immediate post-test, the sixth-grade intervention group showed significant improvement, as indicated by a one-way ANOVA comparing the sixth-, seventh-, and eighth-grade groups [$F(2, 138) = 19.687$, $p < .001$]. The sixth graders achieved nearly identical scores on a delayed post-test administered four months later, indicating that their learning gains were fully maintained. Amongst other subscales, the assessment included sets of items comparing students' performance reading and constructing lengths with conventional versus broken rulers. The sixth-grade intervention group improved on both types of items—but especially on the broken ruler items—from pretest to post-test. Students also made strong gains for items involving fractions.

Discussion

These results indicate that an intervention that consisted primarily of PLM training over the course of about half a dozen class periods yielded dramatic improvements in sixth-grade students' ability to demonstrate competence with linear measurement

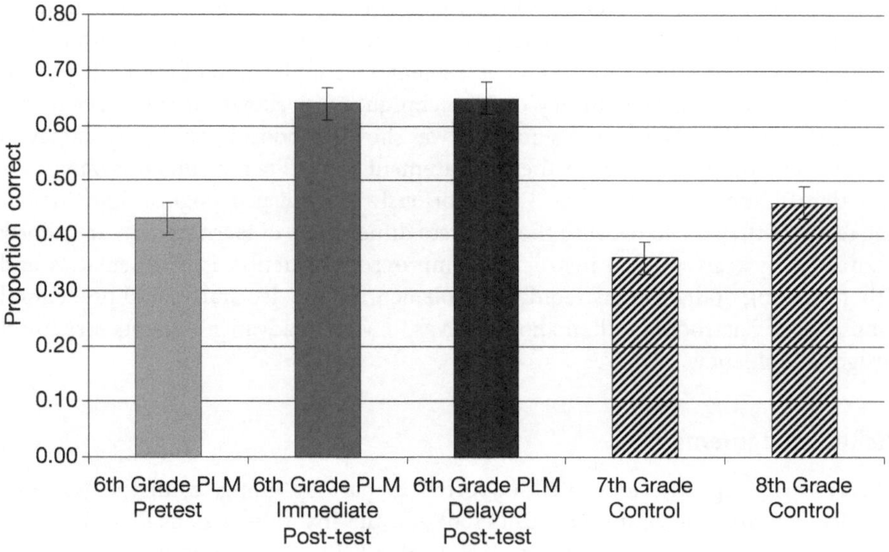

Figure 16.2 Pretest and Post-test Scores on Measurement and Fractions Assessment for Sixth Graders in Intervention Condition in Study 2 Compared with Seventh- and Eighth-Grade Control Groups. Error bars indicate ± one standard error of the mean.

involving both integers and fractions. These results are documented by robust and long-lasting gains on a quantitative assessment, but we also observed qualitative signs of changes in the children's performance. We think it is fair to state that many, perhaps most, of the students at the start of the study had little insight into the structure of linear units of measurement, their organization on a ruler, or the use of fractions to describe partitions of units. At the end of the study, many students had undergone a qualitative change in their understanding, such that they could, literally, perceive the structure and organization of measurement units in rulers and the relations between units and subunits. Their performance on transfer problems indicates that they could also interact with these units and their relations in productive and meaningful ways to solve new problems.

In contrast to the successful learning of the sixth graders in the study, the data obtained from the seventh- and eighth-grade control groups indicate that there otherwise appeared to be no significant learning in this area after two more years of middle-school math instruction. Persisting difficulties in understanding measurement and fractions leave students severely compromised in their ability to work with continuous quantities in meaningful ways, whether in or out of school, and it puts them at significant risk as they move on to more advanced mathematics in high school.

Conclusion

The research described in this chapter suggests that perceptual learning techniques offer clear promise for improving learning in mathematics and other domains. In fraction learning, instruction that focused on recognizing and mapping structural patterns led to clear improvements in students' solving of fraction problems. In measurement, PL interventions led to deep advances in learners' competence with fundamental notions of extent, units of measurement, and application of both integer and fractional quantities. The efficacy of PL techniques was shown in comparison to a no-PLM control in the fraction study, and was shown to boost learning beyond levels found even two grades later in the measurement study. Such results not only indicate that PL methods can succeed in notoriously difficult pedagogical domains but suggest that they do so because they address dimensions of learning that are poorly addressed by conventional instruction. Improved instruction in mathematics and other learning domains may require complementing the declarative and procedural emphases of classrooms with methods such as PLMs that advance students' structural insight and fluency.

Acknowledgments

The research reported here was supported in part by grants from the National Science Foundation (Grant REC-0211826) and the Institute of Education Sciences, U.S. Department of Education (Grant R305H060070). The opinions expressed are those of the authors and do not represent views of the National Science Foundation or the U.S. Department of Education. We gratefully acknowledge Katherine Aguero, Derrick Carpenter, Hye Young Cheong, and Warren Longmire at the University of Pennsylvania and Ji Son and Joel Zucker at UCLA for their able assistance. We also wish to thank the students, teachers, and administrators at our partner schools for their roles in making this research possible.

References

Bell, A., Fischbein, E., & Greer, B. (1984). Choice of operation in verbal arithmetic problems: The effects of number size, problem structure and context. *Educational Studies in Mathematics*, 15, 129–147.

Chase, W. G., & Simon, H. A. (1973). Perception in chess. *Cognitive Psychology*, 4(1), 55–81.

Chi, M. T. H., Feltovich, P., & Glaser, R. (1981). Categorization and representation of physics problems by experts and novices. *Cognitive Science*, 5, 121–152.

DeGroot, A. D. (1965). *Thought and choice in chess*. Amsterdam: Noord-Hollandsche Uitgeversmaatschappij.

Gibson, E. J. (1969). *Principles of perceptual learning and development*. New York: Prentice-Hall.

Gibson, J. J. (1979). *The ecological approach to visual perception*. Boston: Houghton-Mifflin.

Gibson, J. J., & Gibson, E. J. (1955). Perceptual learning: Differentiation or enrichment? *Psychological Review*, 62, 32–41.

Goldstone, R. L. (1998). Perceptual learning. *Annual Review of Psychology*, 49, 585–612.

Hartnett, P. M., & Gelman, R. (1998). Early understandings of number: Paths or barriers to the construction of new understandings? *Learning and Instruction*, 8, 341–374.

Kellman, P. J. (2002). Perceptual learning. In H. Pashler & C. R. Gallistel, (Eds.), *Stevens' handbook of experimental psychology* (3rd edn) (Vol. 3). New York: John Wiley & Sons. pp. 259–299.

Kellman, P. J., & Garrigan, P. (2008). Perceptual learning and human expertise. *Physics of Life Reviews*, 6, 53–84.

Kellman, P. J., & Kaiser, M. K. (1994). Perceptual learning modules in flight training. *Proceedings of the 38th Annual Meeting of the Human Factors and Ergonomics Society*. Nashville, TN: Human Factors and Ergonomics Society, October 24–28, 1994.

Kellman, P. J., Massey, C., Roth, Z., Burke, T., Zucker, J., Saw, A., et al. (2008). Perceptual learning and the technology of expertise: Studies in fraction learning and algebra. *Pragmatics & Cognition, Special Issue on Cognition and Technology*, 16(2), 356–405.

Kundel, H., & Nodine, C. F. (1975). Interpreting chest radiographs without visual search. *Radiology*, 116, 527–532.

Lehrer, R., Jaslow, L., & Curtis, C. (2003). Developing an understanding of measurement in the elementary grades. In D. Clements & G. Bright (Eds.), *Learning and teaching measurement*. Reston, VA: NCTM.

Marr, D. (1982). *Vision*. San Francisco: W. H. Freeman.

National Center for Education Statistics (2008). *The Nation's Report Card*. From http://nces.ed.gov/nationsreportcard/mathematics/

National Mathematics Advisory Panel (2008). *Foundations for success: The final report of the National Mathematics Advisory Panel*. Washington, DC: U.S. Department of Education.

Schmidt, R. A., & Bjork, R. A. (1992). New conceptualizations of practice: Common principles in three paradigms suggest new concepts for training. *Psychological Science*, 3, 207–217.

Schneider, W., & Shiffrin, R. M. (1977). Controlled and automatic human information processing: I. Detection, search, and attention. *Psychological Review*, 84(1), 1–66.

Schwartz, J. (1988). Intensive quantity and referent transforming arithmetic operations. In J. Hiebert & M. Behr (Eds.), *Number concepts and operations in the middle grades*. Reston, VA: NCTM. pp. 41–52.

17 Algebraic Misconceptions

A Test for Teacher (and Researcher) Use for Diagnosing Misconceptions of the Variable

Joan Lucariello and Michele Tine

Learning algebra is the "gatekeeper" to students' future educational and career success (Adelman, 2006; RAND Mathematics Study Panel, 2003; Silver, 1997; U.S. Department of Education, 1999). Given the importance of learning algebra, it is disquieting that its learning proves so challenging. Many students experience difficulty mastering seminal algebraic concepts. For example, a significant proportion of middle-school, high-school, and college students do not have a correct understanding of one of the most foundational elements of algebra, *variables*. Moreover, a substantial number hold significant misconceptions about variables. This situation makes the present goal of developing a teacher-use diagnostic test for variable misconceptions an important one.

Proper understanding of letters as variables includes two components. First, the letter must be interpreted as representing a *varying* quantity (Philipp, 1992) or a range of unspecified values (Kieran, 1992). This is known as the "multiple values" interpretation of literal symbols (Knuth, Alibali, McNeil, Weinberg, & Stephens, 2005). This interpretation has been studied by presenting sixth, seventh, and eighth graders with the problem: "The following question is about this expression '$2n+3$'. What does the symbol (n) stand for?" (Knuth et al., 2005). Responses were correct if they evidenced "multiple values" answers that expressed the idea that the literal symbol could represent more than one value (e.g., "the symbol is a variable, it can stand for anything"; "a number, it could be 7, 59, or even 363.0285"). About 50 percent of sixth graders, 39 percent of seventh graders, and almost 25 percent of eighth graders gave incorrect responses, indicating that they did not possess the correct multiple-values interpretation of variables.

A correct understanding of variables also entails awareness that some kind of relationship exists between letters, as their value changes in a systematic manner (e.g., as b increases, r decreases) (Kuchemann, 1978). Said differently, a correct interpretation of variables entails knowing that related numbers that change together are "variables" (Philipp, 1992). The "which is larger problem" has been relied on to assess this understanding. For example, Kuchemann (1978) presented 3000 secondary-school students with the following problem: "Which is the larger, $2n$ or $n+2$? Explain." Only 6 percent of students were correct and seemingly aware of a "second-order relationship," that the relation between $2n$ and $n+2$ is actually changing with n. The difference between $2n$ and $n+2$ increases as n increases. For example, when $n=2$, the two expressions are equal; when $n=1$, $n+2>2n$. Knuth et al. (2005) also explored this understanding using the "which is larger" problem and showed that students have difficulty grasping the notion that a relationship exists between letters, as their value changes systematically. Only about 18 percent of sixth graders, just over 50 percent of seventh graders, and just over 60 percent of eighth graders evidenced this understanding.

Hence, many students do not understand variables. In addition, however, and more of a problem is that many middle-school, high-school, and college students harbor several significant misconceptions about variables. Accordingly, the problem is not simply absence of correct knowledge, but holding erroneous concepts about variables.

Five significant misconceptions are prevalent. These were originally investigated by Kuchemann (1978). Kuchemann explored student interpretation of variables in 3000 students in their second through fourth years of secondary schools (aged 13, 14, and 15 years, respectively). Students were presented with a 51-item, half-hour paper-and-pencil test of algebraic problems to solve. Subsequently, many investigators have documented one or another of these five misconceptions. These five misconceptions, along with some associated research, are discussed below (see also Kieran, 1992).

1: Letter Evaluated

The letter is assigned a numerical value from the outset; the numerical value of the letter is directly determined by simple trial and error. There is no step at which the letter has to be handled as an unknown. Kuchemann (1978) presented the following problem "If $u = v + 3$ and $v = 1$, $u =$ __". Fourteen percent gave the wrong answer of 2. In this case, the value of u was directly ascertained by simple trial and error and at no point was u handled as an unknown. Only 61 percent of his 14-year-old students were correct in understanding that the literal symbol (u) could represent "multiple values."

2: Letter Not Considered

The letter is ignored or its existence is acknowledged without giving it a meaning. For example, in the Kuchemann (1978) problem "Add 4 onto $n + 5$", only 68 percent of students were correct ($n + 9$). Twenty percent gave the common wrong answer, 9.

3: Letter as a Label for an Object or an Object Itself

The literal symbol is interpreted as a label for an object or as an object itself. This misconception has been studied extensively (Kuchemann, 1978; Macgregor & Stacey, 1997; Stacey & Macgregor, 1997). The Stacey and Macgregor (1997) research entailed more than 2000 students aged 11–15 years. Students were presented with example problems, including the following on "David's Height": "David is 10 cm taller than Connor. Connor is h cm tall. What can you write for David's height?" The correct answer is $10 + h$, wherein 10 is added to the number or quantity denoted by h. Stacey and Macgregor (1997) compiled a list of letter misinterpretations that were written by students at all levels, including those in their third year of algebra.

The misconception of the *letter as a label associated with the name of an object* was found (e.g., $C+10=D$, where C means "Connor's height" and D means "David's height"). They also noted the *abbreviated word* interpretation, which is another indicator of this misconception (e.g., response of D*h* or U*h* where the abbreviation stands for the words David's height or Unknown height, respectively). The abbreviated word interpretation was typical of most children in a class of 11-year-olds who had never been taught algebra.

In explaining the source of this kind of misinterpretation, Stacey and Macgregor (1997) noted that quantities are frequently denoted by the initial letters of their names, as in "p. 6" meaning "page 6" or "cm" meaning "centimeters." Moreover, like Kuchemann (1978) they note a source in teacher talk. "Teachers talk about m as the 'mass' and t as the 'time taken'; they make statements like 'Let C denote the circumference' and 'We'll use C to stand for the cost' " (Stacey & Macgregor, 1997, p. 111).

Letters are frequently interpreted as objects when a problem involving quantities has to be translated into mathematical language and when a mathematical statement has to be interpreted (Kuchemann, 1978; Rosnick, 1981). Most of the research has been based on the "students and professors" problem, which reads as follows:

> Write an equation, using the variables S and P to represent the following statement: At this university there are six times as many students as professors. Use S for the number of students and P for the number of professors.

Rosnick (1981) found that 37 percent of a group of 150 entering engineering students at the University of Massachusetts were unable to write the correct equation, $S=6P$. The most common error was the reversed equation, $6S=P$. This error was also prevalent for the "students and professors" problem among college students (Clement, Lochhead, & Monk, 1981). The source of the problem, according to Rosnick (1981), is students' belief that S is a label standing for *students*, rather than a variable standing for *number of students.*

4: Letter as Specific Unknown

Letter is thought of as a *specific* unknown number, that is, as an unknown number with a fixed value (which can be operated upon without having to be evaluated). Stacey and Macgregor (1997) identified highly specific modes of this variable misinterpretation in their participants. For example, for the David's Height problem, some students used the *letter to represent its alphabetical value* (e.g., response of 18 because H is the eighth letter of the alphabet, therefore 10 more $\{8+10\}$ is the eighteenth; response of R because the tenth letter after H in the alphabet is R). The *alphabetical value* interpretation (e.g., $a=1$; $b=2$) was traced to a parallel in the early Greek numeration system in which each number was denoted by a letter. It was noted also that this interpretation is often used in common puzzles and secret codes and in some textbooks that rely on alphabetical codes in answer keys for "self-correcting" homework assignments.

Other students relied on the misinterpretation that (unless otherwise specified) *letters equal 1* (e.g., response of 11 because $10+h/1=11$). Still other students seemed to think that *letters represent an arbitrary but reasonable value of the quantity described in the problem* (e.g., response of 160 wherein the student has generated a reasonable height for Connor, such as 150 cm, and added 10). I would term this misconception a *real world knowledge* interpretation of variables.

5: Letter as Generalized Number

The letter is taken to represent multiple values, but it is only necessary to think of the letter taking on these values one at a time. For example, in the Kuchemann (1978) problem "What can you say about c if $c+d=10$ and c is less than d?" the most common answer (39 percent) was just a single value for c, usually $c=4$. Only 30 percent gave the correct answer $c<5$ or a systematic list (e.g., 1, 2, 3, 4).

The documentation of student misconceptions of a variable impels the development of a teacher-use diagnostic test for variable misconceptions. Moreover, misconceptions impede learning, making development of a diagnostic test even more pressing. Misconceptions interfere with or distort assimilation of inputs. Furthermore, misconceptions can be entrenched and very resistant to change (Brewer and Chinn, 1991; McNeil & Alibali, 2005). Their grip on reasoning is explained in part by the fact that overcoming misconceptions is a matter of radically reorganizing or replacing student knowledge. Conceptual change or accommodation is necessary for learning to occur (Carey, 1985, 1986; Posner, Strike, Hewson, & Gertzog, 1982; Strike & Posner, 1985, 1992). The need for conceptual change is yet another reason why teachers could benefit from the availability of a diagnostic test. Traditional methods of instruction, such as lectures, labs, discovery learning, or reading text, although effective at achieving conceptual *growth*, are generally ineffective at accomplishing conceptual change (Chinn & Brewer, 1993; Kikas, 1998; Lee, Eichinger, Anderson, Berkheimer, & Blakeslee, 1993; Smith, Maclin, Grosslight, & Davis, 1997). Conceptual change requires specific instructional strategies, such as raising student metacognition and creating cognitive conflict. See Mayer (2008) and Lucariello (in preparation) for a review of these strategies. Accordingly, teachers need to know when such instructional strategies are called for.

For all these reasons, an instrument to enable teachers to diagnose misconceptions would be of great utility. This research develops such an instrument.

Method

Participants

This study included two pools of participants. Pool 1 had 243 participants and Pool 2 had 234 participants. Pool 1 had 104 females and 139 males. Pool 2 had 95 females and 139 males. Subjects ranged from grades 6 to 12. The majority of students were in grades 8 and 9.

Student participants came from classrooms wherein the teacher had enrolled the class to participate in this on-line study of student algebraic reasoning. Teachers were recruited nationally from advertisements with various teacher-membership associations. In return for participation, teachers received feedback information on their students' performance on the algebra ability test (see below).

Data were collected on-line during intervals of the school day designated by the teachers.

Measures

The following measures were administered on-line in the order listed. Once students completed a measure, they no longer had access to the measure.

1. *Student background*. Demographic and background information on students was collected.
2. *Standardized algebra ability*. This test consisted of 20 multiple-choice questions culled from different administration dates of the grade 10 Mathematics Massachusetts Comprehensive Assessment System (MCAS) exam.
3. *Misconception diagnostic test*. This test consisted of 12 multiple-choice items. Items were multiple-choice questions with four answer options: misconception, correct, and two foils. The test was designed to assess four types of misconceptions about variables. Hence, there were three items for each of these types. Below are the test questions for each type of misconception that yielded the highest percentage of misconception responses (for both pools of students).

LETTER EVALUATED (ASSIGNED ONE NUMERICAL VALUE FROM OUTSET)

If $4x+4y=12$, then what is the value of $x+y$?

A 3 (Correct)
B 4
C Cannot be determined. (Misconception)
D 12

LETTER NOT CONSIDERED (IGNORED)

n is a whole number greater than 0 and less than 5. How many values of $3n$ can there be?

A 0
B 3 (Misconception)
C 4 (Correct)
D 5

LETTER INTERPRETED AS A LABEL FOR AN OBJECT OR OBJECT ITSELF

At a university, there are six times as many students as professors. This fact is represented by the equation $S=6P$. In this equation, what does the letter S stand for?

A number of students (Correct)
B professors
C students (Misconception)
D none of the above

Table 17.1 Number of Hummingbirds and Feeders

Number of Feeders (f)	*Number of Hummingbirds (h)*
1	3
2	5

LETTER CONSIDERED A SPECIFIC UNKNOWN NUMBER

Rita put some hummingbird feeders in her backyard. Table 17.1 shows the number of hummingbirds that Rita saw compared to the number of feeders.

Which equation best describes the relationship between h, the number of hummingbirds, and f, the number of feeders?

A $h = 11f$
B $h = 2f + 1$ (Correct)
C $h = f + 2$ (Misconception)
D $h = f + 6$

"Isomorph Near Transfer" Problem Set

Twelve near transfer isomorph problems were constructed, each to match a corresponding item in the diagnostic problem set. Each isomorph problem varied from its match in the original 12 problems by a change only in the actual letters and numbers used. (See sample below.) The isomorph problem set was administered to Pool 1 participants only.

ORIGINAL LEVEL 2 MISCONCEPTION (LETTER NOT CONSIDERED/IGNORED) PROBLEM

Simplify $3m + 5 - 2m + 1$:

A 7 (Misconception)
B 10
C $m + 6$ (Correct)
D $7m + 8$

ISOMORPH (NEAR TRANSFER) PROBLEM

Simplify $4p + 2 - 3p + 7$

A 10 (Misconception)
B 12
C $p + 9$ (Correct)
D $10p + 8$

"Very Far Transfer" Problem Set

Four very far transfer problems were constructed, one each to correspond to the four kinds/levels of variable misconceptions. Each problem was constructed as a variant of one of the three problems at that level. The very far transfer problems varied from their match in the original 12 problems by changes in the sign (positive/negative) and arithmetic operation. Both pools of participants received all four "very far transfer" problems.

"VERY FAR TRANSFER" PROBLEM FOR LEVEL 2 PROBLEM CITED ABOVE

Simplify $-5d-9--6d-2$

A -10 (Misconception)
B -12
C $d-11$ (Correct)
D $-10d-14$

Coding on Diagnostic Test

Performance on the 12-item diagnostic test was used to differentiate students into three student groups:

Knowers—80 percent + Correct responses (10 Correct; 2 Incorrect)
Students under 80 percent correct (at least three and possibly as many as 10 incorrect responses):
Misconceivers—four or more Misconception error responses
Mistakers—three or fewer Misconception error responses and/or presence of other Incorrect (nonmisconception) error responses

Results

Reliability of Diagnostic Test

The reliability of the 12-item diagnostic test was computed, for both Pool 1 and 2 participants, on the basis of correct responding. The Cronbach alpha score for Pool 1 was .77 and for Pool 2 was .78.

Performance on Diagnostic Test: Discrimination of Three Student Groups

The diagnostic test reliably and comparably (across Pools 1 and 2) discriminated students into the three student groups of *Knowers*, *Misconceivers*, and *Mistakers* (see Figure 17.1). The distribution of students into the three groups reliably differs from the expected distribution for Pool 1 [χ^2 (2, $n = 243$) = 13.14, $p < .001$] and Pool 2 [χ^2 (2,

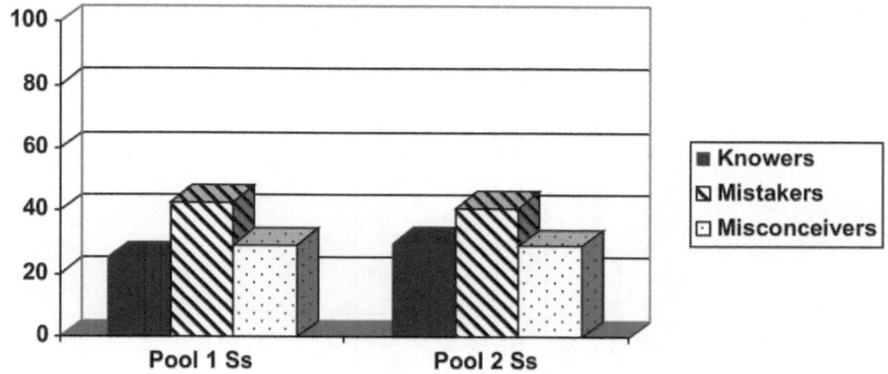

Figure 17.1 Percentage of Students by Group.

$n = 234) = 6.26$, $p < .05$]. For both pools, a significant percentage, about 30 percent, of students were classifiable as Misconceivers. Approximately another 30 percent were found to be Knowers. Just over 40 percent (44 percent Pool 1; 41 percent Pool 2) were classified as Mistakers.

Response-Kinds by Student Group

The mean number of response-kinds on the diagnostic test by student group for Pool 1 and Pool 2 participants was calculated (see Table 17.2). A MANOVA with group (knower, misconceiver, mistaker) as the between-subject factor and grade co-varied was run on the three response types (correct, incorrect, misconception). A main effect for group was found for all three response-types for both pools: *misconception* responses $F(2, 238) = 380.48$, $p < .001$ (Pool 1) and $F(2, 230) = 297.19$, $p < .001$ (Pool 2); *correct* responses $F(2, 238) = 138.66$, $p < .001$ (Pool 1) and $F(2, 230) = 171.33$, $p < .001$ (Pool 2); *incorrect* responses $F(2, 238) = 45.07$, $p < .001$ (Pool 1) and $F(2, 230) = 47.74$, $p < .001$ (Pool 2).

Pairwise comparisons were conducted with LSD (least significant difference) post-hoc test. Knowers correctly answered more questions than Misconceivers ($p < .001$, Pools 1 and 2) and Mistakers ($p < .001$, Pools 1 and 2). Knowers (both Pools 1 and 2) were correct on approximately 10.5 items of the 12 and showed a mean of only one misconception response across the 12 questions. Mistakers also answered more questions correctly than did Misconceivers ($p < .001$, Pools 1 and 2). Misconceivers answered significantly more questions with misconception responses (mean of 4.5 and 4.7 of 12 questions, Pools 1 and 2 respectively) than did Knowers ($p < .001$) and Mistakers ($p < .001$). Knowers also answered fewer questions with misconception responses than did Mistakers ($p < .001$, Pools 1 and 2). Mistakers answered more questions with incorrect responses than did Knowers ($p < .001$, both Pools 1 and 2) and Misconceivers ($p < .05$, Pool 1; $p < .06$, Pool 2). Misconceivers also had more incorrect responses than did Knowers ($p < .001$, Pools 1 and 2).

Table 17.2 Mean Number (of 12) of Response-Kinds by Student Group

	Knowers				*Mistakers*				*Misconceivers*			
	Pool 1		*Pool 2*		*Pool 1*		*Pool 2*		*Pool 1*		*Pool 2*	
Response Kind	*Mean*	*SD*	*Mean*	*SD*	*Mean*	*SD*	*Mean*	*SD*	*Mean*	*SD*	*Mean*	*SD*
Correct	10.35	.25	10.50	.23	6.39	.19	6.45	.19	4.71	.23	4.57	.23
Incorrect	.43	.24	.39	.23	3.29	.18	3.29	.19	2.67	.22	2.72	.23
Misconception	1.01	.10	1.11	.10	2.31	.07	2.26	.09	4.53	.09	4.65	.10

Misconception Responses by Student Group

The kinds of misconceptions exhibited by Misconceivers (Pools 1 and 2) on the diagnostic test were examined. The mean proportion of misconception responses by level of misconception was calculated (number of misconception responses across the three questions per level was tallied with the total and then divided by 3). These data are presented in Figure 17.2. Misconceivers showed all four levels of misconceptions. Chi-squared analyses revealed that the distribution of Misconceivers for each level differed significantly from the expected/chance distribution for both pools of participants (except for Pool 1 Level 1). Pool 1: Level 2 χ^2 (2, $n=73$) = 45.63, $p<.001$; Level 3 χ^2 (2, $n=73$) = 28.43, $p<.001$; Level 4 χ^2 (2, $n=73$) = 16.80, $p<.001$. Pool 2: Level 1 χ^2 (2, $n=68$) = 27.65, $p<.001$; Level 2 χ^2 (2, $n=68$) = 11.85, $p<.01$; Level 3 χ^2 (2, $n=68$) = 38.24, $p<.001$; Level 4 χ^2 (2, $n=68$) = 42.47, $p<.001$.

For both pools of subjects, paired *t*-tests were run to determine if misconception responding was higher for some levels of misconceptions than for others. Level 3 misconceptions were the most prevalent in both pools of participants. For Pool 1, misconception responding was significantly higher for Level 3 than for Level 1 [$t(72)=-3.21$, $p<.01$], Level 2 [$t(72)=-5.30$, $p<.001$], and Level 4 [$t(72)=3.15$, $p<.01$] and for Pool 2 Level 3 responding was higher than that of Level 2 [$t(67)=-3.64$, $p<.001$]. Level 4 was the next highest level of misconception responding for both pools. For both pools, Level 4 responding was significantly higher than Level 2, [Pool 1 $t(72)=-2.41$, $p<.015$, Pool 2 $t(67)=-3.18$, $p<.01$], but not significantly different from Levels 1 and 3. Level 2 misconception responding represented the lowest level of misconception responding for both pools. For Pool 2, misconception responding was lower for Level 2 than Level 1, $t(67)=-2.07$, $p<.05$.

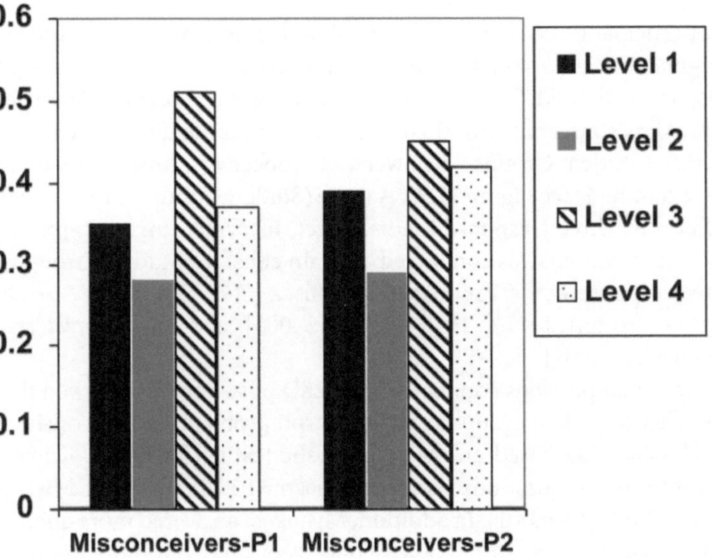

Figure 17.2 Proportion of Misconception Responses by Level for Misconceivers in Pool 1 and Pool 2.

The distribution of Misconceivers by grade level was also examined. For Pool 1, Misconceivers were concentrated in and comparably distributed across grades 8 (27 percent), 9 (26 percent), and 10 (23 percent), with another 14 percent in grade 7. Only 10 percent of Misconceivers fell into grades 11 and 12. Similarly, for Pool 2, Misconceivers were concentrated in grades 8 (18 percent), 9 (34 percent), and 10 (18 percent), particularly in grade 9. Grade 11 had 13 percent of the Misconceivers with another 12 percent in grades 6 and 7. Only 6 percent were in grades 12 and 13. For both pools, the distribution of Misconceivers significantly differed from the chance/expected distribution for Pool 1 [χ^2 (5, $n=73$)$=23.90$, $p<.001$]; and Pool 2 [χ^2 (7, $n=68$)$=42.35$, $p<.001$].

The distribution of Misconceivers by gender was also examined. Both sexes were well represented in the Misconceiver group. In Pool 1, Misconceivers were roughly evenly split across females (52 percent) and males (48 percent). In Pool 2, males accounted for 65 percent of misconceivers and females accounted for 35 percent. Summing across Pools, 44 percent of Misconceivers were female and 56 percent were male.

Construct Validity of Diagnostic Test: Student Group Performance on Other Measures

Student Groups performed differently on the diagnostic test. If these Groups are distinct groups, they should also perform differently on other measures of their algebraic reasoning. Group performance was analyzed on generalization of response measures (near and very far transfer of responses) and on a measure of algebraic ability.

Student Group Performance on 12 Isomorph (Near Transfer) Questions

Pool 1 participants received 12 additional problems, which represented isomorphs to the original set of items on the diagnostic test. If the grouping of students based on the diagnostic test is valid, Student Groups should perform on the isomorph (near transfer) problems consistent with their classification on the diagnostic test. Performance of the three Student Groups—Knowers, Misconceivers, Mistakers—on the isomorph set was examined (see Figure 17.3). A three (Student Group: Knowers, Misconceivers, Mistakers) by three (response kind: correct, incorrect, misconception) MANOVA, with grade co-varied, was conducted. A main effect for Student Group was found for all three response kinds: for *correct* responses [$F(2, 238)=89.85$, $p<.001$]; for *misconception* responses [$F(2, 238)=41.17$, $p<.001$]; and for *incorrect* responses [$F(2, 238)=40.58$, $p<.001$].

Pairwise comparisons conducted with LSD post-hoc tests showed that the performance of each student group on the isomorph problem set was consistent with their group classification based on their diagnostic test performance. Knowers answered significantly more questions correctly (mean$=10.46$) than did Misconceivers and Mistakers (both at $p<.001$). In addition, Mistakers answered more questions correctly than did Misconceivers ($p<.05$). Misconceivers answered more questions with misconception answers than did Knowers ($p<.001$) and Mistakers ($p<.05$). In addition, Mistakers chose more misconception answers than Knowers ($p<.001$). Mistakers answered more questions incorrectly than Knowers ($p<.001$) and Misconceivers as

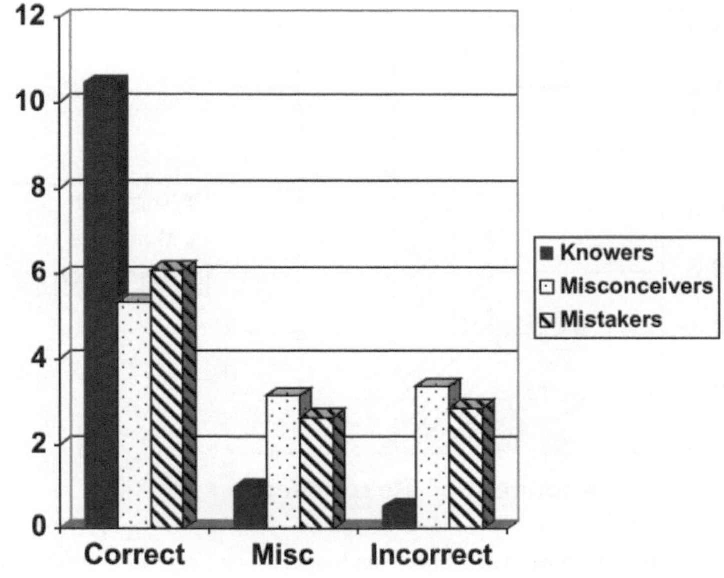

Figure 17.3 Near Transfer Performance on Isomorph Problem Set by Student Group and Response Kind.

well ($p < .08$), albeit at a marginally significant level. Misconceivers also had more incorrect responses than did Knowers ($p < .001$).

Student Group Performance on Very Far Transfer Problems

All Pool 2 participants received four very far transfer problems. A univariate ANOVA with Student Group (Knowers, Misconceivers, Mistakers) as the between-subjects factor, and grade as the covariate was conducted on the number of questions (out of four) answered with the misconception answer choice (see Figure 17.4). A main effect for group was found [$F(2,223) = 20.57$, $p < .001$]. Pairwise comparisons with LSD post-hoc tests revealed that Misconceivers answered more of the "very far transfer" questions with misconception responses than did Knowers and Mistakers ($p < .001$). Hence, Misconceivers are more likely than the other student groups to generalize misconception responding. In addition, Mistakers answered more very far transfer questions with misconception responses than did Knowers ($p < .001$).

Student Group Performance on Algebra Ability Test

Students' proportion of correct responses on 19 of the 20 ability test questions was calculated (one item was omitted owing to administration problems). The reliability of the 19 test items (based on correct responses) on the algebra ability test was calculated. For Pool 1 Cronbach's alpha = .72 and for Pool 2 Cronbach's alpha = .75.

The performance of the three student groups on the algebra ability test was calculated for each pool (see Figure 17.5). Univariate ANOVAs on the proportion of correct responses on the algebra ability test scores by Student Group (Knowers, Mistakers,

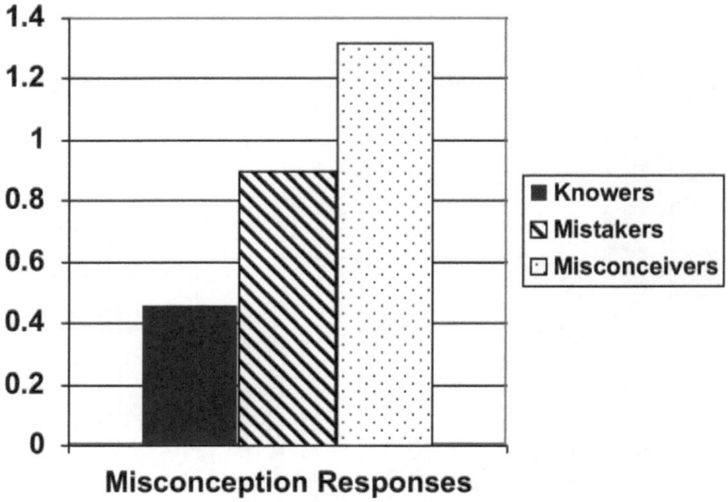

Figure 17.4 Mean Number of Misconception Responses on Very Far Transfer Problems by Student Group (Pool 2 Students).

Misconceivers), with grade co-varied, showed a main effect for group [Pool 1: $F(3,196) = 26.50$, $p < .001$; Pool 2: $F(3,193) = 34.04$, $p < .001$]. Post-hoc tests revealed that Knowers performed better on the algebra ability test than Misconceivers and Mistakers (Pool 1 and Pool 2: $p < .001$). Moreover, Misconceivers performed more poorly than Mistakers (Pool 1 and Pool 2: $p < .05$).

To explore the contribution of student group to algebra ability test performance, hierarchical, separate entry regression analyses were conducted (for both Pools) with proportion of correct responses on the algebra test as the dependent variable. Grade was entered in the first step, followed by student group (see Table 17.3). Student Group explained an additional 15 percent and 18 percent of the variance in ability test scores over and above grade, respectively for Pool 1 and 2 participants [Pool 1: $F(1, 197) = 37.21$, $p < .001$, Pool 2: $F(1, 194) = 47.77$, $p < .001$].

Discussion

The aim of the present research was to develop a diagnostic test for teacher use that would provide teachers with information on the kinds of variable misconceptions that impede their students in learning algebra. A diagnostic test was successfully developed.

The 12-item multiple-choice instrument exhibited excellent reliability as a scale. Moreover, it differentiated students into three distinct groups: *Misconceivers*, *Knowers*, and *Mistakers*. Misconceivers were found to represent a significant proportion of students, approximately 30 percent. Furthermore, students classified into these three distinct groups, on the basis of the diagnostic test, performed differently from each other on other measures of their algebraic reasoning and consistent with their classification on the diagnostic test.

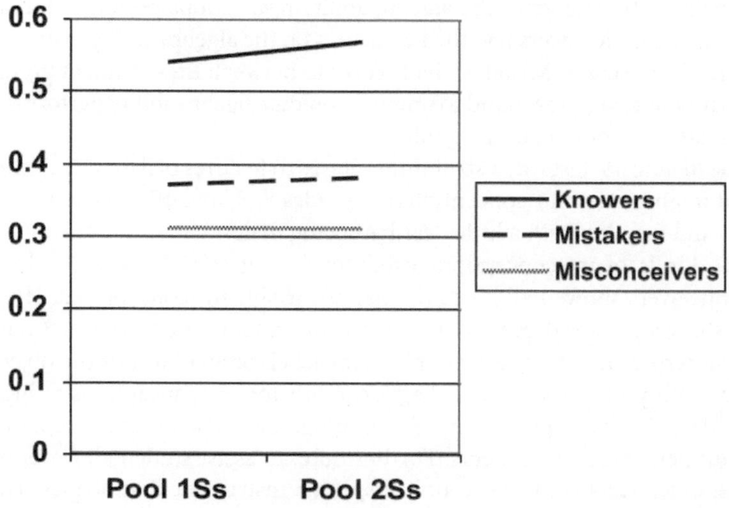

Figure 17.5 Performance on Algebra Ability Test by Student Group.

Table 17.3 Hierarchical Regression Analysis of Variables Predicting the Proportion of Correct Responses on Algebra Ability Test (Pool 1 and Pool 2 Students)

Variable	ΔR^2	B(*SE*)	β	R^2
Pool 1				
Step 1: Grade[a]	.05***	–.02(.01)	–.15*	.05
Step 2: Student Group	.15***	–.06(.01)	–.42***	.20
Pool 2				
Step 1: Grade[a]	.08***	–.02(.01)	–.15*	.08
Step 2: Student Group	.18***	–.08(.01)	–.45***	.26

Betas are for final step in the model.
[a]Grade is 7–12.
*$p < .05$
***$p < .001$

With respect to their generalization behavior, on the set of isomorph (near transfer) problems, student groups performed consistently with their diagnostic category. Knowers answered significantly more questions correctly than any other group. Misconceivers answered more questions with misconception answers than any other group. Finally, Mistakers answered more questions with incorrect (non-misconception variety) responses than did Knowers and with a tendency to do so relative to Misconceivers. Also with respect to generalization, the three student groups performed consonant with their diagnosis on the "very far transfer" problems. Misconceivers were more likely than any other group to select misconception responses to these problems, whose structure was most divergent from the original diagnostic problems.

Finally, with respect to the algebra ability test, groups performed in accord with their diagnosis. Knowers had the best scores on the algebra ability test; Misconceivers had the worst scores. Mistakers had scores in between these student groups. In addition, student group was found to uniquely predict algebra ability performance beyond the variance accounted for by grade.

The diagnostic test was also illuminating as to features of Misconceivers. They were found in all grades but concentrated in grades 8, 9, and 10. Misconceivers included males and females, with a little over half being males.

The kinds of misconceptions exhibited by Misconceivers were also examined. Misconceivers showed all kinds (levels) of variable misconceptions. They did, however, show a greater degree of misconception responding for Level 3 questions that entail interpreting the letter/variable as the label for an object or the object itself.

The utility of a teacher-use diagnostic test for misconceptions is highlighted in these data. Misconceptions impede learning, and Misconceivers represented a significant percentage of students. Furthermore, to assist students in overcoming misconceptions, teachers need to utilize specific instructional strategies. However, the tests used by most teachers and state-mandated testing programs provide little new information that teachers can use to modify their instructional practices to increase the learning of individual students. A shortcoming in testing's current technology stems from a narrow focus on estimating student ability within a given domain, by examining the percentage of items to which a student responds correctly. Such assessments provide information largely redundant with what teachers already know about their students' knowledge and little insight into *why* students may not be performing well within a content area. A diagnostic test provides teachers with this information, which then can be used to adapt instruction.

The diagnostic test developed can also be used in research. For example, the diagnostic test can be used to categorize experimental subjects into misconceiver and nonmisconceiver groups and study behavior across them. It can also be used to test the effectiveness of various interventions by using the diagnostic test (and isomorph problem set) as pre- and/or post-test measures. Hence the diagnostic test is multipurpose, having both classroom and laboratory/scientific uses.

Acknowledgement

This research was supported by Grant No. R305H040099 from the U.S. Department of Education, Institute for Education Science, awarded to the first author.

References

Adelman, C. (2006). *The toolbox revisited: Paths to degree completion from high school through college*. Washington, DC: U.S. Department of Education.

Brewer, W. F., & Chinn, C. A. (1991). Entrenched beliefs, inconsistent information, and knowledge change. In L. Birnbaum (Ed.), *The international conference of the learning sciences: Proceedings of the 1991 conference*. Charlottesville: Association for the Advancement of Computing in Education. pp. 67–73.

Carey, S. (1985). *Conceptual change in childhood*. Cambridge, MA: MIT Press.

Carey, S. (1986). Cognitive science and science education. *American Psychologist*, 41, 1123–1130.

Chinn, C. A., & Brewer, W. F. (1993). The role of anomalous data in knowledge acquisition: A theoretical framework and implications for science instruction. *Review of Educational Research*, 63, 1–49.

Clement, J., Lochhead, J., & Monk, G. (1981). Translation difficulties in learning mathematics. *American Mathematical Monthly*, 88, 286–290.

Kieran, C. (1992). The learning and teaching of school algebra. In D. Grouws (Ed.), *Handbook of research on mathematics teaching and learning*. New York: Macmillan Publishing Co. pp. 390–419.

Kikas, E. (1998). The impact of teaching on students' definitions and explanations of astronomical phenomena. *Learning and Instruction*, 8, 439–454.

Kuchemann, D. (1978). Children's understanding of numerical variables. *Mathematics in School*, 7, 23–26.

Knuth, E. J., Alibali, M. W., McNeil, N. M., Weinberg, A., & Stephens, A. C. (2005). Middle school students' understanding of core algebraic concepts: Equivalence & variable. *Zentralblatt fur Didaktik der Mathematik (International Reviews on Mathematical Education)*, 37, 1–9.

Lee, O., Eichinger, D. C., Anderson, C. W., Berkheimer, G. D., & Blakeslee, T. D. (1993). Changing middle school students' conceptions of matter and molecules. *Journal of Research in Science Teaching*, 30, 249–270.

Lucariello, J. (in preparation). Instructional strategies for addressing student misconceptions: Achieving conceptual change. Online. Available at: http://www.apa.org/education/k12/curricular-materials.aspx [How do I Get my Students Over Their Alternative Conceptions (Misconceptions) for Learning?].

Mayer, R. E. (2008). *Learning and instruction*. Upper Saddle River, NJ: Pearson.

Macgregor, M., & Stacey, K. (1997). Students' understanding of algebraic notation: 11–15. *Educational Studies in Mathematics*, 33, 1–19.

McNeil, N. M., & Alibali, M. W. (2005). Why won't you change your mind? Knowledge of operational patterns hinders learning and performance on equations. *Child Development*, 76, 883–899.

Philipp, R. A. (1992). A study of algebraic variables: Beyond the student–professor problem. *Journal of Mathematical Behavior*, 11, 161–176.

Posner, G. J., Strike, K. A., Hewson, P. W., & Gertzog, W. A. (1982). Accommodation of a scientific conception: Toward a theory of conceptual change. *Science Education*, 66, 211–227.

RAND Mathematics Study Panel (2003). *Mathematical proficiency for all students: Toward a strategic research and development program in mathematics education*. Santa Monica, CA: RAND Mathematics Study Panel.

Rosnick, P. (1981). Some misconceptions concerning the concept of variable. *Mathematics Teacher*, 74, 418–420.

Silver, E. A. (1997). Algebra for all—Increasing students' access to algebraic ideas, not just algebra courses. *Mathematics Teaching in the Middle School*, 2(4), 204–207.

Smith, C., Maclin, D., Grosslight, L., & Davis, H. (1997). Teaching for understanding: A study of students' pre-instruction theories of matter and a comparison of the effectiveness of two approaches to teaching about matter and density. *Cognition and Instruction*, 15, 317–393.

Stacey, K., & Macgregor, M. (1997). Ideas about symbolism that students bring to algebra. *The Mathematics Teacher*, 90, 110–113.

Strike, K. A., & Posner, G. J. (1985). A conceptual change view of learning and understanding. In L. H. T. West & A. L. Pines (Eds.), *Cognitive structure and conceptual change*. Orlando: Academic Press. pp. 211–231.

Strike, K. A., & Posner, G. J. (1992). A revisionist theory of conceptual change. In R. Duschl & R. Hamilton (Eds.), *Philosophy of science, cognitive psychology, and educational theory and practice*. Albany: SUNY. pp. 147–176.
U.S. Department of Education (1999). Do gatekeeper courses expand education opportunities? White paper prepared for U.S. Secretary of Education Richard W. Riley. Washington, DC.

18 Towards Instructional Design for Grounded Mathematics Learning

The Case of the Binomial

Dor Abrahamson

The Problem

Consider a penny. It is flipped four times. Now consider two possible outcomes of this experiment:

(a) Heads, Heads, Heads, Heads.
(b) Heads, Heads, Tails, Tails.

Is one of these two outcomes more likely than the other, or are they equally likely?

This item targets basic knowledge of probability. Namely, it aims to evoke the phenomenon of a random generator (e.g., coins, dice, spinners, etc.) as a context for eliciting and gauging an understanding of randomness, independence, and distribution. Solving this item does not demand any numerical reasoning or arithmetical calculation—one need only grasp the logic of the situation so as to determine the appropriate response. We might thus hope that graduates of the U.S. school system, who have studied at least basic probability concepts, fare well on this simple item. But do they?

According to probability theory, this penny-flipping situation describes a *compound-event* problem, because each trial is defined as a *conjunction* of two or more *singleton* events. Here, each trial is a conjunction of four singleton coin flips. Also, this is a *binomial* situation, because each singleton-event variable—that is, each coin flip—yields one of exactly two possible values (Heads or Tails). In this particular compound-event binomial problem, the two hypothetical experimental outcomes, HHHH and HHTT, are equiprobable, because the conjunction consists of a sequence of *independent events*. That is, the coin has no memory, so each of the four singleton flips is not affected by the result of a preceding flip nor does it affect the result of a subsequent flip. And yet an overwhelming proportion of the adult population chooses option (b), HHTT, as more likely, arguing that it appears to better capture the structure and function of the random generator, that is, its equal numbers of Heads and Tails better represent the essence of a two-sided coin (Tversky & Kahneman, 1974). These findings have been robustly replicated in numerous studies (Jones, Langrall, & Mooney, 2007). Moreover, there is reason to believe that the findings reflect early or even innate reasoning mechanisms. For example, Xu and Vashti (2008) showed eight-month-olds a tub full of balls of two colors from which there issued into a narrow tube a sample of several balls that were thus arranged sequentially. The infants' reactions to different types of samples suggested that they found more interesting those samples whose ratio composition was less representative of the population; for example, a sample of four green balls was more interesting than a sample of two green balls and two blue balls, irrespective of their order.[1]

What are the pedagogical implications of these empirical findings of flawed probabilistic reasoning across the ages? Should we conclude that the subject matter of probability is inherently counterintuitive and that therefore students can at best develop basic fluency in the rote execution of probability solution procedures?

Overview of the Chapter

In the following, I will attempt to cast a more optimistic light on the prospects of a grounded probability education, in which students can make sense of fundamental concepts such as the binomial. I begin by building the following case: (1) There are at least two ways of interpreting inscribed strings such as HHTT; (2) the natural human inclination is to ignore the order of singleton events in conjunctions, for example, to construe a comparison of HHHH and HHTT as though it were a comparison of "four Heads" (hence, 4H) and "two Heads and two Tails in any order" (hence, 2H2T); (3) ignoring the order is the source of participants' canonical error on this item; (4) the inclination to ignore the order results from a tacit perceptual constraint imposed naturally by innate or very early cognitive mechanisms; and (5) once we appreciate that respondents are construing orderless compound events, we may acknowledge the mathematical truth of stating that HHTT is more likely than HHHH (see below).

Yet where would such an exoneration of human intuition leave us in terms of instructional goals? Can students advance in their mathematical learning of probability if they persist to ignore the dimension of order within compound events? Probably not. Attending to the order of singleton events within compound events is absolutely vital for performing *combinatorial analysis*, a key classicist-probability procedure used for establishing anticipated frequencies of random events. Combinatorial analysis is performed by listing all the possible outcomes—that is, building the *sample space* (event space, probability space) of the experiment—and calculating the proportion of *favorable events* out of the entire sample space. For example, there are 16 different ways of arranging four singleton binomial events—TTTT, TTTH, TTHT, THTT, HTTT, TTHH, THTH, HTTH, THHT, HTHT, HHTT, THHH, HTHH, HHTH, HHHT, HHHH (hence, the *elemental events*)—and six of these satisfy the criterion of having two Heads and two Tails in any order (hence, the *aggregate event* 2H2T). Therefore, there is a 6/16 probability (.375) that an experimental trial will result in 2H2T. However, if students are inclined to aggregate the singletons of compound-event outcomes, how are they to learn the notions of combinatorial analysis and sample space?

This juxtaposition of naïve and professional *orientations of view* (Stevens & Hall, 1998) toward compound-event outcomes—as either aggregate or elemental events—suggests that pedagogical efforts could focus on helping students see mathematical signs such as HHTT in the culturally intended manner. At the same time, the *constructivist* educational perspective (von Glasersfeld, 1987) warns us against forsaking students' natural ways of constructing the world, lest they fail to make sense of mathematical concepts. The current chapter addresses this pedagogical tension between tacit and cultural constructions of mathematical inscriptions. I describe the instructional design for the binomial that we have created in light of this research problem as well as preliminary empirical work for evaluating the potential of our design to foster a deep understanding of the binomial. In turn, I use the empirical results to comment on issues of mathematical epistemology, cognition, and pedagogy.

Background and Motivation: Toward Bridging Tacit and Mathematical Views

A broadly accepted fundamental hypothesis of cognitive development researchers working on the origins and nature of mathematical reasoning is that humans posses domain-specific, ecologically adapted cerebral mechanisms for processing quantitative information (Dahaene, 1997; Devlin, 1999; Gelman & Williams, 1998; Gigerenzer & Brighton, 2009). Specifically, the "ecological hypothesis" implies that in some natural contexts, which mathematicians would model as binomial, attending to outcomes as aggregate, rather than elemental events, is advantageous to human survival. Our agenda has been to identify and simulate these contexts, so that we can build instruction that embraces learners' intuitive reasoning yet steers it toward mathematically normative understandings (Abrahamson, 2009a,b; Clement, Brown, & Zeitsman, 1989; Fischbein, 1975; Smith, diSessa, & Roschelle, 1993; Wilensky, 1997). Thus, students may be able to make sense of the notion of probability rather than just memorize formulas with little if any retention, let alone understanding. We have focused on binomial situations such as the coin-flipping scenario because these situations are fundamental to the study of probability (Weisstein, 2006) and their learning is well researched (for a review, see Jones et al., 2007). By working in the reciprocal tracks of scholarship and design, the study potentially contributes to developing theoretical models of learning as well as viable instructional materials.

Our emerging conjecture is that, although humans possess an ecologically powerful cognitive mechanism that could potentially ground the binomial, its compatibility with mathematical notions depends on the inscriptional forms, discursive framings, and pragmatics that encapsulate formal constructions of mathematical information (see Bamberger, 1996; Borovcnik & Bentz, 1991; Gelman & Williams, 1998; Gigerenzer, 1998; diSessa & Wagner, 2005). That is, mathematical semiotic formulations of source phenomena do not necessarily cater to naturalistic constructions of these same phenomena (Bamberger & diSessa, 2003). Further encumbering the adoption of mathematical views, gestalt perception of inscriptions operates temporally a priori to analytic reasoning and is very slow to become entrained (Navon, 1977; Stroop, 1935; Van Dooren, De Bock, Weyers, & Verschaffel, 2004).

We thus face the following pedagogical dilemma. In order to trigger students' tacit knowledge that is relevant to the study of a mathematical concept, we are to create instructional materials aligned with students' ecologically viable perceptions—under such conditions, students' initial judgments are expected to be mathematically correct, if approximate. However, in order to advance from these qualitative judgments to mathematical formulations of the phenomenon, students are to adopt a new, mediated view of the phenomenon, yet such a demand is liable to defeat the entire rationale of recruiting students' tacit knowledge. How, then, might we help students coordinate the tacit and cultural ways of perceiving a certain class of phenomena, so that they can sustain their intuitive views even as they adopt the mathematical formulations inherent to modeling this phenomenon? That is, how might students build meaning for a semiotic artifact, when doing so demands a counterintuitive view of the world?

A Proposed Design for the Binomial

We maintain that students should begin by working with a source phenomenon rather than its semiotic reformulation. Thus, the mathematical validity of students' initial inference regarding quantitative properties of the phenomenon would be objective, not only contingent on interpreting the inference vis-à-vis tacit perceptual construction. Subsequently, students should interact with the mathematical reformulation of the phenomena. Through negotiating tacit and cultural perceptions of these inscriptions, students are to recognize their own perceptual biases. Our agenda has been to engender and then examine students' struggle to coordinate unmediated and mediated perceptions of mathematical situations. We thus wished to articulate epistemological and cognitive aspects of mathematical learning, a developmental process that we perceive as a reconciling of tacit and cultural resources (Abrahamson, 2003, 2004, 2008a,b, 2009a,b; Abrahamson & Cendak, 2006; Schön, 1981; diSessa, Hammer, Sherin, & Kolpakowski, 1991; Wilson, 1998).

Design-Based Research of Grounded Mathematical Learning

Our scholarly focus on developing theory through developing instructional materials has suggested the suitability of the design-based research approach (Brown, 1992; Cobb, Confrey, diSessa, Lehrer, & Schauble, 2003; Collins, 1992; Confrey, 2005; Gravemeijer, 1994; Sandoval & Bell, 2004). In this methodological approach, theory and design are mutually informative through iterated cycles of research and development. This relatively new approach is itself still under development and is thus still fraught with methodological challenges relative to more established research practices of the cognitive and learning sciences represented in this volume (Kelly, 2003, 2004). Yet any methodological shortcomings of this approach appear to trade off favorably with the emergence of new theoretical constructs as well as instructional material with potential educational impact (Barab et al., 2007; Edelson, 2002; diSessa & Cobb, 2004). Accordingly, both the theory and the design discussed in this chapter are in progress, yet these tentative results may nevertheless be of value to theoreticians, designers, and instructors.

We have thus designed and built a set of mixed-media instructional materials for the binomial and interviewed grade 4–6 students as they engaged in activities based on these materials (Abrahamson, 2009b; Abrahamson & Cendak, 2006; Abrahamson, Janusz, & Wilensky, 2006; Abrahamson & Wilensky, 2005a,b).

A Design for the Binomial

The learning materials revolved around a single mathematical object, the *4-block*: a 2-by-2 grid of four location "variables" that each can take on one of two color "values," green or blue (see Figure 18.1). The 4-block is instantiated in both substantive

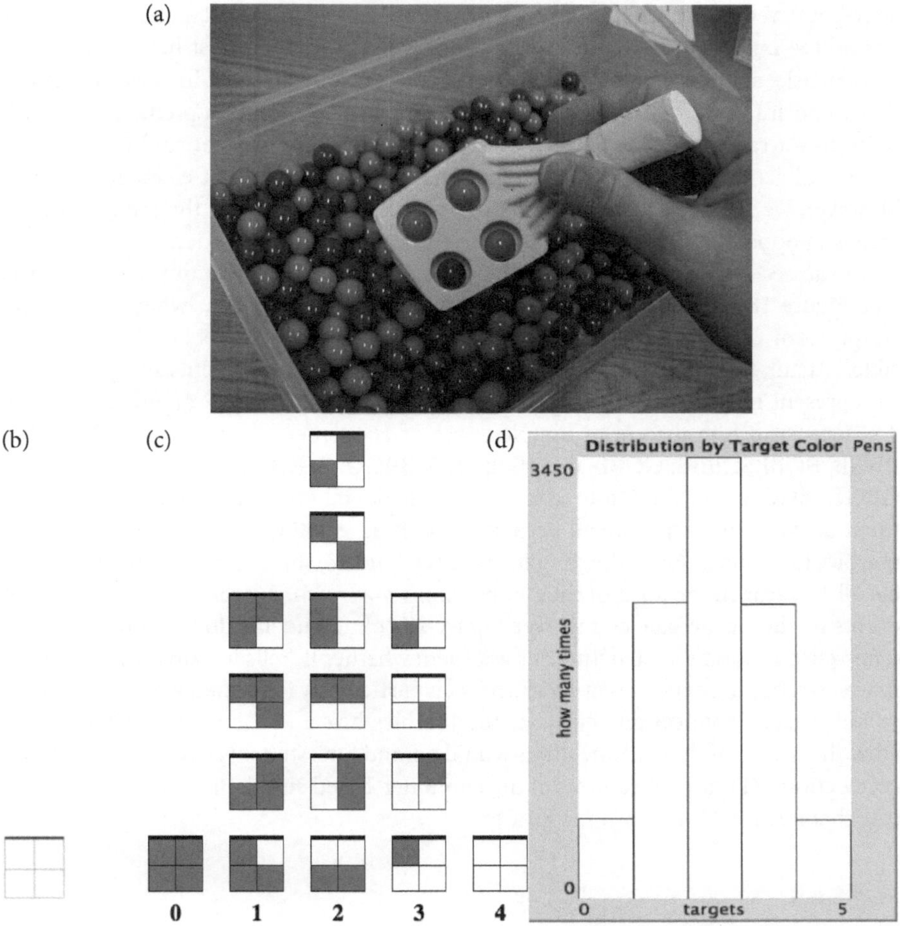

Figure 18.1 (a) The Marble Scooper Bearing a Sample of Four Marbles Drawn from a Box Containing an Equal Number of Green and Blue Marbles. (b) A 4-Block Template for Conducting Combinatorial Analysis of the Experiment. (c) The Combinations Tower Is the Distributed Sample Space, Assembled by Categories into a Pictograph Bar Chart. (d) Histogram from a Computer-Based Simulation of the Marbles Experiment.

and virtual environments and is used in both theoretical probability contexts (figuring out what can be obtained and how often) and empirical probability contexts (running real or simulated experiments while monitoring cumulative results).

In the first activity (see Figure 18.1a), the student is shown a tub full of hundreds of green and blue marbles of equal numbers. An interviewer demonstrates the operation of a 4-block *marbles scooper*, a utensil for drawing out samples of four marbles. Next, the interviewer asks the participant to predict what will happen when we scoop. This question is phrased ambiguously, so as not to specify interpretation along three dimensions: (a) the "probable" (anticipating the outcome distribution) or the "possible" (creating the sample space); (b) the next scoop or the long term; or (c) an aggregate event, for example, 3g1b, or an elemental event, for example, a scoop that has a blue marble in the bottom right-hand corner, with the rest being green. If a student asks about the color composition of the marbles box, the interviewer replies that there are equal numbers of each color. Once students offer a prediction, we ask them to warrant it. We predicted that these students, who had not studied probability, would articulate their expectations in terms of aggregate, not elemental, events. However, we also hoped that the explicit spatial configuration of the scooper would create an opportunity for the dyad to discuss the dimension of order.

In the second activity, students are provided with a pile of empty 4-block cards (see Figure 18.1b) and a pair of crayons and are asked to create "what we can get" (to perform combinatorial analysis). We thus shift from the "probable" to the "possible." Again, the instructions do not specify whether or not students should attempt to represent only the five aggregate events—no green, 1 green, 2 green, 3 green, or 4 green—or create the entire sample space of 16 elemental events—BBBB; BBBG, BBGB, BGBB, GBBB; GGBB, GBGB, BGBG, BBGG, BGGB, GBBG; GGGB, GGBG, GBGG, BGGG; GGGG. (Note that these 16 elemental events are equiprobable, as in the four-coin-flips experiment, because there is an equal number of green and blue marbles in the box.) Once the sample space is complete, students are steered to assemble all 16 cards in the form of the *combinations tower*, which organizes the elemental events by their aggregate counts (see Figure 18.1c). If students do not spontaneously offer reflections on this structure, we ask them whether it "tells us" anything about the marbles-scooping experiment. We have been particularly interested whether students would discern connections between the marbles box and the combinations tower, what the nature of these connections would be, and how students would express these connections. (Figure 18.1d hints at the computer-based simulations not discussed in this chapter; see Abrahamson, 2009a.)

Methods

Participants

Forty-six grade 4–6 students from a private K–8 school in the San Francisco Bay Area (33 percent on financial aid; 10 percent minority students) volunteered to participate in the study. The two grade 4–6 mathematics teachers reported that none of these students had had formal instruction in probability. The teachers collaboratively assigned the pool of volunteering students to three achievement levels ("high," "middle," and "low") within each grade level, on the basis of the students' performance on assessments. From these 3-by-3 grade/performance groups, we selected roughly equal numbers of students, balancing for gender, so that we worked with 28 participants. We preferred selecting students whom the teachers had indicated as typically more disposed to express their thoughts verbally. These more verbose students were distributed roughly uniformly across grade, gender, and achievement level. Although this selection as well as the demographics of the school may create methodological blind spots, this initial stage of the research required articulate feedback from the students as they interacted with the designed learning tools. The research team included the author—an experienced interviewer—and a novice researcher under training.

Materials

The study was conducted in the form of semistructured clinical interviews (Ginsburg, 1997; diSessa, 2007). Therefore, in addition to the learning tools (see above), we created a protocol designed to step the participant through a set of key questions, while leaving room for extemporized follow-up questions, based on student response. The protocol was developed iteratively over several pilot interviews that are not reported in this chapter.

Procedure

Over a span of three weeks, the researchers visited the school a total of nine days so as to complete the set of 28 one-to-one interviews. The interviews took place in quiet rooms within the school facilities and lasted on average 56 minutes (SD = 12 minutes). An audio/video camera recorded all the sessions, and this footage as well as physical and virtual artifacts that the students created and the interviewers' field notes and daily debriefs were all subsequently used for data analysis. At the end of each interview, the interviewer explained to the participant the mathematical formulations of the situations and answered any remaining questions with respect to the content or the interview.

Data Analysis

Audio and video data were studied collaboratively by the design-research team using *microgenetic* qualitative analysis (Schoenfeld, Smith, & Arcavi, 1991; Siegler & Crowley, 1991) as well as *grounded theory* techniques (Glaser & Strauss, 1967). Through the analysis, we attempted to (a) clarify and interpret students' utterances and multimodal actions; (b) understand student difficulty in coordinating aggregate- and elemental-event constructions of the sample space; (c) capture and articulate the nature of prompts that helped students overcome these difficulties; (d) identify connections students were making between the activities; (e) reveal patterns across students; (f) characterize these patterns in terms of our emergent design framework (Abrahamson & Wilensky, 2007); and (g) identify necessary improvement to the design of objects, activities, and facilitation. As patterns in the data were discerned, we used these as lenses to reexamine earlier data, until we were satisfied that we had built a coherent description of typical learning trajectories through this particular design and that all student confusions were identified and defined.

Results and Discussion

In this section, we evaluate, through the lens of our data, the prospects of creating pedagogical contexts that enable students to coordinate tacit and mathematical constructions of binomial situations.

Eliciting Students' Informal Inference for Binomial Outcome Distribution

All but three of the 28 students working with the marbles box and scooper predicted that the most likely 4-block should be 2g2b. Of these three, a single student (grade 5 "middle" girl) did not offer any guess, and two students (grade 4 "high" and "middle" girls) expected the plurality of scoops to have exactly three marbles of either of the two colors—a mathematically correct inference (8/16) that was possibly based on a single sample they had each scooped as they operated the device to learn its mechanism. For the rest of the students, there does not appear to be any definitive relation between the several samples they took and their guesses.

Of the 25 students who initially guessed 2g2b, 23 supported their statement by pointing to the apparently equal distribution of marbles in the bin, which they described as "half–half" or "even," whereas two students initially could not explain their guess. When we pressed students to explain how or why the "half–half"-ness of the marbles bore on the "half–half"-ness of the 4-block, only one student (grade 6 "high" boy) elaborated, albeit he, too, could not explain *why* the half–half distributive property should carry over from the bin to the scooper (see also Abrahamson, 2007). Unsurprisingly, not a single student generated or even attempted to generate the sample space to support the intuitive inference that 2g2b would be the most frequent outcome.

We concluded that, in line with our design rationale, the marbles-scooping random generator creates for the dyad opportunities for discursive exchanges wherein students can articulate their naïve expectation of outcome frequencies in verbal forms that agree unconditionally with probability theory. Yet, as the following section demonstrates, our participants' verbatim judgments were objectively true only because the protocol had not yet required of them to engage semiotic forms wherein order is relevant.

Insight—Appropriating a Mathematical Artifact to Warrant a Tacit Inference

Asked to build the sample space, each of the participants initially created no more than five cards, which included some permutation on each of 4b, 1g3b, 2g2b, 3g1b, and 4g. As they built this aggregate-event-based sample space, the participants did not refer to the particular location of the singleton colorations within the 4-block, and not one participant suggested how this activity might support their earlier guess as to the most likely event in the scooping experiment. The participants' apparent obliviousness to the particular configuration of the green and blue values in the 4-block variable is striking, given that the 4-block cards do not enable the representation of an aggregate event that is objectively orderless. That is, any coloring-in of, say, 3g1b necessarily instantiates an elemental event—one of the four particular permutations on the 3g1b combination. Yet, just as people tend to compare HHHH and HHTT as aggregate events despite the "glaring" order in these inscriptions, so our participants ignored the 4-blocks' spatial configurations.

By and large, it was only when we guided the participants to expand the sample space so as to include all 16 elemental events that the dimension of order was first mentioned. Students appeared confused as to why we wished them to build the additional 11 elemental events. Some students argued that they had already shown us all the possible events, so that the cards they were now asked to create were perforce redundant. It was as though we had asked our participants to measure the diameter of a coin or smell it as a means of anticipating its outcomes—the dimension of permutation was perceived as entirely irrelevant to the task at hand.

Yet, once the sample space was completed and, in any case, once the combinations tower was erected, all but one student—the same student who had not initially inferred that the most likely experimental outcome would be 2g2b—eventually offered the insight that 2g2b is the most likely aggregate event because it contains more elemental events than the others. So doing, the participants used the sample space, a mathematical artifact, to support their tacit inference. Yet how was this key insight achieved? Did this insight signal that the students fully and/or stably understood the binomial? Did they map their five-object aggregate view of likelihood onto the five *sets* of elemental events?

In Abrahamson (2009b), I build on cognitivist, socio-cultural, and educational-semiotic perspectives in an attempt to explain the participants' insight. In particular, I investigate how the participants came to appropriate the combinations tower, which contains five sets of elemental events, as a *semiotic means of objectifying* (Radford, 2003) their presymbolic sense of outcome distribution, which is couched in five aggregate events. So doing, I draw on *conceptual-blending* theory (Fauconnier & Turner, 2002; Hutchins, 2005), *intuitive-rules* studies (Stavy & Tirosh, 1996), Wilson's (1998) theory of creativity as conjoining disparate ideas, and Peirce's (1931–1958) construct of *generative abduction*. I there argue that students' insight can be characterized as a *semiotic leap*.

A semiotic leap is the discursive action of identifying an available artifact as enabling the objectification of a presymbolic multimodal notion for a phenomenon previously experienced, even though the artifact initially does not appear to couch the phenomenon in the appropriate perceptual unit. The leap is nevertheless motivated by a felt need to communicate an inference, such as because of the pragmatics of a clinical interview.

The semiotic leap does not demarcate complete understanding. On the contrary, the learner has yet to engage in abductive and inductive reasoning in order to rationalize the semiotic artifact as a logical discursive tool. Indeed, when we subsequently asked participants to compare the likelihoods of two 4-block cards, one each from the 3g1b set and the 4g set (see Figure 18.2), the participants still demonstrated great difficulty in reconciling the elemental-event perspective, by which the two cards are equiprobable, and the aggregate-event perspective, by which they are heteroprobable. We interpreted this difficulty as marking that the students had not yet adequately recognized their own tendency to shift between elemental and aggregate views of individual 4-blocks: participants thus believed that their inferences were mutually exclusive whereas they were in fact complementary. In a sense, disambiguating the 4-block can be viewed as the pivotal reflective activity in our design. Because in so doing students first appreciate differences between naïve and scientific formulation of compound events and can thus begin to reconcile these formulations.

Conclusions

Learners' explicit commitment to a particular way of seeing the world is essential to the process of building meaning through semiotic objectification (Radford, 2003). Yet, as psychological experiments demonstrate (Tversky & Kahneman, 1974; Xu & Vashti, 2008), once a phenomenon is reified, its tacit and semiotic formulations are liable to be discordant (see Bamberger & diSessa, 2003; Sfard, 2007). It is precisely such

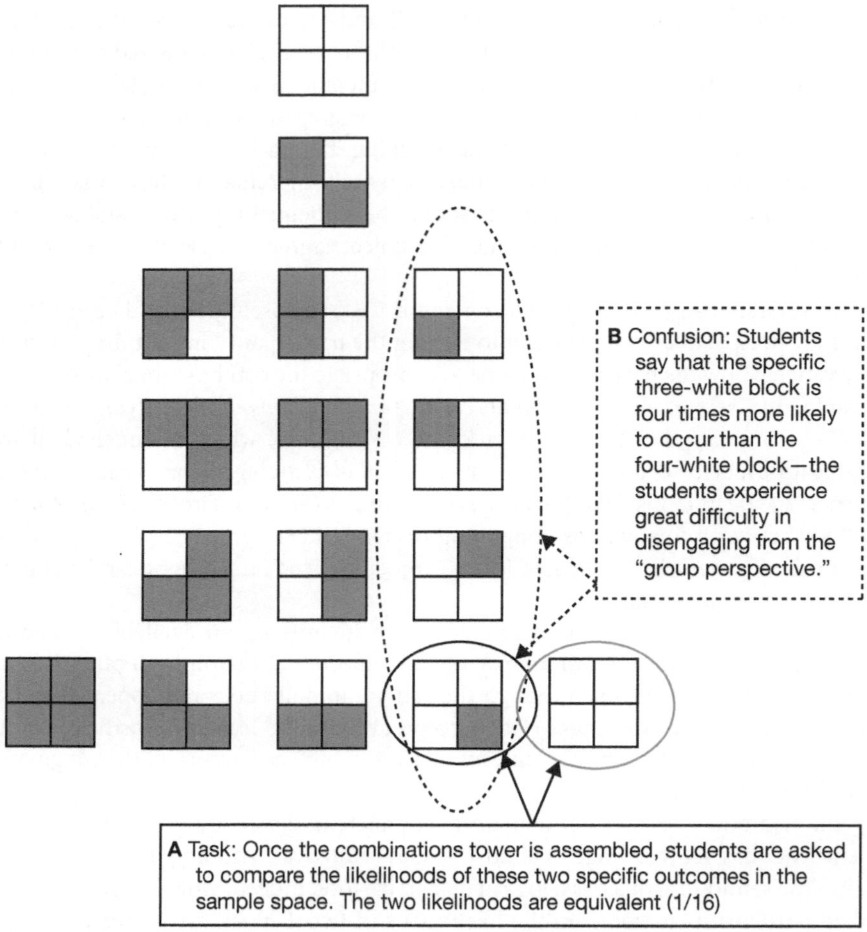

Figure 18.2 Students Confuse Aggregate and Elemental Constructions of a Compound Event.

discord that educational designers and practitioners should help students breach. So doing, students need to make explicit their implicit construction of stimuli.

This process of naming not what one *infers* when one perceives a phenomenon but *how one is perceiving it* is difficult—especially so when privileged forms of perceptions are engaged, because they are inaccessible to cognition (they are "cognitively impenetrable"; Pylyshyn, 1973). Thus, for a person to explain how they are seeing the world is contingent on that person realizing *that* they are seeing the world in a particular way rather than another way (Gopnik & Rosati, 2001; Wittgenstein, 1953). And yet the equipmentality of perception is covert and is revealed only in breakdowns (Dreyfus, 1990; Heidegger, 1962; Koschmann, Kuuti, & Hickman, 1998). To bring about such breakdown, a teacher may wish to foster a semiotic conflict, which is revealed to the student as a conflict of inferences drawn for a single referent that is covertly assigned two alternative senses.

Finally, coordinating tacit and mathematical formulations of phenomena involves certain discontinuous logical progressions, wherein multimodal imagistic reasoning precedes logical rationalization. Given this, students may make sense of formal mathematical procedures only retroactively, once the germane conceptual images have been blended and anchored into the semiotic artifacts.

Final Remark and Future Work

What might be realistic objectives for the teaching and learning of probability? Fischbein (1987) maintains that certain mathematical concepts cannot be grasped with *primary intuition*, so that students have to develop alternative, *secondary intuition* that includes the heuristic of inhibiting the primary intuition. Cobb (1989), commenting on Fischbein, thus dubs primary intuition a "double-edged sword." One might assume that attending to order in compound events is a case of secondary intuition that must inhibit the primary intuition of ignoring the order. Yet how might such secondary intuition be fostered?

For "radical constructivists" (e.g., von Glasersfeld, 1987) who would provide students only minimal guidance, the prospect of directly nurturing secondary intuition is anathema. For socio-cultural theorists (Vérillon & Rabardel, 1995; Vygotsky, 1978/1930), however, mediating students' construction of knowledge through guiding their operation of cultural artifacts is humans' fundamental modus operandi, such that acculturating students to perform procedures that foster effective performance is a natural and desirable practice. Somewhere between the constructivists and socio-culturalists are adherents of "realistic mathematics education" (Freudenthal, 1986; Gravemeijer, 1994), who seek to design contexts for students' *guided reinvention* of mathematical concepts. Despite these differences in pedagogical perspectives, though, mathematics-education researchers agree on the merits of learning environments that encourage students to be reflective as they learn to operate mathematical procedures (NCTM, 2006).

Working in an emerging design framework (Abrahamson & Wilensky, 2007), we have been creating instructional materials that leverage students' concept-specific reasoning and then surface productive semiotic conflict between the student and teacher. As they participate in these activities, students experience structured opportunities to acknowledge and then reconcile phenomenologically immediate and semiotically mediated perceptions of situations wherein quantitative properties and relations have been problematized (Abrahamson, 2009c). Optimally, students need not substitute one intuition for another—rather than inhibit their primary intuition, they may sustain and qualify it ("tune it toward expertise"; diSessa, 2008).

Equipped with an apparently effective means of demonstrating to students the utility of combinatorial analysis, we have still to demonstrate that future participants in our activities carry on to apply and extend this procedure correctly in situations they identify as cases of the binomial function. Design-based research projects can be very long journeys. Having been entrenched in data analysis for several years, we are now prepared to spiral through new empirical studies of the residual effects our improved activities may engender for student understanding of fundamental notions of probability.

Acknowledgement

I wish to thank members of the Embodied Design Research Laboratory who participated in the implementation and analysis of the empirical study, and Blair Lehman, Nancy Stein, and Betina Zolkower for their comments on previous drafts.

Note

1 Strictly speaking, this and our own proposed probability problem, discussed in this chapter, below, are considered without-replacement hypogeometric approximations of the binomial, because the four singleton events are not truly independent.

References

Abrahamson, D. (2003). A situational-representational didactic design for fostering conceptual understanding of mathematical content: The case of ratio and proportion. Paper presented at the annual meeting of the American Educational Research Association, Chicago, IL, April 21–25, 2003.

Abrahamson, D. (2004). Embodied spatial articulation: A gesture perspective on student negotiation between kinesthetic schemas and epistemic forms in learning mathematics. In D. E. McDougall & J. A. Ross (Eds.), *Proceedings of the twenty sixth annual meeting of the North American Chapter of the International Group for the Psychology of Mathematics Education* (Vol. 2). Toronto: Preney. pp. 791–797.

Abrahamson, D. (2007). Handling problems: Embodied reasoning in situated mathematics. In T. Lamberg & L. Wiest (Eds.), *Proceedings of the twenty ninth annual meeting of the North American Chapter of the International Group for the Psychology of Mathematics Education.* Stateline (Lake Tahoe): University of Nevada, Reno. pp. 219–226.

Abrahamson, D. (2008a). The abduction of Peirce: The missing link between perceptual judgment and mathematical reasoning? Presentation at the Townsend Working Group in Neuroscience and Philosophy (A. Rokem, J. Stazicker, & A. Noë, Organizers), UC Berkeley. Online. Available at: http://edrl.berkeley.edu/content/abduction-peirce-missing-link-between-perceptual-judgment-and-mathematical-reasoning (video). and http://neurophilosophy.berkeley.edu/AbrahamsonJLS-EmbCogSI-2008-10-21.pdf (paper) (accessed August 12, 2010).

Abrahamson, D. (2008b). Writes of passage: From phenomenology to semiosis in mathematical learning. In T. Rikakis & A. Keliiher (Eds.), *Proceedings of the Creative IT 2008 Workshop: Success factors in fostering creativity in it research and education.* Tempe: Arizona State University. Online. Available at: http://l3dswiki.cs.colorado.edu:3232/CreativeIT/247 (accessed August 12, 2010).

Abrahamson, D. (2009a). Embodied design: Constructing means for constructing meaning. *Educational Studies in Mathematics*, 70(1), 27–47.

Abrahamson, D. (2009b). Orchestrating semiotic leaps from tacit to cultural quantitative reasoning: The case of anticipating experimental outcomes of a quasi-binomial random generator. *Cognition and Instruction*, 27(3), 175–224.

Abrahamson, D. (2009c). A student's synthesis of tacit and mathematical knowledge as a researcher's lens on bridging learning theory. In M. Borovcnik & R. Kapadia (Eds.), Research and developments in probability education (special issue). *International Electronic Journal of Mathematics Education*, 4(3), 195_226. Online. Available at: http://www.iejme.com/032009/main.htm (accessed August 12, 2010).

Abrahamson, D., & Cendak, R. M. (2006). The odds of understanding the Law of Large Numbers: A design for grounding intuitive probability in combinatorial analysis. In J. Novotná, H. Moraová, M. Krátká, & N. Stehlíková (Eds.), *Proceedings of the thirtieth conference of the International Group for the Psychology of Mathematics Education* (Vol. 2). Charles University: PME. pp. 1–8.

Abrahamson, D., Janusz, R., & Wilensky, U. (2006). There once was a 9-Block . . . : A middle-school design for probability and statistics [electronic version]. *Journal of Statistics Education*, 14(1). Online. Available at: http://www.amstat.org/publications/jse/v14n1/abrahamson.html (accessed August 12, 2010).

Abrahamson, D., & Wilensky, U. (2005a). ProbLab goes to school: Design, teaching, and learning of probability with multi-agent interactive computer models. In M. Bosch (Ed.), *Proceedings of the Fourth Congress of the European Society for Research in Mathematics Education*. Universitat Ramon Llull, Catalonia: FUNDEMI IQS. pp. 570–579.

Abrahamson, D., & Wilensky, U. (2005b). Understanding chance: From student voice to learning supports in a design experiment in the domain of probability. In G. M. Lloyd, M. Wilson, J. L. M. Wilkins, & S. L. Behm (Eds.), *27th annual meeting of the North American Chapter of the International Group for the Psychology of Mathematics Education* (Vol. 7). Roanoke, VA: PME-NA. pp. 1–8.

Abrahamson, D., & Wilensky, U. (2007). Learning axes and bridging tools in a technology-based design for statistics. *International Journal of Computers for Mathematical Learning*, 12(1), 23–55.

Bamberger, J. (1996). Turning music theory on its ear: Do we hear what we see; Do we see what we say? *International Journal of Computers for Mathematical Learning*, 1(1), 33–55.

Bamberger, J., & diSessa, A. A. (2003). Music as embodied mathematics: A study of a mutually informing affinity. *International Journal of Computers for Mathematical Learning*, 8(2), 123–160.

Barab, S., Zuiker, S., Warren, S., Hickey, D., Ingram-Goble, A., Kwon, E.-J., et al. (2007). Situationally embodied curriculum: Relating formalisms and contexts. *Science Education*, 91, 750–782.

Borovcnik, M., & Bentz, H.-J. (1991). Empirical research in understanding probability. In R. Kapadia & M. Borovcnik (Eds.), *Chance encounters: Probability in education*. Dordrecht: Kluwer. pp. 73–105.

Brown, A. (1992). Design experiments: Theoretical and methodological challenges in creating complex interventions in classroom settings. *Journal of the Learning Sciences*, 2(2), 141–178.

Clement, J., Brown, B., and Zietsman, A. (1989). Not all preconceptions are misconceptions: Finding "anchoring conceptions" for grounding instruction on student's intuitions. *International Journal of Science Education*, 11, 554–565.

Cobb, P. (1989). Review: A double-edged sword. *Journal for Research in Mathematics Education*, 20(2), 213–218.

Cobb, P., Confrey, J., diSessa, A., Lehrer, R., & Schauble, L. (2003). Design experiments in educational research. *Educational Researcher*, 32(1), 9–13.

Collins, A. (1992). Towards a design science of education. In E. Scanlon & T. O'Shea (Eds.), *New directions in educational technology*. Berlin: Springer. pp. 15–22.

Confrey, J. (2005). The evolution of design studies as methodology. In R. K. Sawyer (Ed.), *The Cambridge handbook of the learning sciences*. Cambridge: Cambridge University Press. pp. 135–151.

Dahaene, S. (1997). *The number sense*. Oxford: Oxford University Press.

Devlin, K. (1999). *The math gene: How mathematical thinking evolved and why numbers are like gossip*. New York: W. H. Freeman and Company.

Dreyfus, H. L. (1990). *Being-in-the-world: A commentary on Heidegger's Being and Time, Division I*. Cambridge, MA: MIT Press.

Edelson, D. C. (2002). Design research: What we learn when we engage in design. *Journal of the Learning Sciences*, 11(1), 105–121.
Fauconnier, G., & Turner, M. (2002). *The way we think: Conceptual blending and the mind's hidden complexities*. New York: Basic Books.
Fischbein, E. (1975). *The intuitive sources of probabilistic thinking in children*. London: Reidel.
Fischbein, E. (1987). *Intuition in science and mathematics*. Dordrecht: D. Reidel.
Freudenthal, H. (1986). *Didactical phenomenology of mathematical structures*. Dordrecht: Kluwer Academic Publishers.
Gelman, R., & Williams, E. (1998). Enabling constraints for cognitive development and learning: Domain specificity and epigenesis. In D. Kuhn & R. Siegler (Eds.), *Cognition, perception and language* (5th edn., Vol. 2). New York: John Wiley and Sons. pp. 575–630.
Gigerenzer, G. (1998). Ecological intelligence: An adaptation for frequencies. In D. D. Cummins & C. Allen (Eds.), *The evolution of mind*. Oxford: Oxford University Press. pp. 9–29.
Gigerenzer, G., & Brighton, H. (2009). *Homo heuristicus*: Why biased minds make better inferences. *Topics in Cognitive Science*, 1(1), 107–144.
Ginsburg, H. P. (1997). *Entering the child's mind*. New York: Cambridge University Press.
Glaser, B. G., & Strauss, A. L. (1967). *The discovery of grounded theory: Strategies for qualitative research*. Chicago: Aldine Publishing Company.
von Glasersfeld, E. (1987). Learning as a constructive activity. In C. Janvier (Ed.), *Problems of representation in the teaching and learning of mathematics*. Hillsdale, NJ: LEA. pp. 3–18.
Gopnik, A., & Rosati, A. (2001). Duck or rabbit? Reversing ambiguous figures and understanding ambiguous representations. *Developmental Science*, 4(2), 175–183.
Gravemeijer, K. P. E. (1994). Educational development and developmental research in mathematics education. *Journal for Research in Mathematics Education*, 25(5), 443–471.
Heidegger, M. (1962). *Being and time* (J. Macquarrie & E. Robinson, Trans.). New York: Harper & Row. (Original work published 1927.)
Hutchins, E. (2005). Material anchors for conceptual blends. *Journal of Pragmatics*, 37(10), 1555–1577.
Jones, G. A., Langrall, C. W., & Mooney, E. S. (2007). Research in probability: Responding to classroom realities. In F. Lester (Ed.), *Second handbook of research on mathematics teaching and learning*. Charlotte, NC: Information Age Publishing. pp. 909–955.
Kelly, A. E. (2003). Research as design. Theme issue: The role of design in educational research. *Educational Researcher*, 32(1), 3–4.
Kelly, A. E. (2004). Design research in education: Yes, but is it methodological? *Journal of the Learning Sciences*, 13(1), 115–128.
Koschmann, T., Kuuti, K., & Hickman, L. (1998). The concept of breakdown in Heidegger, Leont'ev, and Dewey and its implications for education. *Mind, Culture, and Activity*, 5(1), 25–41.
Navon, E. (1977). Forest before trees: The precedence of global features in visual perception. *Cognitive Psychology*, 9, 353–383.
NCTM (2006). *Curriculum focal points for prekindergarten through Grade 8 mathematics: A quest for coherence*. Reston, VA: NCTM.
Peirce, C. S. (1931–1958). *Collected papers of Charles Sanders Peirce*. C. Hartshorne & P. Weiss (Eds.). Cambridge, MA: Harvard University Press.
Pylyshyn, Z. (1973). What the mind's eye tells the mind's brain: A critique of mental imagery. *Psychological Bulletin*, 80(1), 1–24.
Radford, L. (2003). Gestures, speech, and the sprouting of signs: A semiotic-cultural approach to students' types of generalization. *Mathematical Thinking and Learning*, 5(1), 37–70.
Sandoval, W. A., & Bell, P. E. (2004). Special issue on design-based research methods for studying learning in context. *Educational Psychologist*, 39(4), 199–201.

Schoenfeld, A. H., Smith, J. P., & Arcavi, A. (1991). Learning: The microgenetic analysis of one student's evolving understanding of a complex subject matter domain. In R. Glaser (Ed.), *Advances in instructional psychology*. Hillsdale, NJ: LEA. pp. 55–175.

Schön, D. A. (1981). Intuitive thinking? A metaphor underlying some ideas of educational reform (Working Paper 8): Division for Study and Research, MIT.

diSessa, A. A. (2007). An interactional analysis of clinical interviewing. *Cognition and Instruction*, 25(4), 523–565.

diSessa, A. A. (2008). A bird's-eye view of the "pieces" vs. "coherence" controversy. In S. Vosniadou (Ed.), *Handbook of research on conceptual change*. New York: Routledge. pp. 35–60.

diSessa, A. A., & Cobb, P. (2004). Ontological innovation and the role of theory in design experiments. *Journal of the Learning Sciences*, 13(1), 77–103.

diSessa, A. A., Hammer, D., Sherin, B., & Kolpakowski, T. (1991). Inventing graphing: Meta-representational expertise in children. *Journal of Mathematical Behavior*, 10(2), 117–160.

diSessa, A. A., & Wagner, J. F. (2005). What coordination has to say about transfer. In J. Mestre (Ed.), *Transfer of learning from a modern multi-disciplinary perspective*. Greenwich: Information age. pp. 121–154.

Sfard, A. (2007). When the rules of discourse change, but nobody tells you: Making sense of mathematics learning from commognitive standpoint. *Journal of Learning Sciences*, 16(4), 567–615.

Siegler, R. S., & Crowley, K. (1991). The microgenetic method: A direct means for studying cognitive development. *American Psychologist*, 46(6), 606–620.

Smith, J. P., diSessa, A. A., & Roschelle, J. (1993). Misconceptions reconceived: A constructivist analysis of knowledge in transition. *Journal of the Learning Sciences*, 3(2), 115–163.

Stavy, R., & Tirosh, D. (1996). Intuitive rules in science and mathematics: The case of "more of A—more of B." *International Journal of Science Education*, 18(6), 653–668.

Stevens, R., & Hall, R. (1998). Disciplined perception: Learning to see in technoscience. In M. Lampert & M. L. Blunk (Eds.), *Talking mathematics in school: Studies of teaching and learning*. New York: Cambridge University Press. pp. 107–149.

Stroop, J. R. (1935). Studies of interference in serial verbal reactions. *Journal of Experimental Psychology*, 18, 643–662.

Tversky, A., & Kahneman, D. (1974). Judgment under uncertainty: Heuristics and biases. *Science*, 185(4157), 1124–1131.

Van Dooren, W., De Bock, D., Weyers, D., & Verschaffel, L. (2004). The predictive power of intuitive rules: A critical analysis of the impact of "More A—More B" and "Same A—Same B." *Educational Studies in Mathematics*, 56, 179–207.

Vérillon, P., & Rabardel, P. (1995). Cognition and artifacts: A contribution to the study of thought in relation to instrumented activity. *European Journal of Psychology of Education*, 10(1), 77–101.

Vygotsky, L. S. (1978/1930). *Mind in society: The development of higher psychological processes*. Cambridge, MA: Harvard University Press.

Weisstein, E. W. (2006). Probability. *MathWorld: A Wolfram Web Resource*. Online. Available at: http://mathworld.wolfram.com/Probability.html (accessed August 12, 2010).

Wilensky, U. (1997). What is normal anyway?: Therapy for epistemological anxiety. *Educational Studies in Mathematics*, 33(2), 171–202.

Wilson, E. O. (1998). *Consilience: The unity of knowledge*. New York: Vintage Books.

Wittgenstein, L. (1953). *Philosophical investigations* (G. E. M. Anscombe, Trans.). Upper Saddle River, NJ: Prentice Hall.

Xu, F., & Vashti, G. (2008). Intuitive statistics by 8-month-old infants. *Proceedings of the National Academy of Sciences of the U.S.A.*, 105(13), 5012–5015.

Part IV

Theoretical and Methodological Concerns

19 Linking Cognitive and Developmental Research and Theory to Problems of Educational Practice

A Consideration of Agendas and Issues

James W. Pellegrino

The chapters in this volume consider some of what we know about the nature of student cognition, its development, and the possible implications of that work for educational practice. The goal of doing so is to improve the conditions of teaching and student learning and, ultimately, the overall outcomes of the educational enterprise. With regard to these educational outcomes, we know from various policy reports examining student performance on national and international assessments, such as National Assessment of Educational Progress (NAEP), Trends in International Math and Science Studies (TIMSS), and Program for International Student Assessment (PISA), that there is much room for improvement in all areas of curriculum, especially mathematics and science (see for example Phillips, 2007).

The individual and collective work reported in this volume is representative of the considerable body of cognitive, developmental, and educational psychological research currently being done on questions related to the teaching and learning of important subject-matter content. On what basis might the work be judged? One potentially useful "evaluative" context is to see how the work fits within frameworks that have been offered by other groups with similar concerns about the state of knowledge for "linking research with practice." In that spirit, the goal of this chapter is to present a perspective on important issues in this arena of research and development (R&D) and offer that as a lens to help the reader of this volume consider its individual and collective contributions. It is explicitly not the intent of this chapter to provide an analysis and critique of the collective papers included in the volume. Such an appraisal is far beyond the scope of what might be possible in the space available.

The perspective that is presented is based on work—done by a panel assembled by the National Research Council—to consider what a comprehensive R&D agenda for learning and instruction might look like, especially when set in the context of a larger orchestrated effort known as the Strategic Education Research Partnership (SERP) (see Donovan, Wigdor, & Snow, 2003). For present purposes, we are concerned not with the feasibility or policy wisdom of something like SERP, but rather with the logic behind the framework offered by the SERP Panel on Learning and Instruction regarding important questions and possible directions for research (see Donovan & Pellegrino, 2004). This serves as a context to situate the work reported in this volume.

Given the goals articulated above, the remainder of this chapter consists of two sections. In the first, we consider aspects of the rationale guiding the R&D framework developed and offered by the SERP Learning and Instruction Panel. In so doing, we include discussion of some overarching issues about relating research and practice.

We then describe the elements of the proposed framework and their application to the areas of reading, mathematics, and science instruction. This allows us to examine what was argued then as important concerns for research, and use that as a basis for judging progress given the work reported in this volume. In the second section, we briefly discuss an area of research that the SERP Panel felt needed considerable attention across all domains of instruction: the assessment of student learning. We offer this discussion as a context for understanding the importance of work on student cognition and its development as reported in this volume, and also as a means to further exhort researchers, including the contributors to this volume, to consider applications of their theory and research for addressing important problems of assessment design and use. Such problems are fundamental and lie at the core of linking cognitive and developmental research and theory with instructional practice.

Framing an Agenda for Research and Development on Learning and Instruction: Making the Problems of Practice Central

The goal of the SERP Panel was to develop a prototype agenda for an R&D program on learning and instruction that would have the potential to *significantly improve and support practice*. The Committee felt that their design must reflect hypotheses about why research has thus far failed to achieve that end. Building on ideas about research and practice linkages from *How People Learn: Bridging research and practice* (Donovan, Bransford, & Pellegrino, 1999), they proposed three closely related general shortcomings of the existing R&D enterprise (to which there are, of course, exceptions):

1 Research has not placed the problems teachers face in their educational practice on center stage.

Powerful incentives discourage research on the problems of practice, including a reward structure for university researchers that devalues research on the "applied problems of practice," the transaction costs involved in gaining access to school settings for research, and the disparate cultures of the classroom and university. Further complicating matters is that many researchers lack intimate knowledge of K–12 teaching practice. Thus, even when they conceptualize research intended to inform practice, their assumptions about the nature of practice are often mismatched with the realities of schools and classrooms.

2 Research from the variety of disciplines related to education has not been coordinated in a sustained R&D program aimed at improving educational practice.

Pockets of excellent research exist in the study of learning, pedagogy, measurement and testing, and subject-matter domains. But for these pockets of research to influence practice requires that they be developed with a grounded understanding of the needs of practitioners. They must be incorporated into instructional programs and materials that are developed and tested in classrooms—in small- and large-scale settings—and with the variety of students for whom they are intended.

3 Too little attention has been paid to the interdependence of student learning, teacher knowledge and skill, and organizational environment.

Even when careful R&D efforts are carried through to the production of curricula or instructional programs that are "usable" and that show positive results on student achievement, this initial success does not guarantee that they will be "used" persistently or more widely. Whether R&D efforts have an impact is likely to be influenced by two factors: (a) whether teachers have the knowledge and skill that allows them to understand the purpose of the program or curriculum and effectively incorporate it into their practice; and (b) whether the organizational environment of the school supports and facilitates teachers' efforts and learning that is entailed in the adoption of the materials.

Core Questions behind an Agenda for R&D

We can view a teacher's practice as organized around a set of core questions about human learning and instruction. These questions focus on the normal course of development and learning, as well as on diagnosing and responding to student problems in mastering new concepts and acquiring new skills. The questions provide a schema for approaching practice, as well as a set of dimensions on which knowledge must be supported through professional education. They are:

- What do we want students to know or be able to do?
- What are the typical understandings and preconceptions these students hold on this topic at the outset?
- What is the expected progression of understanding, and what are the predictable points of difficulty that must be overcome?
- What interventions can move students along a path from their initial understanding to the desired outcome?
- What general and discipline-specific norms and practices best constitute and support student learning?
- Finally, how can the individual student's understanding and progress be monitored, and how can the student become engaged in the instructional activities he or she needs to take the next step toward understanding?

The above questions are often answered without reference to a research base. The "what" and "how" might, for example, be answered by the adoption of a textbook that reflects little of the research knowledge on student learning. And individual student differences in understanding are often overlooked entirely because textbooks rarely provide the tools for formative assessment and guidance on instructional responses to student difficulties. As accumulating evidence points to the complexity of the underlying processes, the need for research-based knowledge to support professional practice becomes ever more apparent.

The elements in the schema for educational practice are informed by several very different knowledge bases, reviewed more thoroughly in *How People Learn: Brain, Mind, Experience, and School* (Bransford, Brown, Cocking, Donovan, & Pellegrino, 2000) and *Knowing What Students Know* (Pellegrino, Chudowsky, & Glaser, 2001), as well as other domain-specific synthesis reports covering areas of reading, mathematics, and science (Snow, Burns, & Griffin, 1998; Kilpatrick, Swafford, & Findell, 2001; Donovan & Bransford, 2005; Duschl, Schweingruber, & Shouse, 2007). These and other explorations of the knowledge base on learning and instruction (National Reading Panel, 2000; Snow, 2002; Ball, 2003; Pashler et al., 2007) provide major contributions to understanding what is currently known from theory and research and what it might mean for educational practice. Such knowledge helps considerably in the framing of an agenda regarding what we still need to know in the way of research and theory development.

- *What students should know or be able to do* in an area is informed by disciplinary expertise. It requires an understanding of the core concepts around which the discipline is organized, characteristic methods of reasoning and problem solving, language, and patterns of discourse. What to teach becomes a matter not only of the information and skills considered desirable for students to possess, but of helping the student build the conceptual framework that transforms information into understandings.
- *Knowledge of common student conceptions* of a topic, and the expected progression of student thinking, requires careful research on the typical trajectory of understanding. In part, this research attempts to identify the nature and limits of children's changing cognitive abilities with age and instruction. It also attempts to uncover common understandings that can either support learning or undermine it. Research findings suggest that students' everyday understandings are resilient, even after specific instruction to the contrary. That resilience highlights the importance of using a carefully designed research program to inform and support teaching in order to achieve conceptual change. Research of this sort is often done by cognitive scientists and education researchers, though the knowledge may emerge from the experience of teachers and the observation of exemplary practice.
- *Educative experiences intended to move students along a learning path* constitute the core of "instruction." These experiences might be designed by curriculum developers, teachers, or researchers. But, regardless of the source, the contributors to skill development, knowledge acquisition, and conceptual change are themselves a research agenda, and the effectiveness of the instructional approach a matter for empirical testing.
- *General and discipline-specific norms and practices* define the rules for interaction in the classroom. Learning takes place in classrooms, which are themselves communities. Every community is distinguished by norms for work and interactions, ranging from when and how people collaborate to how they speak with one another. Some norms are general, rooted in our understanding of schools in a democratic society; others are discipline specific, for what it means to do mathematics differs from what it means to do literary analysis or chemistry. In all cases, the relationships between particular norms and learning outcomes are a matter for empirical investigation.
- *Assessing an individual student's understanding* is the task of R&D on methods and systems of assessment. This knowledge base can be developed and tested in the context of curriculum R&D, but it may also draw on more fundamental research—such as research on the nature and measurement of "comprehension." Again, whether the assessment serves the intended purpose is a matter for research.

Typically, these various knowledge bases are not adequately developed to support teaching practice, and more targeted, instruction-relevant research would help (Hiebert, Gallimore, & Stigler, 2002). As an example, take research on the misconceptions students harbor about aspects of physics. In the course of everyday experience, we develop understandings or "models" of how the physical world works. As we move closer to a heat source, temperature rises. We then draw inferences based on our experiences (e.g., in the summer, the earth must be closer to the sun) that are scientifically incorrect. A persuasive body of evidence suggests that the models of physical principles that students deduce incorrectly from everyday experience are powerful and resilient (e.g., DiSessa, 1982; Vosniadou & Brewer, 1994). While students may learn physics in the classroom and even perform quite well on tests, outside the classroom they revert to their untrained model. If teachers had materials that were designed to address student conceptions and habits, instruction would be improved.

The principle at work in physics is at play in all disciplines (Donovan et al., 1999; Bransford et al., 2000), undermining the effectiveness of the educational process. In history, for example, everyday notions of "telling the truth" interfere with historical notions of truth. But this "relevant" knowledge regarding the tenacity of everyday understandings becomes usable for a teacher only when it is applied to the subject that he or she teaches and incorporated into instructional activities that draw out and effectively work with students' conceptions. These activities must have potential for being incorporated into existing instructional practice—or be an acceptable replacement for that practice—if they are to be a viable substitute for the status quo. Only the most extraordinary teachers will be able to undertake such a task on their own. Furthermore, only the most inefficient profession would require the design of individual solutions to a general problem.

To effectively "bridge" research and practice, then, an R&D program must generate and/or draw on existing, relevant knowledge from a variety of disciplines, elaborate that knowledge so that it is usable in practice, and then incorporate it into carefully tested tools and programs (directed both at student and at teacher learning) that allow it to be used. The point of departure in designing such a program must be the questions that lie at the core of instructional practice. Grounding research in the problems of instructional practice necessarily means that the research and practice relationship is neither linear nor unidirectional. Instead, they must interact in meaningful, progressively more sophisticated ways. Research is not neatly packaged and sent out to teachers to be implemented. Instead, researchers and teachers need to work together to both identify meaningful problems and to conceptualize relevant research to inform those problems.

When research findings are compelling, sustained attention is required to assure that independent replication of research and evaluation results in the range of environments of intended use for an educational intervention. Because education is a complex enterprise in which any outcome is influenced by a variety of factors, the conditions that support success in one setting may not be understood until it is attempted in other settings where conditions differ. Moreover, evaluations are often conducted only by the designers of the intervention or their "critical friends." Although many research findings are criticized for this reason, research attention in education is rarely directed at independent replication. Too often, promising outcomes mark the end point of a research endeavor rather than the beginning of a research effort aimed at replication and scientific generalization (see, for example, Shavelson & Towne, 2002).

Finally, to be broadly useful to practice, attention must be sustained through careful study of the scaling up of successful interventions (see, for example, Dede, Honan, & Peters, 2005; Schneider & McDonald, 2006). When researchers design and study educational interventions, the process of study itself entails involvement by the researcher. The tacit and explicit knowledge of the researcher may support the teacher's effective use of the intervention. Similarly, a teacher's tacit knowledge and skill may be crucial to an intervention's success. But when the original researcher and teachers are no longer present, the innovation is often much harder to implement and sustain. Thus, the requirements for effective, sustainable use of an innovation must themselves be the subject of study.

Interdependence: Student Learning, Teacher Knowledge, and Organizational Environment

Practice is not embodied solely in the tools and protocols of the trade. Rather, these work in tandem with both the practitioner's knowledge and skill and the organizational environment. Similarly, instructional tools, teacher knowledge, and the organization of the school are interdependent. Although each is, in a sense, independent, the effectiveness of any one in supporting student achievement depends on the strength of the other two. Focusing on the strength of one without attending to the other two is a strategy that holds little promise for success. A program of research, then, must coordinate not only the knowledge from different fields on effective instructional programs, but also research on the knowledge requirements for teachers to carry out instruction effectively, and requirements of the environment to support effective instruction.

Teacher knowledge and skill matter a great deal in student learning. Ferguson (1991) analyzed data from 900 Texas school districts and found that teacher licensing exam scores, master's degrees, and experience accounted for over 40 percent of the variance in students' reading and math achievement scores after controlling for socio-economic status. Other studies suggest a similarly powerful effect (Ferguson & Ladd, 1996). Yet, despite its importance, the research base on teacher learning is relatively undeveloped.

As we have already noted, the schematic questions that frame student learning apply just as aptly to teacher learning. Take our first question: "What do we want the teacher to know and be able to do?" We want teachers to understand the learning process of the student well enough to assess and guide it, and teach the content well. We want teachers to guide the pace and direction of instruction, and flexibly respond to student questions and thoughts. Teachers should know the curriculum materials well enough to use them as a means to an end (rather than as the end itself), the norms and practices that constitute effective practice well enough to create a supportive learning environment, and assessments well enough to interpret the outcomes and respond appropriately.

What are not well defined are the forms and levels of knowledge a teacher must master in order to reach that end. What mathematics must he or she know, and what pedagogical knowledge does he or she need to make and implement appropriate decisions about the next best instructional steps to develop student thinking? Though these questions are central to effective practice, little research has been done to provide answers.

The typical learning trajectory for teachers, and how it changes, also requires empirical investigation. Much that teachers need to know cannot be learned apart from practice. This raises several questions: Under what conditions can teachers best learn while engaged in practice? What knowledge and skill must teachers acquire at the beginning of their careers? What knowledge and skill is best acquired once they enter the profession? What resources are necessary to support and sustain teacher learning over time?

Widespread adoption and use of improved instructional methods is often hampered by institutional barriers that prevent or frustrate efforts to change: problems of organizational structure, incentive structures, organizational culture, career patterns of teachers and administrators, and financial constraints. Even when presented with demonstrably effective instructional reforms, school systems are frequently incapable of moving the organizational machinery to achieve systemwide adoption (Briars & Resnick, 2000). Effective changes, instituted with the help of program developers, are frequently not sustained when the developer departs.

Schools are certainly not unique in resisting change. Change is effortful; it imposes uncertainty and requires risk of failure. One would therefore expect to see change more often where uncertainty can be reduced by effective supports for success, where risk is minimized by careful evaluation of candidate changes, and where effort and risk taking are rewarded. However change is not, in and of itself, a desirable end; there are certainly changes for the worse. What is desirable is an organization that can learn: that can systematically assess its own performance, evaluate the potential of alternative approaches, monitor the effect of change, and alter course to improve outcomes.

If advances in instructional programs and teacher knowledge are to have sustained impact on student learning, the organizational structure of schools must support that change. Because the success of each component (instructional program, teacher knowledge, and organizational structure) in contributing to improving student learning depends on the success of the others, all three must be integrated in a comprehensive research program.

A compelling example is the Algebra Tutor developed by researchers at Carnegie Mellon University in cooperation with algebra teachers in Pittsburgh schools. When the curriculum was used in schools, the student achievement outcomes were below those in research trials. Upon investigation, the researchers discovered that the amount of instructional time for mathematics in many of the city's schools was 65 hours a year—far below expectation (Anderson, Corbett, Koedinger, & Pelletier, 1995). In this case, organizational and instructional issues were critically intertwined.

Framework for a Research Agenda

The coordinated program of research that is required to effectively support educational practice can be mapped onto the schematic questions posed above. The SERP Panel used a travel metaphor to illuminate certain features of these schematic questions. This metaphor parallels learning as a journey. For any subject matter taught, we can assess the quality of the research base as it relates to practice by asking about the guidance it provides in answering each of the schematic questions. The full set of questions—for students, teachers, and organizations—is posed in Table 19.1. The charge to the Panel was to focus on the first two of the three columns: student- and

Table 19.1 Schematic Questions for Research on Teaching and Learning

	Learner		
Metaphor	*Student*	*Teacher*	*Organization*
Destination	What should he or she know and/or be able to do (regarding the discipline or topic)?	What should he or she know about the discipline or topic? What should he or she know about student learning of and the teaching of the discipline or topic?	How should the organization be functioning to support student and teacher learning?
Point of Departure	What are the typical preconceptions and informal understandings that students bring to the topic?	What are the teacher's existing understandings about the topic and about student learning?	What are the operative rules in the current environment, and what purpose do they serve?
Route	What is the expected progression of understanding, and what are the predicatable points of difficulty or hurdles?	What are the typical preservice and in-service learning trajectories and what difficulties are likely to be encountered?	What are the normal steps toward change and what barriers are typically encountered?
Vehicle	What curriculum, pedagogy, and classroom norms and practices facilitate learning?	What factors and/or experiences facilitate learning?	What factors and/or processes facilitate change?
Checkpoints/Course Corrections	How can individual student progress be monitored and instructional activities matched to current understanding?	How can progress be monitored and instructional activities matched to current understanding?	How can consequences of change be monitored and course corrections be made to facilitate progress?

Reprinted with permission from Donovan and Pellegrino (2004), Courtesy of the National Academies Press, Washington, DC.

teacher-as-learner. Suffice to say, however, that policy, professional, and local organizational issues all serve to support or constrain opportunities for the improvement of teaching and learning. To avoid mounting a research program that could otherwise be narrow and insular, it is essential that these issues be systematically included in a comprehensive and context-sensitive program of research.

In the long run, providing research-based knowledge to support answers to the schematic questions for every subject taught in schools is a desirable end. Yet the reality of the limited resources devoted to education research, and the existing capacity to conduct that research, suggests the need for focus on a limited set of subjects in order to assure that work can be carried through all stages necessary for usability. As a knowledge base is consolidated in some areas, attention can be devoted to new areas as well.

The areas in which R&D could be useful to educational practice are innumerable. To narrow the focus, the Panel asked two questions:

1. Are there examples of rigorous R&D efforts that already show impressive gains in student achievement in research trials? These would provide the starting point for R&D and move it forward by following promising program outcomes with research on the circumstances under which the results are obtained, the feasibility of the intervention in the classroom and school context, the teacher knowledge and support required for success, and the organizational factors that influence outcomes. The Panel referred to these as "downstream" cases because the work in these areas has already traveled some distance.
2. Are there pervasive problems of practice widely recognized as critical, but where the knowledge base is too weak to guide instructional interventions? The Panel referred to these as the "upstream" cases, since the work is still in early stages. Since upstream research will need to travel the full course through all phases of R&D, the impact on practice is likely to require more time.

Whereas Question 1 is driven by the opportunities provided by the current knowledge base, Question 2 is driven by the needs of practitioners. Precisely because the knowledge base is weak, R&D on these questions would draw more heavily on the study of practice, and the extension of research from the social and behavioral sciences to strengthen the knowledge base.

Focal Areas for Research

The Panel recommended three instructional areas for focus: reading, mathematics, and science. With regard to "downstream research" potential, in both mathematics and reading, the strongest research bases address learning in the early-elementary years. In contrast, for science the Panel chose physics—a subject generally taught in high school. Interestingly enough, there are strong parallels between the explicit cases discussed in the Panel's report and the areas of research reported in this volume. Many of the questions outlined in Table 19.1 with regard to student learning remain the focus of study and we continue to show progress in developing a research base with implications for practice. Noticeably, there is far less of a focus on teacher-as-learner in the collective work reported in this volume.

All three instructional domains were recommended for "upstream" research as well, but the reasons were not uniform. First, practitioners on the Panel emphasized the central role standardized tests play in the lives of students, particularly as they reach upper grades. Although there is considerable disagreement regarding the desirability of these tests, since performances in mathematics and in reading comprehension are major contributors to test results, these are seen as important areas for improving student outcomes. The choice of these areas was further supported by the need for proficiency in order to successfully meet the demands of modern life.

Further, reading comprehension features prominently in the teaching of most subject areas, including English, social studies, mathematics, and science. Teachers of these subjects must decide to what extent reading comprehension instruction is required as part of their effort to teach the subject matter. There is currently little guidance in this regard. Research on reading comprehension therefore has the potential to provide benefits in virtually all subject areas. Finally, reading comprehension is poorly measured. Current tests emphasize the speed of reading and short-term recall of factual information. But existing research suggests that both speed and short-term recall are weak predictors of the construction of understanding from text that comprehension requires. Given the importance of reading comprehension in standardized tests, making progress on test measurement issues would have substantial potential to influence practice. It is significant that some of the research reported in this volume has begun to attend to several of these concerns. This is exemplified in research being done on the comprehension of various forms of "digital" content and in the assessment of student comprehension processes.

In contrast to math and reading comprehension, science was identified for upstream research because the development of science curricula, particularly for the elementary- and middle-school years, has been remarkably weak. The American Association for the Advancement of Science reviewed widely used textbooks in middle-school science. "The study probed beyond the usual superficial alignment by topic heading and examined each text's quality of instruction aimed specifically at the key ideas, using criteria drawn from the best available research about how students learn" (Roseman, Kesidou, Stern, & Caldwell, 1999). Not one of the middle-grades science texts evaluated by the project was rated satisfactory. High-school biology texts scored slightly higher than the middle-school texts, but the evaluation found serious shortcomings in their content coverage and instructional design (Budiansky, 2001). Textbooks across the grades were characterized as "overstuffed and undernourished," with presentation of a great many facts, and too few opportunities to provide the concepts that make those facts meaningful (Budiansky, 2001). In contrast to mathematics, there is little agreement in science as to the sequence and content of study, or even when science education should begin. It is noteworthy then that chapters in this volume report interesting work on key issues in science teaching and learning across the K–12 spectrum.

The Role of Assessment in a Comprehensive Program of R&D

Although many themes are woven into the research agenda proposed by the SERP Panel, *assessment* of the outcomes of learning and instruction merits special attention. High-quality evidence that permits practitioners, researchers, and policymakers to ask and answer critical questions about these outcomes—what students know and are able to do—is critical to advancing an R&D agenda in any domain. At least three different purposes for educational assessment play distinct roles in the agenda:

1 *Formative assessment for use in the classroom to assist learning.* As mentioned, teachers need assessment data on their individual students to guide instructional process.
2 *Summative assessment for use at the classroom, school or district level* to determine student attainment levels. These include tests at the end of a unit or school year to determine individual student achievement.
3 *Assessment for program evaluation* to make comparisons across classrooms and/or schools. These include standardized tests to determine outcomes from different instructional programs, or to compare performance across schools.

The first requirement for developing quality assessments is that the concepts and skills that signal progress toward mastery of a subject be understood and specified. In various areas of the curriculum—such as early reading, early mathematics, and high-school physics—substantial work has already been done in this regard (see the various reports cited earlier). In some cases, researchers have capitalized on such knowledge to develop the elements of an assessment strategy, though that work has generally concentrated on the development of materials for formative assessment (e.g., Wilson & Sloane, 2000; Minstrell & Kraus, 2005; Wiliam, 2007). In contrast to these cases, we have yet to fully capitalize on research and theory to develop similarly valid assessment tools for many other areas of mathematics, reading comprehension, and numerous aspects of elementary- and middle-school science. These assessment design and development activities constitute part of the prospective R&D agendas outlined for each separate content domain.

Assessment of the overall outcomes of instruction—summative assessment—is important to the R&D agenda because it allows for a test of program effectiveness. But it is more broadly important because the content of those assessments can drive instructional practice for better or worse. The deployment of the Force Concept Inventory in Physics illustrates a case in which a summative assessment tool based on cognitive and instructional research has had a powerful, positive impact on the redesign of instruction, serving as an evaluation tool to determine the effectiveness of new instructional methods (e.g., Hake, 1998). In most areas of the curriculum, however, little progress has been made regarding how to assess student learning in ways that capture what is known about the nature of expertise and its development. Related issues concern assessment of the types of knowledge and skill implied in documents such as the mathematics and science standards.

Assessing the impact of long-term programs of R&D is also important. The public and policymakers need ways to determine the return on investing in research in cognition and the learning sciences. Such use of assessment for program evaluation purposes needs measures that are sensitive to learning and the impact of good-quality teaching.

Unfortunately, the educational system currently lacks sophisticated approaches to addressing these three levels. There has been a general failure to recognize the different assessment needs of formative, summative, and program evaluation purposes

(Pellegrino et al., 2001). This is coupled with a failure to design systematic ways to organize assessments that coordinate within and between these levels, and a failure to incorporate into the educational system provisions for the flow of information from one level to the other (in both directions). Current large-scale standardized tests, used by most states to assess academic achievement, will not suffice (Pellegrino et al., 2001; National Research Council, 2003; Wilson & Bertenthal, 2005). The models of learning and measurement underlying such tests are rather shallow, raising doubts about the quality of the evidence they can provide about student learning or the impact of instructional programs. Such concerns about the use of standardized test data are also likely to apply to the tests in reading, math, and science, yet to be developed under the Elementary and Secondary Education Act and No Child Left Behind legislation.

Researchers, educators, and policymakers must therefore concern themselves with supporting R&D on effective and appropriate assessment procedures that can serve multiple purposes. The elements of such an agenda are many. Here we mention just a few, emphasizing components described in the NRC report on assessment (Pellegrino et al., 2001).

Elements of an Assessment R&D Agenda

Although we distinguish three purposes of assessment, quality instruments for each depend on the same three components: (a) theories and data about content-based cognition that indicate the knowledge and skills that should be tested; (b) tasks and observations that provide information on whether the student has mastered the knowledge and skills of interest; and (c) qualitative and quantitative techniques for scoring student performance that capture fairly the differences in knowledge and skill among assessed students. Research related to each of these components is needed for assessments to provide reliable indicators of student achievement. But so too is research on how to integrate these three components into educational settings. For example, researchers have developed sophisticated models of student cognition in various areas of the curriculum, but in many cases this has not been translated into sets of tasks that can be used for assessment. Even in subject domains for which characteristics of expertise have been identified, a detailed understanding of patterns of growth that would enable one to identify progression toward competence is often lacking.

To develop assessments that are fair (comparably valid across different groups of students), it is crucial that patterns of learning for different student populations be studied. Much of the development of cognitive theories has been conducted with restricted groups of students (i.e., mostly middle-class whites). In many cases, it is not clear whether current theories of developing knowledge and expertise apply equally well with diverse populations of students, including under-represented minority students, English language learners, and students with disabilities. Although there are typical learning pathways, often there is not a *single* pathway to competence. Furthermore, students will not necessarily respond in similar ways to assessment probes designed to diagnose knowledge and understanding. These kinds of natural variations among individuals need to be better understood through empirical study and incorporated into assessment design.

Sophisticated models of learning by themselves do not produce high-quality assessment information. Also needed are methods and tools both for eliciting appropriate and relevant data and for interpreting the data collected about student performance. As described elsewhere (Pellegrino et al., 2001), current measurement methods enable a much broader range of inferences to be drawn about student competence than many people realize. In particular, it is now possible to characterize student achievement in terms of multiple aspects of proficiency, rather than a single score; chart students' progress over time, instead of simply measuring performance at a particular point in time; deal with multiple paths or alternative methods of valued performance; model, monitor, and improve judgments based on informed evaluations; and model performance not only at the level of students, but also at the levels of groups, classes, schools, and states. But research is needed to (a) investigate the relative usefulness of existing statistical models for capturing critical aspects of learning specified in cognitive theories; (b) develop tools that make it easier for those who have professional interest, but not the full range of psychometric expertise to use these measurement approaches; and also (c) develop tools that allow educational professionals, including teachers and policymakers, to use the *results* from these measurement models.

Beyond issues of the knowledge base needed to sustain effective assessment design, there are important issues about assessment use that also need to be pursued. Research needs to explore (a) how new forms of assessment can be made accessible for teachers, and practical for classroom use; (b) how they can be made efficient for use in large-scale testing contexts; (c) how we can reliably combine information from the classroom-level assessments and large-scale assessments for use in addressing educational problems; and (d) how various new assessment forms affect student learning, teacher practice, and educational decision making. Research must explore ways that teachers can be assisted in integrating new forms of assessment into their instructional practices and how they can best make use of information from these (see, for example, Black, Harrison, Lee, Marshall, & Wiliam, 2003). It is particularly important that such work be done in close collaboration with practicing teachers who have varying backgrounds and teaching experience. Also to be studied are ways that school structures (e.g., length of class time, class size, and opportunities for teachers to work together) impact the feasibility and effectiveness of implementing new types of assessments.

It is easy to see that the work needing to be done to link research and theory to issues of educational practice is vast. It can be somewhat foreboding given the scope of the issues that must be addressed, and assessment is one such issue given its significance in the worlds of policy and practice. The positive side of things is that we can see in the research reported in this volume much positive movement in addressing fundamental questions, including those related to the assessment of student learning. As a field, we continue to make progress, although at times it can be hard to see where and how things fit together and where we need to go next. It is to be hoped that the discussion of the research-practice issues and the framework for R&D described in this chapter provide some useful perspectives on progress in achieving important educational goals.

References

Anderson, J. R., Corbett, A., Koedinger, K., & Pelletier, R. (1995). Cognitive tutors: lessons learned. *Journal of the Learning Sciences*, 4, 167–207.

Ball, D. L. (2003). *Mathematical proficiency for all students: Toward a strategic R&D program in mathematics education*. Santa Monica, CA: RAND, RAND Mathematics Study Panel.

Black, P., Harrison, C., Lee, C., Marshall, B., & Wiliam, D. (2003). *Assessment for learning: Putting it into practice*. Buckingham: Open University Press.

Bransford, J. D., Brown, A. L., Cocking, R. R., Donovan, S., & Pellegrino, J. W. (Eds.) (2000). *How people learn: Brain, mind, experience, and school* (Expanded edition). Washington, DC: National Academies Press.

Briars, D., & Resnick, L. (2000). *Standards, assessment—and what else? The essential elements of standards-based school improvement*. Los Angeles: University of California at Los Angeles, Center for the Study of Evaluation. Online. Available at: http://www.cse.ucla.edu/products/Reports/TECH528.pdf.

Budiansky, S. (2001). The trouble with textbooks. *Prism* (February). Online. Available at: http://www.prism-magazine.org/feb01/index.htm.

Dede, C., Honan, J. P., & Peters, L. C. (Eds.) (2005). *Scaling up success: Lessons learned from technology-based educational improvement*. New York: Jossey-Bass.

DiSessa, A. (1982). Unlearning Aristotelian physics: A study of knowledge-based learning. *Cognitive Science*, 6, 37–75.

Donovan, M. S., & Bransford, J. D. (Eds.) (2005). *How students learn: History, mathematics, and science in the classroom*. Washington, DC: National Academies Press.

Donovan, M. S., Bransford, J. D., & Pellegrino, J. W. (Eds.) (1999). *How people learn: Bridging research and practice*. Washington, DC: National Academies Press.

Donovan, M. S., & Pellegrino, J. W. (Eds.) (2004). *Learning and instruction: A SERP research agenda*. Washington, DC: National Academies Press.

Donovan, S., Wigdor, A. K., & Snow, C. (Eds.) (2003). *Strategic education research partnership*. Washington, DC: National Academies Press.

Duschl, R. A., Schweingruber, H. A., & Shouse, A. W. (Eds.) (2007). *Taking science to school: Learning and teaching science in grades K–8*. Washington, DC: National Academies Press.

Ferguson, R. (1991). Paying for public education: New evidence on how and why money matters. *Harvard Journal on Legislation*, 28, 465–498.

Ferguson, R., & Ladd, H. F. (1996). How and why money matters: An analysis of Alabama schools. In H. Ladd (Ed.), *Holding schools accountable: Performance-based reform in education*. Washington, DC: Brookings Institution. pp. 265–298.

Hake, R. R. (1998). Interactive-engagement versus traditional methods: A six-thousand-student survey of mechanics test data for introductory physics courses. *American Journal of Physics*, 66, 64–74.

Hiebert, J., Gallimore, R., & Stigler, J. W. (2002). A knowledge base for the teaching profession: What would it look like and how can we get one? *Educational Researcher*, 31(5), 3–15.

Kilpatrick, J., Swafford, J., & Findell, B. (Eds.) (2001). *Adding it up: Helping children learn mathematics*. Washington, DC: National Academies Press.

Minstrell, J., & Kraus, P. (2005). Guided inquiry in the science classroom. In M. S. Donovan & J. Bransford (Eds.), *How students learn: Science in the classroom*). Washington, DC: National Academies Press. pp. 475–514.

National Reading Panel (2000). *Teaching children to read: An evidence-based assessment of the scientific research literature on reading and its implications for reading instruction* (NIH Pub. No. 00-4769). Washington, DC: National Institute of Child Health and Human Development.

National Research Council (2003). *Assessment in support of learning and instruction: Bridging the gap between large-scale and classroom assessment.* Committee on Assessment in Support of Instruction and Learning. Board on Testing and Assessment, Committee on Science Education K–12, Mathematical Sciences Education Board, Center for Education, Division of Behavioral and Social Sciences and Education. Washington, DC: National Academies Press.

Pashler, H., Bain, P., Bottge, B., Graesser, A., Koedinger, K., McDaniel, M., & Metcalfe, J. (2007). *Organizing instruction and study to improve student learning* (NCER 2007-2004). Washington, DC: U.S. Department of Education, National Center for Education Research, Institute of Education Sciences. Online. Available at: http://ncer.ed.gov.

Pellegrino, J. W., Chudowsky, N., & Glaser, R. (Eds.) (2001). *Knowing what students know: The science and design of educational assessment.* Washington, DC: National Academies Press.

Phillips, G. (2007). *Chance favors the prepared mind: Mathematics and science indicators for comparing states and nations.* Washington, DC: American Institutes for Research.

Roseman, J. E., Kesidou, S., Stern, L., & Caldwell, A. (1999). Heavy books light on learning: AAAS Project 2061 evaluates middle grades science textbooks. *Science Books & Films,* 35(6). Online. Available at: http://www.project2061.org/newsinfo/research/textbook/articles/heavy.htm (accessed March 8, 2000).

Schneider, B., & McDonald, S.-K. (Eds.) (2006). *Scaling up educational reform: Vol. 2: Issues in practice.* New York: Rowman & Littlefield.

Shavelson, R., & Towne, L. (Eds.) (2002). *Scientific research in education.* Washington, DC: National Academies Press.

Snow, C. (2002). *Reading for understanding: Toward an R&D program in reading comprehension.* Santa Monica, CA: RAND, RAND Reading Study Group.

Snow, C., Burns, M., & Griffin, P. (Eds.) (1998). *Preventing reading difficulties in young children.* Washington, DC: National Academies Press.

Vosniadou, S., & Brewer, W. F. (1994). Mental models of the day/night cycle. *Cognitive Science,* 18, 123–182.

Wiliam, D. (2007). Keeping learning on track: Formative assessment and the regulation of learning. In F. K. Lester Jr. (Ed.), *Second handbook of research on mathematics teaching and learning* (2nd ed.). Greenwich: Information Age Publishing. pp. 1053–1098.

Wilson, M., & Bertenthal, M. (Eds.) (2005). *Systems for state science assessment.* Washington, DC: National Academies Press.

Wilson, M., & Sloane, K. (2000). From principles to practice: An embedded assessment system. *Applied Measurement in Education,* 13, 181–208.

20 The Evolution of Head Start

Why the Combination of Politics and Science Changed Program Management More than Program Design

Thomas D. Cook, Manyee Wong, and Vivian C. Wong

This chapter describes the historical process whereby senior Head Start staff at first sought hard-headed social science evidence about their program's effectiveness but then retreated from this commitment, burned by biased evaluation results that were used against them in the political process. But political calls to learn about Head Start's effectiveness could not be silenced. Initially, Head Start officials answered these calls by sponsoring small, local studies of limited technical quality that could not provide either definitive or general causal answers. Such studies were later meta-analyzed but the technical quality of individual studies still precluded clear answers about the program's causal impact. The officials also redirected research efforts toward developing and monitoring program compliance in order to reduce local variation and to create a more professional-appearing program without avoiding the large-scale summative evaluations whose results might threaten the program's budget and political support. These strategies postponed Congressional pressure to obtain broad-based causal results about program impact, but they did not still it.

Fortunately, methods eventually became available that permitted a study of Head Start's effectiveness with high internal and external validity that would not alienate the many parts of the Head Start political constituency, which, over the years, had become wary of scientific research. In this context of joint political pressure and technical method advance, several truly scientific studies of Head Start's implementation and impact were conducted, lending more credence to earlier evaluation findings that showed short-term cognitive gains.

But science is not just about method. It is also about substantive theories and the critically appraised evidence for them. So we ask: How has Head Start used theories and findings from developmental science? The program was launched in great haste, and its origins were a mish-mash of generalities from health, socio-emotional development, early education, parent involvement, parent employment, and community development. Little theory was invoked to justify the program content other than to claim it should be holistic, focusing on many outcomes. Applied early childhood researchers were later involved as grantees and program consultants as officials sought to improve Head Start by introducing components from other more successful theory-based preschool programs, such as Perry Preschool (Lazar et al., 1982; Ramey & Campbell, 1984; Schweinhart & Weikart, 1980). However, developmental science as a whole has not contributed much to program renewal through, for example, the specification of new standards, the identification of new possible program components, or close theoretical examination of the many Head Start purposes and how they can be sharpened and combined. Head Start's program theory is still largely a legacy of holistic child development theory of the 1960s, leavened by dated sociological theory about poverty and community development.

In recent years, Head Start has made efforts to respond to the shifting needs of immigrants and of families with working mothers. These attempts speak more to modifying outreach activities than to scientific advancement in the conceptual and operational theory of the program. Can a national program improve itself enough merely by sponsoring more scientific evaluations and borrowing program details from other preschool programs? Are there fruitful roles that the developmental sciences could play in program improvement but have not yet done so? What responsibility, if any, do developmental scientists have for helping improve program design? These are the questions we pose as we seek to add more and better scientific input into the content of Head Start.

Head Start Past and Present

Head Start was launched in 1965 as part of Lyndon B. Johnson's War on Poverty program. It began as a national eight-week summer program. Program proponents hoped that Head Start would counter poverty's adverse effects by enhancing poor children's cognitive development and school attitudes early in life (Bloom, 1964). So a Planning Committee of scholars from medicine, social work, psychology, and education was formed and charged to draft recommendations for the new program's goals. With such diverse disciplinary representation, significant pressure emerged to focus on various domains of child development (e.g., socio-emotional development, parental involvement, and nutritional and health services). Thus, Head Start was charged with promoting all aspects of children's development through the provision of comprehensive services.

Head Start was initially funded by the Office of Economic Opportunity (OEO). At the time, OEO emphasized community action and encouraged parental involvement in program decision making and the establishment of partnerships with local social service organizations and community groups. By the program's fourth year, it had grown to serve over 2 million children with an annual operating budget of $350 million (White, 1970).

Today, Head Start is a year-round program with federal appropriation of nearly $7 billion and, in February 2009, Congress passed the stimulus-spending bill that provides an additional $2.1 billion for Head Start and Early Head Start. The program serves mostly children (51 percent of four-year-olds, 36 percent of three-year-olds, and 3 percent of five-year-olds) from families at or below 100 percent of the poverty level, but centers can enroll up to 10 percent of families with higher incomes, and full services are available to disabled children regardless of income. At least 10 percent of Head Start's enrollment must be of children with special needs. Head Start still aims to promote the school readiness of low-income children by enhancing their social, cognitive, and physical development through the delivery of comprehensive services but now prioritizes on academic achievement.

The Earliest Evaluation Studies

Whether Head Start should be evaluated at all has been a point of contention since its inception. Physicians on the Planning Committee argued that, if poor children were receiving prescribed meals and health services, there was nothing to evaluate, while others argued that valid educational performance measures were not available to assess poor, minority populations. Nonetheless, an evaluation of the 1965 summer program was conducted. Zigler and Muenchow (1992) describe the effort as a disaster owing to the lack of valid measures, the wide range of outcomes assessed (e.g., demography, health, and social competency), and the desire to include all enrolled children in the study. The result was a deluge of survey data that could not be processed. Moreover, no IQ measures were included, drawing criticisms from both members of Congress and researchers. The experience led Zigler to conclude that evaluation efforts needed to be more local and intensive (Zigler & Muenchow, 1992). As a result, 14 regional centers were created for the research and evaluation of Head Start. Some of these centers provided data that would be used in two evaluation efforts that we discuss later: the Consortium Report (Lazar et al., 1982) on early childhood programs and the meta-analysis of Head Start (McKey et al., 1985).

Local evaluations of Head Start summer programs yielded some useful results. Eisenberg and Conners (1966) claimed that a Baltimore summer program led to IQ gains of eight to ten points, and OEO officials enthusiastically publicized this result. But Wolff and Stein (1966) examined four New York City kindergartens and found that the benefits of the summer program faded after six months. These early evaluations were methodologically very weak, most being narrative accounts of presumed local successes and failures. They had the misfortune to appear almost simultaneously with the Coleman Report (1966), which concluded that educational spending had little effect on students' achievement when compared with their socio-economic background. This also raised doubts about whether compensatory programs could reverse poverty's deleterious effects at all. Taken together, these early studies prompted concerns about Head Start's long-term ability to raise academic achievement, whatever its short-term benefits.

Mounting political pressure to prove that its portfolio of programs worked led OEO to establish an Office of Research, Planning, Programs and Evaluation (RPP&E). RRP&E was to be an independent scientific auditing arm to examine the cost-effectiveness of every program so they could be prioritized given limited funding (Zigler & Muenchow, 1992). Staffed mostly by quantitatively trained evaluators and economists, RRP&E was not impressed by the technical quality of the 1965 evaluation or by the myriad of outcomes examined. Because Head Start was primarily a program aimed to improve children's learning, RPP&E chose to prioritize assessments of participants' cognitive development. In spring 1968, it proposed a national evaluation of Head Start. The purpose was to examine the effects of the summer and full-year programs for 1965, 1966, and 1967 using data collected in disadvantaged neighborhoods with Head Start centers and from children in the first three years of school in order to assess the cognitive and affective status of children. The study was to be retrospective, with a comparison group set up three years after the program had begun; and it was to be completed by spring 1969, less than a year after awarding the contract.

The Head Start Steering Committee (renamed from Planning Committee) had misgivings about the project. Its experience was that state-of-the-art methods were not yet suitable for a rigorous national evaluation and that such an evaluation could not be responsive to the tremendous variation in quality across sites and so would miss positive effects attributable to higher-quality programs. Members also argued that it was unfair to evaluate the first summer, when program operations had been chaotic given the great haste with which Head Start was launched and the speed with which it grew. They also argued that the study should assess program impacts on health, nutrition, and community involvement outcomes as well as on cognitive gains; and that no comparable control group could be achieved three years after the program had been launched. The Planning Committee proposed an alternative design, a longitudinal study. But OEO proceeded with its planned summative study, awarding the evaluation contract to Westinghouse/Ohio State.

The Westinghouse/Ohio State Report

The National Impact Study examined how Head Start children in the first, second, and third grades differed in their intellectual and socio-emotional development from comparison children who had not been in the program. The sample consisted of 104 Head Start centers randomly selected from a national list and reflected the program's proportion of summer (70 percent) and full-year programs (30 percent) at the time. Children in the same grades at public schools near selected Head Start centers were identified as the comparison group and were matched on race, sex, and kindergarten attendance. Despite attempts to match children within neighborhoods and schools, comparison children were more materially advantaged than Head Start children.

Assessments were conducted on cognitive and affective outcomes as well as environmental factors that might influence children's achievement in school. The study found that comparison children outperformed summer Head Start students in all three grades, implying that the summer Head Start program was harmful to those it served. The full-year program, however, produced small but reliable gains in cognitive development among treatment children relative to controls in grades 1 and 2, but not for those in grade 3, where the sample size was small. No improvements in children's affective development were found for the full-year program. Effects seemed to be larger in centers that enrolled mostly African-American children, but, even so, Head Start children performed below national norms in language development and scholastic achievement, approaching them in school readiness only by first grade. However, it was clear that Head Start parents strongly supported the program and reported substantial participation in it (White, 1970). Despite evidence that indicated immediate short-term gains for the full-year Head Start program and the larger effects found for African-American children, the final report concluded that the benefits of Head Start were unsatisfactory given the substantial efforts and resources used to remediate the effects of poverty on disadvantaged children (Cicirelli, Evans, & Schiller, 1969).

The Westinghouse Report had its share of critics (Campbell & Erlebacher, 1970; Smith & Bissell, 1970). But it was released close to when Jensen (1969) published his influential monograph emphasizing the inheritability of IQ and implying the futility of compensatory education efforts aimed at raising IQ. The confluence of the Jensen and Coleman Reports and the Westinghouse/Ohio State Study probably helped end the first era when it was widely believed that compensatory education programs could overcome genetic and environmental limitations. Indeed, the Nixon administration planned to phase out Head Start in three years, and it took no small effort from Elliot Richardson, then Secretary of Health, Education and Welfare, to convince Nixon to keep it in place. In spring 1969, Nixon created the Office of Child Development (OCD) and assigned Head Start to it, all while reducing the program's summer efforts and capping its total budget.

Improvement and Innovation

Burned by the Westinghouse Study, Edward Zigler (head of OCD) sought to keep Head Start relevant by embarking on a period of experimentation in order to identify improvements in program design and delivery. In the early 1970s, he launched the Improvement and Innovation Initiative (I&I) that would become a blueprint for Head Start's operations for the next 30 years. The initiative sought solutions to two seemingly contradictory program challenges. First, management needed to be tightened to reduce disparities in program quality across sites; but, paradoxically, programs also needed greater latitude to innovate and discover new ways of doing things.

In late 1971, OCD began developing new performance standards for services that full-year center-based programs were required to provide. The goal was to identify state-of-the-art practices that had been shown to improve healthy child development and could be implemented in Head Start centers. For some fields, such as health and nutrition, empirical evidence and professional consensus provided clear guidance for identifying state-of-the-art practices. However, for other fields, including early childhood education, the evidence was less clear, so OCD set standards that were generally believed to be good practices though not empirically tested approaches to early childhood education.

Aside from setting performance standards for specific services, OCD also established a monitoring system that remained in effect until the 1980s. Head Start grantees were required to evaluate their performance annually relative to both the performance standards and a community needs assessment. In addition, regional Health, Education, and Welfare (HEW) officials conducted annual visits to centers, and a multidisciplinary monitoring team conducted full-scale evaluations every three years. This was to improve management and services, and to demonstrate to Congress and the administration that OCD was operating Head Start cost-effectively and professionally (Zigler & Muenchow, 1992).

The "innovation" side of I&I was quite different. It involved launching a number of experimental programs that moved Head Start beyond its center-based model. Zigler viewed Head Start as a national laboratory for modeling the best and most innovative programs in child and family services. Head Start had earlier attempted several experimental programs, such as Follow Through (aimed to support children's cognitive gains during the early school years) and a curriculum development projects. Later demonstration projects included Health Start, which aimed to expand health-screening services to poor children not enrolled in Head Start; Home Start, which delivered Head Start–like services in children's homes; and the Child Development Associate Program, which credentialed preschool workers with relatively little formal education. These projects sought to gain additional funding for experiments to identify what works.

Report from the Consortium for Longitudinal Studies

Despite the no-growth budget, the Westinghouse Study's political fallout, and the new monitoring and local experimentation priorities, OCD maintained a small residual interest in assessing Head Start's impact not shared by the broader Head Start constituency. In 1971, Zigler asked Urie Bronfenbrenner to review all the available longitudinal data on the effectiveness of early intervention programs. When Bronfenbrenner (1974) released results corroborating the claims of Wolff and Stein (1966) and the Westinghouse Report, which showed initial cognitive gains followed by diminished performance in later years, it almost permanently soured the Head Start community against impact evaluations (Zigler & Muenchow, 1992). The report was a greater threat to Head Start than the Westinghouse Study and, in response, Zigler (who had by then left OCD but still worked with it) encouraged several researchers to launch a new study on the long-term effects of preschool in general.

The Consortium for Longitudinal Studies synthesizes the results of all studies of the long-term effects of preschool programs. Led by Lazar and Darlington, its goals were to estimate the overall effectiveness of early childhood programs and to probe whether the size of any effects varied by the type of child. Eleven independent demonstration projects were examined. All had treatment and control groups and all were moderately well implemented, most with close involvement of the program developer. But they varied in many specifics, including project location, age of children, length of intervention exposure, and racial composition of the groups served. None of these 11 studies was of Head Start itself, though two were related to the program.

Overall, the study found that children who attended preschool programs were more likely to be competent performers in school. They were less likely to be in special education classes or held back a grade. They scored higher on intelligence tests several years after leaving preschool and achieved at a higher level in later grades. They were more likely to cite achievement-related reasons for being proud of themselves, and they rated their own school performance higher than controls did. Parents also cited higher aspirations for their children than did controls.

The stakes for this study were quite high. It was the first rigorous longitudinal study of preschool education and its results could be seen as helping or hindering not just Head Start but the whole concept of early childhood education (Zigler & Muenchow, 1992). The gamble paid off but many commentators made the mistake of generalizing the study's positive results to Head Start (Hacsi, 2002; Hood, 1992; Woodhead, 1988; Zigler & Styfco, 1993). In particular, many reporters cited the work as evidence that Head Start works. In reality, although the results may have some indirect implications for Head Start, Head Start was not among the programs studied. Nonetheless, Head Start officials attached themselves to the study's coat tails and were not conspicuous in denying its applicability to their own program, perhaps seeing it as a demonstration of Head Start's potential if not its achievements. The long-term results of some of these non-Head Start demonstrations are still widely used to support the idea that early childhood intervention is a proven cost-beneficial commodity worth even more public investment (Heckman & Masterov, 2005). So it serves Head Start politically to be (mis)construed as one of these successful demonstrations of efficacy studies (small-scale and carefully controlled) rather than of effectiveness studies (scaled up with little implementation oversight).

One important and concrete consequence occurred because of Head Start reaction to the positive synthesis of preschool programs. The Perry Preschool/High Scope intervention showed up very well in the Consortium Study and became widely known for its long-term effects on adult education, employment, marriage, and incarceration outcomes. So Head Start authorities featured it prominently when recommending educational curricula from which individual Head Start grantees or centers could choose. The rationale was simple: If the common criticism is that Head Start does not raise achievement on a long-term basis, then why not ask centers to adopt those educational materials that have been best demonstrated to work long term? Here, developmental science and specific evaluation findings were used to improve the program by adopting components not actually identified and developed within Head Start.

The Meta-analysis of Head Start Effects

Shortly after the Consortium Report, the Chief of Evaluation at OCD, Dennis Deloria, began receiving constant inquiries from members of Congress about Head Start's effectiveness. This led Deloria to pull together all the research on Head Start's effects and to publish an annotated bibliography (Mann, Harrell, & Hurt, 1978). The effort was not funded with OCD evaluation monies because of fear that the results might be negative and would have to be released to Congress. Fortunately, the qualitative review seemed to promise positive results, paving the way for a formal meta-analysis of all Head Start studies dealing with the program's impact on child health, cognitive, and socio-emotional development. The main intent was to learn about child outcomes over a longer period than in the past and to use the new knowledge to help guide policy.

The meta-analytic results (McKey et al., 1985) again showed that Head Start had an immediate but short-term impact on cognitive development but also that it positively affected grade retention and special education placement. The study also claimed overall positive results for the socio-emotional outcomes, with Head Start students having higher self-esteem, greater achievement motivation, and more positive social behavior. The study also found that program impacts were greater after the 1970s than earlier, suggesting that the ongoing programmatic changes favoring full-day preschool, greater technical assistance, and enforced performance standards may have been collectively or individually beneficial.

Unfortunately, these results are based on very few studies and, of all studies used in the synthesis, half used some form of treatment/control design whereas the others used pre-/post-test designs without a control group. So the technical quality of the studies in the meta-analysis was not high, even by the standards of the day. Gamble and Zigler (1989) argued that meta-analysis is summative evaluation that should be done very carefully and only in extraordinary circumstances because the stakes are those of a single large-scale evaluation. Schweinhart and Weikart (1980) echoed this, further noting that many of the included studies had noncomparable treatment and control groups likely to cause biased estimates, that the studies were not formally representative of the national Head Start program, and that many conclusions depended on small numbers of studies with few children evaluated.

Despite Deloria's claim that politics did not guide much of the impetus for the meta-analysis other than in response to calls from Congress to learn about impacts, Zigler's own account suggests otherwise (Zigler & Muenchow, 1992). Although there was widespread support from the public and many legislators, key players in OCD could begin to see the erosion of support for Head Start, in part because of mixed evaluation findings. According to Ray Collins (personal communication, October 1, 2007), support for the study was definitely a move by the Head Start "people above" to challenge the politically threatening claim of no long-term effects on children. Indeed, the impetus to use research to improve program management, to identify new program practices, to generate ammunition about long-term effects in other preschool programs, and to make meta-analysis conditional on an informal qualitative assessment of its likely impacts can be seen as partial responses to fears about national research into Head Start's accomplishments. In any event, the meta-analysis provided positive results for program lobbyists who no longer thought they needed to maintain the status quo. The National Head Start Association could now launch a successful campaign to increase Head Start funding and, in 1978, the program received a $150 million increase, a magnitude not seen since 1967 (Zigler & Muenchow, 1992).

Improving Performance

The Clinton 1994 reauthorization of Head Start expanded the program, but also instituted mandates for revising the program's performance standards (Hacsi, 2002). Released in November 1996, the revised standards covered new services for Early Head Start, eliminated unnecessarily prescriptive rules (e.g., mandates for child-size utensils), and required programs to implement curricula consistent with the new standards (U.S. Department of Health and Human Services, Administration for Children and Families, Head Start Bureau, 1996). The training and technical assistant (T/TA) system, dismantled during the Reagan era, was also renewed and a program of research and evaluation to improve Head Start and disseminate new ideas was required.

However, questions about the program's overall effectiveness remained. Although Health and Human Services (HHS) and Head Start administrators and advocates adamantly asserted that Head Start works, lingering doubts remained. A 1997 U.S. General Accounting Office report criticized the Consortium Report and the 1985 meta-analysis, claiming they were based on studies that were 20 years old, had questionable research methodologies, and were not nationally representative. Only a handful of the included studies explicitly examined Head Start, and most of these were limited to the program's cognitive benefits. Critics also pointed to what they interpreted to be the growing achievement gap by family income level, using it as evidence to contend that Head Start was obviously failing to prepare poor children for school.

The 1998 reauthorization of Head Start limited the program's expansion and refocused its efforts on quality improvements and impact evaluations. Congress required HHS to develop new education performance measures to monitor what children and families should attain from participating in Head Start (see Head Start Act, 1998). HHS was also to provide grantees with technical assistance to assure school readiness, promote family literacy, encourage larger salaries for staff with higher levels of education, and institute new teacher and staff educational requirements. Head Start agencies were also required to coordinate with local education agencies and community schools, and to develop procedures to help children transition to kindergarten. Finally, the 1998 reauthorization reserved funds for national impact evaluations of Head Start. What we see here is a growing political interest in directing Head Start's focus toward children's cognitive development to prepare them for reading and math in schools.

George W. Bush's administration was charged with responsibility for implementing the mandates of the 1998 Head Start reauthorization, and he proposed moving Head Start from HHS to the Department of Education in order to prioritize the goal of intellectual development over Head Start's many other goals for children (Haskins & Sawhill, 2003). Although the proposed move did not happen, a new set of accountability measures called the National Reporting System (NRS) did. Under NRS, Head Start children are tested twice a year in math and preliteracy skills. A new T/TA program called Steps to Success (STEP) was also established to provide Head Start staff and administrators with training in strategies to support literacy skills.

Head Start advocates voiced strong opposition to the Bush administration's emphasis on cognitive development. They also opposed the NRS, claiming the system would have little practical benefit since it provided low-quality data, was burdensome, and was even culturally inappropriate in some aspects. They also criticized the Bush administration for dismantling the old T/TA system in favor of a far less effective assistance network (National Head Start Association, 2007). However, the NRS did not require legislative approval and, although both House and Senate versions of the Head Start reauthorization bills have included amendments to suspend it until the National Academy of Sciences can develop appropriate and valid assessments, it was eliminated in the 2007 Head Start reauthorization.

These political skirmishes go to the heart of Head Start. How focused should it be? How closely should it be tied to professional versus local community development norms? How tightly integrated should it be to schools and their kindergarten curricula? How important is it as a health program? We have been struck by how little developmental theory and findings have been used in these debates, now brought to a peak but hardly novel in the history of Head Start. Where is the research, or even the grounded speculations, about whether children's life chances are better promoted by raising their test scores at age four as opposed to achieving gains in other cognitive domains, or in socio-emotional, physical, or health status? Economists and lawyers are central in the current political debates, and their adherence to human capital theory predisposes them to value cognitive gains over all others, but developmentalists should be disposed to consider a more holistic framework than achievement test scores. We have uncovered little evidence that they have been used in debates about Head Start performance standards and measures, currently the major mechanisms for driving program improvement.

Also striking in the debates about performance standards is how research funds are to be used. Although there is a call for impact study, the main emphasis seems to be on setting up nationwide and expensive monitoring systems. The monitoring systems are about improving implementation in the service of reaching at least some clear goals. New and better teachers are needed; existing teachers need more and better professional development; sites need to be sampled to check that these things are happening and to learn what teachers are actually working to implement at the classroom level. Monitoring is important and central to most formative evaluation systems, but its implementation is expensive and so summative evaluation may take a backseat. Of course, when total funds for Head Start increase, so do funds for research and evaluation, potentially permitting both formative and summative evaluation goals to be explored together. But, as funds expanded at the time, the major beneficiary was a single large monitoring study to describe Head Start as a whole called FACES.

Family and Child Experience Survey (FACES)

Launched in 1997 (West et al., 2007), FACES is a longitudinal study of a nationally representative sample of Head Start children, parents, teachers, and classrooms. Its purpose is to develop performance measures for each of these entities and to examine how program quality is related to child outcomes. FACES provides descriptions of the cognitive, socio-emotional, and physical development of Head Start children; of the characteristics and well-being of Head Start families; of the observed quality of Head Start classrooms; and of the characteristics, needs, and opinions of Head Start teachers and program staff.

Results from the FACES 2003 report (Zill, Sorongon, Kim, Clark, & Woolverton, 2003) showed that Head Start children enter with skills well below the national averages in early literacy and math. Despite statistically significant gains in vocabulary, early writing, and math skills among four-year-olds (but not three-year-olds), Head Start children remained below national norms in all areas. The average number of letters that Head Start children know increased from 7.2 in 1997 to 10.1 in 2001 (Head Start Bureau, U.S. Department of Health and Human Services, Administration for Children and Families, 2001), thus reaching the 1998 Congressional goal. In addition, Head Start children showed significant growth in such social skills as more cooperative and less withdrawn behavior. In kindergarten, Head Start children made improvements toward national averages in vocabulary, early writing, and math, but did not meet these goals. Regression techniques were used to examine associations between child outcomes and teacher quality and curriculum choice, holding certain background characteristics constant. The results indicated that higher teacher salaries and the use of an integrated curriculum such as High/Scope or Creative Curriculum were associated both with larger gains in letter knowledge and with reductions in problem and hyperactive behaviors. Higher teacher education was also linked to greater gains in children's writing skills, whereas longer classroom days were associated with greater gains in both letter knowledge and early writing. However, neither observed classroom quality nor adult-to-child ratios were related to child outcomes.

FACES is a descriptive longitudinal study that provides a comprehensive and scientifically valid description of the Head Start program, not of its causal effects. It illustrated the levels at which Head Start children were performing and how they related to national norms. When the FACES results were released, many Head Start program officials were shocked to see how disadvantaged Head Start children were compared with middle-class samples. Ron Haskins characterized results from FACES as being at best modestly encouraging. For him, the worst news was that Head Start children entered the program scoring only slightly above the twentieth percentile in many subjects and, though they made significant gains by the end of the year, they could not attain national norms (Haskins & Sawhill, 2003). That would be quite an accomplishment, however, since most middle-class four-year-olds are themselves engaged in formal and informal educational activities in child-care arrangements and homes.

The Head Start Impact Study

A comprehensive descriptive study such as FACES did not temper Congress's demand for a good-quality study of program effectiveness. In the 1998 Head Start reauthorization, Congress also mandated the national Head Start Impact Study (HSIS) and required the use of a stratified cluster random sampling design to examine how Head Start affects children's cognitive, socio-emotional, and health status. The sample was specified to be nationally representative and to include three- and four-year-olds, who would be randomly assigned to the program or not and who would be followed through the spring of first grade. Four broad outcome domains were to be examined: cognitive development, socio-emotional development, health status and access to health care, and parenting practices. Thus, the study included both random selection (of Head Start sites) and random assignment (of individual children) with assessments in multiple outcomes, the ultimate in experimental design, an aspiration more often envisaged than attained.

It was carried out despite many Head Start advocates' fear and distrust of the science behind it and the expectation that many sites would refuse to collaborate since they would be required to deny program services to some children. So why was this ambitious and state-of-the-art summative evaluation done? Part of the reason has to do with Congress's specification of the study's design and HHS acceptance and experience in the use of random assignment in program areas such as job training. But the most likely reason is that the design of the study required so few children at each site that a vast majority of the centers could recruit the few controls needed without turning away eligible children. Also, center directors recognized that their own centers could not be evaluated because of each site's small sample sizes. The evaluation was not about each center's accountability; it was about the national program's accountability.

Study findings showed small to moderate one-year positive impacts in most cognitive tests for both three- and four-year-olds. The study also showed a modest reduction in problem behaviors for three-year-olds but not four-year-olds. However, the results for other socio-emotional outcomes were weak and rarely reliable. These results depended on data from parents, but future analyses will include information from teachers and other caregivers. As to health, there was a small to moderate increase in access to dental care for three- and four-year-olds and an improvement in health status for three-year-olds only. In terms of parenting, there is more use of educational activities in homes and less use of physical discipline by the parents of three-year-olds. All these results are based only on one year of program exposure. Two-year results are due out soon.

Conclusions

This chapter is about Head Start as a case study in the use of science as a method for program improvement. We have described a remarkable growth and acceptance of science in the Head Start community: from early origins that supported quality methods, to a reaction against these methods, to avoiding them through the use of local monitoring and outcome studies of little resonance and low technical quality, to a national monitoring system that provides formative feedback to sites for self-improvement, to a national program of formative evaluation designed to identify implementation needs and child performance levels relative to national norms, and to a national summative evaluation characterized by random selection followed by random assignment—a seldom achieved duo—that is designed to describe the program's overall achievement in health, socio-emotional development and preschool academic performance. Few federal agencies can boast a system of evaluation as cutting edge and comprehensive as Head Start's.

But it would be wrong to see this system as emerging easily. Politics often played a crucial role. Head Start's origins are in local community development and its priorities are on local actions, with research and evaluation a distant priority. So the push and pull between Head Start program officers who view evaluation as a potential threat and members of Congress who continually demand demonstration of effectiveness played a crucial role in how the program's evaluation evolution. Program officers pushed for formative evaluation aimed to improve program functions while Congressional members pushed for impact evaluation aimed to provide accountability. The Head Start Impact Study would not have happened had it not been for the Congressional push and specification of scientific standards. It would also probably not have been possible without science advancing to create a design that assessed only the national program.

Science is not only about method choices but also about substantive theory and its evidence base. Social scientists have been widely used in Head Start over the last 40 years, including experts on child development, but their task has rarely been to reconfigure the basic program theory or to offer advice about its components and their relationship to each other. To be sure, applied developmentalists have been involved with Head Start in a number of different ways as described in this chapter, but we are struck by the relative paucity of input from developmentalists with a basic theoretical orientation and especially by experts on specific aspects of development (self-regulation, for instance). Should they not be more involved in the nitty-gritty of program design and modification? It is always difficult in science to pontificate about what is missing, and so it is only our impression that theoretical developmentalists are under-represented in Head Start. Nevertheless, the most pressing problem currently facing Head Start is to improve its design to be more competitive with other preschool programs. We argue that Head Start can make its objectives clearer and more interdependent and can improve its content with more basic developmental research. We suspect that most developmentalists are happy doing basic research in their labs and see Head Start as an applied interest of little value in the discipline. Their research tends to be on micro-processes within specific research domains rather than on the whole child across multiple life domains. But, unless efforts are put in to bridge the cultural and structural divide, Head Start will fail to take advantage of the most advanced thinking in cognitive, emotional, and social development to help with their effort to improve children's cognitive and social outcomes. So this chapter is a *cri de coeur* to readers of this volume, who we assume are much more likely to be scientists from the field of child development than from other disciplines. Help put as much science into the design of Head Start activities as there is in its current evaluation activities. The politics of academic life and its rewards and punishments may be a more severe limitation to Head Start than the politics of Washington.

Acknowledgments

This work was supported by Institute of Education Sciences Grant no. R305U07003. Thanks go to Ben Allen, Ray Collins, Lois-Ellin Datta, Dennis Deloria, Ron Haskin, Blair Lehman, Ruth McKey, Nancy Stein, Louisa Tarullo, Mary Bruce Webb, and Edward Zigler for providing background information and feedback. All errors are the authors'.

References

Bloom, B. (1964). *Stability and change in human characteristics.* New York: John Wiley & Sons.

Bronfenbrenner, U. (1974). *A report on longitudinal evaluations of preschool programs. Vol. II: Is early intervention effective?* Washington, DC: Office of Child Development.

Campbell, D. T., & Erlebacher, A. (1970). How regression artifacts in quasi-experimental evaluations can mistakenly make compensatory education look harmful. In J. Hellmuth (Ed.), *Disadvantaged child* (Vol. 3). New York: Brunner/Mazel, Inc. pp. 195–210.

Cicirelli, V. G., Evans, I. W., & Schiller, T. S. (1969). *The impact of Head Start: An evaluation of the effects of Head Start on children's cognitive and affective development.* Athens: Westinghouse Learning Corporation, Ohio University.

Coleman, J. S. (1966). *Equality of educational opportunity study.* Washington, DC: U.S. Department of Health, Education, and Welfare.

Eisenberg, L., & Conners, C. K. (1966). *The effect of Head Start on developmental processes.* Washington, DC: Department of Health, Education and Welfare; Office of Economic Opportunity, OEO-510.

Gamble, T. J., & Zigler, E. (1989). The Head Start synthesis project: A critique. *Journal of Applied Developmental Psychology*, 10, 267–274.

Hacsi, T. (2002). *Children as pawns: The politics of educational reform.* Cambridge, MA: Harvard University Press.

Haskins, R., & Sawhill, I. (2003). The future of Head Start. *The Brookings Institute: Policy Brief on Welfare and Beyond*, 27, 1–8.

Head Start Act (1998). Community opportunities, accountability, and training and education services act of 1998. Public Law, 105–285.

Head Start Bureau, U.S. Department of Health and Human Services, Administration for Children and Families (2001). *Head Start program performance standards & other regulations.* Washington, DC: Author.

Heckman, J., & Masterov, D. V. (2005). *The productivity argument for investing in young children.* Washington, DC: Committee on Economic Development.

Hood, J. (1992). *Caveat emptor: The Head Start scam.* (Policy Analysis No. 87). Washington, DC: Cato Institute.

Jensen, A. R. (1969). How much can we boost IQ and scholastic achievement? *Harvard Educational Review*, 39(1), 1–123.

Lazar, I., Darlington, R., Murray, H., Royce, J., Snipper, A., & Ramey, C. T. (1982). Lasting effects of early education: A report from the Consortium for Longitudinal Studies. *Monographs for the Society for Research in Child Development*, 47(2/3), 1–151.

Mann, A. J., Harrell, A., & Hurt, M., Jr. (1978). *A review of Head Start research since 1969 and an annotated bibliography*. Washington, DC: Department of Health, Education, and Welfare, Office of Human Development Services, Administration for Children, Youth and Families.

McKey, R. H., Condelli, L., Ganson, H., Barrett, B. J., McConkey, C., & Plantz, M. C. (1985). *The impact of Head Start on children, families, and communities: Final report of the Head Start evaluation, synthesis, and utilization project*. Washington, DC: U.S. Department of Health and Human Services, Office of Human Development Services, Administration for Children, Youth, and Families, Head Start Bureau.

National Head Start Association. (2007). *National Head Start Association 2007 policy agenda*. Online. Available at: http://www.nhsa.org/advocacy/advocacy_agenda.htm (accessed October 9, 2007).

Ramey, C.T., & Campbell, F. A. (1984). Prevention of intellectual disabilities: Early intervention to improve cognitive development. *Preventive Medicine*, 27, 224–232.

Schweinhart, L. J., & Weikart, D. P. (1980). *Young children grow up: The effects of the Perry preschool program on youths through age 15*. Ypsilanti, MI: High Scope Press.

Smith, M. S., & Bissell, J. S. (1970). Report analysis: The impact of Head Start. *Harvard Educational Review*, 40, 51–104.

U.S. Department of Health and Human Services, Administration for Children and Families, Head Start Bureau (1996). *Fact sheet: Improving quality in the Head Start program: New performance standards*. Online. Available at: http://www.os.dhhs.gov/news/press/1996pres/961105.html.

U.S. General Accounting Office (1997). *Head Start: Research provides little information on impact of current program*. (Report Number HEHS-97-59). Washington, DC: Author.

West, J., Tarullo, L. B., Aikens, N., Sprachman, S., Ross, C., & Carlson, B. L. (2007). *FACES 2006 study design*. Washington, DC: Mathematica Policy Research, Inc.

White, S. H. (1970). National impact study of Head Start. In J. Hellmuth (Ed.), *Disadvantaged child* (Vol. 3). New York: Brunner/Mazel, Inc. pp. 163–184.

Wolff, M., & Stein, A. (1966). Head Start six months later. *Phi Delta Kappan*, 48(7), 349–350.

Woodhead, M. (1988). When psychology informs public policy. *American Psychologist*, 43(6), 443–454.

Zigler, E., & Muenchow, S. (1992). *Head Start: The inside story of America's most successful educational experiment*. New York: Basic Books.

Zigler, E., & Styfco, S. J. (1993). Using research and theory to justify and inform Head Start expansion. *Social Policy Report: Society for Research in Child Development*, 12(2), 1–22.

Zill, N., Sorongon, A., Kim, K., Clark, C., & Woolverton, M. (2003). *FACES 2003 research brief: Children's outcomes and program quality in Head Start*. Washington, DC: Administration for Children and Families.

21 Connecting Developmental Science to Educational Policy by Studying Classroom Instruction

Stephen W. Raudenbush

Educational policy research has historically focused on the costs and consequences of actions available to policymakers such as providing resources, introducing incentives, modifying school governance, and regulating credentials. The first argument of this chapter is that this approach to policy research is incapable of supporting substantial improvements in teaching and learning. My alternative view roots educational policy research in the study of well-conceived instructional regimes. As the chapters of the current volume illustrate, powerful ideas for improving instruction, rooted in basic research on children's learning, are increasingly available, and a concerted attempt to transform classroom instruction in order to take advantage of these ideas will reveal crucial questions for educational policy. Thus reframed, educational policy research can become a significant contributor to educational improvement. However, to realize this potential also requires the recognition that research on instruction faces challenging methodological problems that I shall consider.

Research on Resources

For at least half a century, policy researchers have studied how resources such as facilities, teacher qualifications, class size reduction, and instructional materials affect student achievement. However, Cohen, Raudenbush, and Ball's (2003) review suggests that the yield for policy has been weak. To be sure, meaningful classroom learning without well-prepared teachers, usable facilities, comprehensible instructional materials, or manageable class sizes is hard to imagine, yet an oft-repeated summary of the research is Hanushek's (1989, 1996) proposition that "money doesn't matter." Such a summary strikes many practitioners as nonsensical, generating a crisis in the credibility of educational research.

Hanushek's claim is, of course, open to dispute. Policy research has revealed evidence of some effects of resources on student learning (Hedges, Laine, & Greenwald, 1994) and particularly in regard to class size reduction (Finn & Achilles, 1990; Krueger &Whitmore, 2001; Nye, Hedges, & Konstantopoulos, 2000). More recently, analysis of large-scale longitudinal data collected for school accountability has revealed some effects of teacher qualifications (Clotfelter, Ladd, & Vigdor, 2007). Yet most studies of teacher characteristics, teacher training, school facilities, and per pupil expenditures report remarkably small effects. Thus, despite its centrality in educational policy research, research on resources has shed few insights into how to improve student learning, and it is hard to argue that this research has had more than a marginal impact on the quality of classroom learning in the United States.

Educators use resources in such diverse ways for such varying purposes and with such varying effects that the potential benefits of spending money on schools are hard to discern. The implication is that we must look more deeply into how successful practitioners use resources in classroom instruction to understand how and when investments in resources come to have effects on student learning.

Past research on educational resources is implicitly based on a simple causal model, in which investments in what Cohen et al. (2003) call a "conventional resource" (e.g., per pupil expenditures, teacher credentials, physical facilities, or class size) are the direct causes of a student outcome, typically an achievement test score (see Figure 21.1a). The causal connection between such conventional resources and student outcomes has been found in many studies to be small (Clotfelter et al., 2007; Hanushek, 1989, 1996; Hedges et al., 1994). This sobering experience implies that investing in conventional resources cannot generally be relied upon as a strategy for substantially improving the cognitive skill of the nation's youth.

An alternative reaction to these findings is to criticize the simplicity of the causal model itself. Schooling serves multiple purposes. Increases in a school's budget might support new athletic facilities, new art or music programs, or new approaches to foster student health or social development. The impact of such uses of resources may be positive but is unlikely to be captured in measures of student cognitive skill. To understand the impact of resources, one must therefore consider how educators use those resources in pursuing specific aims.

This reasoning leads to an elaborated model (Figure 21.1b). According to this model, any impact of investing in conventional resources is mediated by how those resources are used. If the model is well specified, there should be no direct connection between conventional resources and instruction. Rather, resources come to affect student outcomes only through specific uses that can be observed and statistically related to the outcome.

The more elaborate model that attempts to "explain" the link between resources and student outcomes has motivated substantial research on school climate, organization, and classroom processes (Cohen et al., 2003). Resources are used for so many purposes, with such varied methods and such varied skill, however, that the task of "explaining" how resources come to affect student outcomes is overwhelming. Moreover, studies of resource use have generally been fraught with selection bias. Practitioners who appear to use resources skillfully are likely to be skilled in many ways, and they may be assigned to comparatively advantaged students.

Instruction as the Proximal Cause of Student Learning

A more straightforward approach is to formulate a model in which the clinical practice of education—classroom instruction—is the immediate cause of student learning (Figure 21.1c). The idea is to test the impact of well-formulated systems of instruction that I shall label "instructional regimes" on student learning. In this setting, one can ask whether enhancing or constraining resources will modify the impact of the regime (Figure 21.1d). Suppose that a well-conceived approach to promoting reading comprehension in second grade is found to be reliably effective across a range

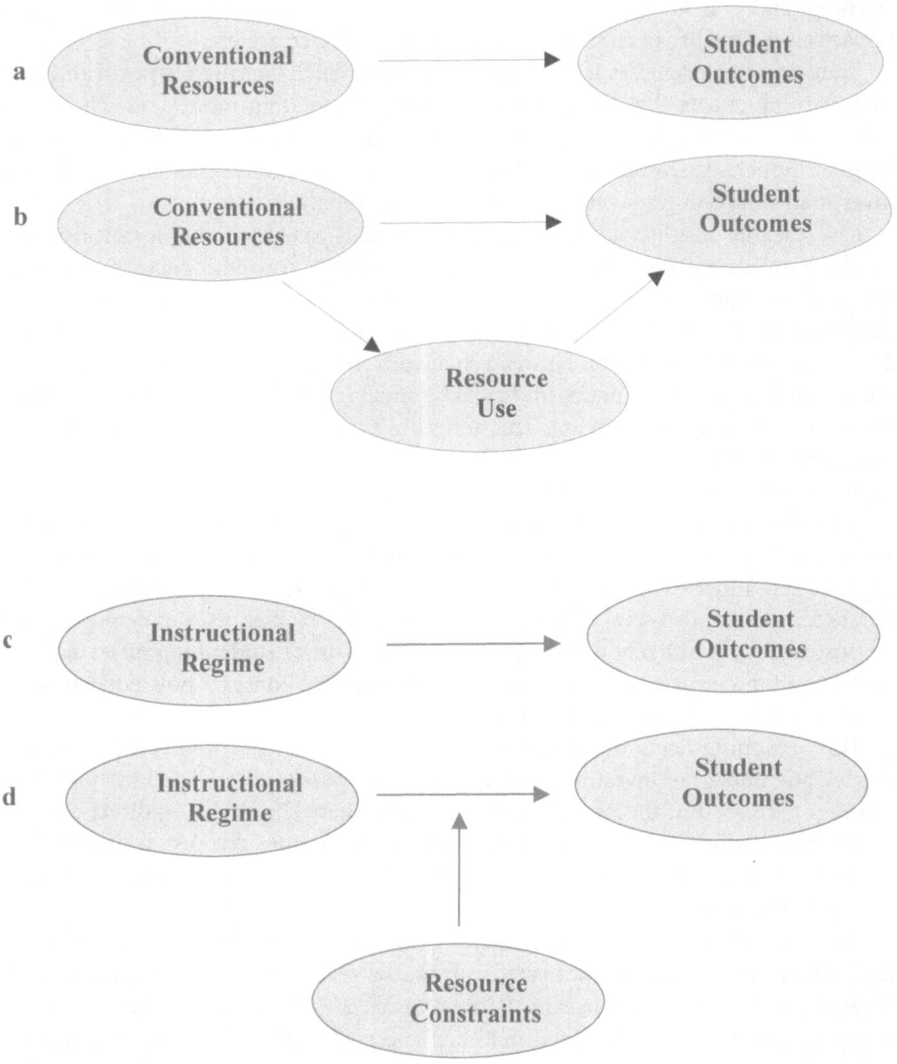

Figure 21.1 Models for Resources, Instruction, and Outcomes.

of school settings. Resource questions immediately follow: What kinds of preservice and in-service teacher preparation are required to enact the program? How small must class sizes be to make the program work? Are alternative regimes that require fewer resources equally effective?

Instructional Regimes

The notion of an instructional regime is founded on a picture of teaching as a continuous interplay between assessment and instruction. An effective teacher is ever attentive to the current level of student skill and knowledge. The interplay is dynamic: The appropriate next step in teaching a concept or skill is predicated on the observable success in having taught the prior step. Assessments can be formal and large-scale (e.g., a districtwide test) or informal and small-scale (e.g., evaluating a student response to a question during classroom discussion).

I therefore define an instructional regime as a more or less explicit set of rules for repeatedly assessing student skill and then tailoring instructional activities in response to the assessment in order to achieve more or less explicit aims for student learning. This process of assessment and tailored instruction produces, over time, considerable experience about a child's progress and the success of attempts to improve that student's learning. This cumulative knowledge, in principle, increases the effectiveness of subsequent efforts to engage the child in instruction.

In some cases, the classroom instructional regime is quite explicit. For example, in algorithm-guided individualized literacy instruction (Connor, Morrison, Fishman, Schatschneider, & Underwood, 2007; see chapter 2), frequent student assessments on specific early literacy skills generate computer-produced recommendations for next instructional steps. In a similar vein, the "Success for All" program (Borman et al., 2005) prescribes periodic assessment of student literacy skills followed by a regrouping of students, who are then targeted for specified modes of instruction. This process of assessing, regrouping, and instruction continues over time. The University of Chicago Charter Elementary Schools use a "balanced" literacy curriculum according to which every student is assessed three times per year (Kerbow, 2006). Each assessment generates a new plan of instruction for the next period. The record of assessments and instructional interventions provides evidence for staff discussions about how each student can reach explicit instructional aims.

In other cases, the interplay between assessment and instruction may not be formalized and may in fact be under the control of essentially autonomous classroom teachers. However, the assumption that effective teachers will engage in something like a regime remains intact even if the essential features of that regime are not readily open to public inspection. The decision to allow autonomous teachers to construct their own regimes might itself be regarded as a regime and evaluated as such. What appears to be a "nonregime" is in fact a hidden regime or a mix of hidden regimes.

The Far-Reaching Implications of Developmental Science for Educational Policy

Developmental science ought to be central to the conception and study of instructional regimes. It potentially prescribes what we teach, when we teach it, how we assess it, and how we differentiate instruction based on the results of the assessment. These considerations, in turn, clarify levels of teacher expertise, and conceptions of teacher learning and career advancement. They have strong implications for the recruitment, training, and induction of new teachers as well as for how the leadership of the school should manage and support the work of teaching. These conditions, together, sharply constrain the framing of questions for policy research.

What We Teach

Several chapters in this volume make an eloquent case for a prominent role for science education in preschool and early elementary education. Romance and Vitale (chapter 9) provide evidence that the process of eliciting, representing, organizing, and analyzing knowledge about the physical world helps children develop vocabulary and reading comprehension. Gelman and Brenneman (chapter 8) make a similar case for preschool children. With this, powerful science instruction becomes a central and explicit feature of the language arts curriculum. This recommendation has strong implications for the role of disciplinary knowledge in the preservice and in-service training of elementary school teachers.

As Stein and her colleagues (chapter 7) emphasize, however, the representation of the scientific content to be taught within this paradigm is the foundation of any success that can follow. What might be called a minimally sufficient representation of the core concepts (for example in early modules on physics) is a starting point, along with explicit representation of the relations among those concepts. This view provides a perspective for critically evaluating current science texts, and the results are sobering.

When We Teach It

Developmental science challenges broadly held notions of when children are able to learn things. Stein and her colleagues (chapter 7) claim that children can learn seemingly advanced topics in physics early—if the core concepts and their relations are represented explicitly, if the instruction strategically uncovers and corrects misconceptions, if children's knowledge is systematically assessed, and if the children receive timely feedback. Gelman and Brenneman (chapter 8) similarly find that preschool children are capable of drawing correct inferences from scientific investigations, generalizing from these, and transferring their knowledge to related domains. They claim that, by learning to make predictions, observe results, and check predictions, these children begin to develop powerful skills in academic language.

Huttenlocher and colleagues (chapter 12) systematically study the emergence of very young children's notions of quantity (e.g., length), with implications for parent and preschool teacher input. Levine and her colleagues show whether and how caregivers' talk about numbers affects early development of the notion of a cardinal set, as well as aspects of spatial reasoning. As this work unfolds, it will have implications for the training of preschool teachers and the assessment of early emergence of cognitive skills that may be essential for later conceptual learning. Early elementary educators will want to know how to capitalize on the results of effective early input as well as how to compensate when that input is missing.

How We Teach It

Morrison and Connor (chapter 2) have tested a developmental paradigm for dynamically assessing literacy skills in the early grades and for tailoring the next instructional steps based on those assessments. This dynamic interplay between assessment and intervention within a coherent developmental framework is an essential feature of a dynamic treatment regime in any discipline, including medicine (Murphy, 2003) or education. A crucial distinction between clinical practice in medicine and in education, however, is that patients are assessed and treated individually whereas in school settings a teacher faces a collection of children with varied instructional histories and varied skills. In that setting, the teacher must attempt to approximate what each child needs in a group setting. The social nature of instruction makes teaching remarkably complex, and arguably more challenging than other forms of clinical practice. Morrison and Connor (chapter 2) aim not only to assess kids' skill and prescribe what they need, but also to supply notions of how the class should be socially organized so that the instruction comes as close as possible to meeting the differentiated literacy needs of the students. Similarly, the value of large-scale randomized trials, such as those discussed by Romance and Vitale (chapter 9), is to show that a developmentally coherent interplay between instruction and assessment can be enacted across a wide range of field settings in which a diverse array of teachers must master the social organization of the instruction.

Implications for Teaching, School Organization, and Leadership

As developmental scientists codify their findings into coherent instructional regimes and take those regimes to scale, we will inevitably encounter challenges to conventional notions of teacher expertise, teacher learning, school organization, and school leadership.

Expertise

Regime-specific teacher expertise must be an essential ingredient to success. At the lowest level, the teacher must understand the basic concepts to be taught and the relations among them, as discussed by Stein and her colleagues. They must also comprehend the developmental logic of the assessments, the feedback, and next instruction steps. A higher level of understanding would entail automaticity in the basic process of assessing and instructing, with working memory freed up to respond creatively to the idiosyncrasies of student responses, questions, and misconceptions. Higher levels would also involve knowing how to teach other teachers to enact the regime effectively.

Levels of Expertise

A differentiated notion of teacher expertise within a regime or across regimes confronts us with questions about the professional recognition and advancement of teachers. To the extent differences in expertise become publicly recognizable and to the extent all teachers in a school are committed to each other's advancement, it becomes in the interest of every teacher to have access to regime-specific instructional leadership. The school principal or instructional leaders among the teacher staff or outside coaches might provide this leadership. One question, then, is whether the teaching profession should become more differentiated than it currently is, such that growth in expertise and the responsibility to teach other teachers is associated with increased pay and status. However, this scenario implies that encouraging competition within the school building—by linking pay to individual classroom performance—would undercut the notion of teachers collaborating to improve each other's performance (Neal, 2008).

School Organization

If an instructional regime is conceived as highly scripted and requiring low levels of teacher knowledge, it may follow that the most efficient mode of school organization would be a centralized, bureaucratic approach, with little scope for teacher collaboration or collective control. On the other hand, if enactment of the regime requires considerable expertise and confronts the teacher with highly varied challenges and considerable uncertainty, the appropriate mode of school organization would emphasize time and support for staff collaboration and a role for collective teacher control over the conditions of instruction (Rowan, 1990).

My reading of most of the chapters in this volume suggests that enacting the best ideas emerging from developmental science generally places higher than usual demands on teacher knowledge and skill. For example, elementary teachers will have to learn the physics they have to teach. Knowledge about teaching reading comprehension must infuse science teaching. Those teaching early reading may focus on word decoding much of the time, but they must be aware that if children are going to be able "read to learn" by grade 3, early language development and vocabulary are essential and therefore must also be emphasized prior to grade 3, likely within the teaching of subject-matter domains.

A related policy question that may emerge is whether the knowledge requirements for new regimes are sufficiently varied and demanding such that it makes sense to implement teacher specialization in the early grades. For example, it may be that specialists in science and math will remove a burden from those who are working hard to develop expertise in teaching word decoding. If specialization is the right choice, an implication is that greater collaboration will be needed. For example, specialists in science and in language arts would need to closely collaborate if the key premise is that science teaching boosts vocabulary, syntax, and comprehension. If we reject specialization, then we need a strategy to promote new learning in multiple subject domains among the generalist teachers.

School Leadership

An emphasis on enacting ambitious regimes in the subject areas transforms the role of the school principal from building manager to instructional leader. The varied subject-matter expertise among the teachers creates opportunities for teacher leadership, but the school principal must have excellent assessments of the varied expertise and of the potential roles teachers can play in advancing the instructional mission of the school.

Implications for Research on Policy

A theme in this discussion is that well-articulated notions of effective instructional regimes, related levels of teacher expertise, and corresponding conceptions of school leadership and instruction frame new questions for policy and render some old questions obsolete. In this view, the causal model that relates the regime to student outcomes drives questions about the appropriate resources, regulations, incentives, and governance structures that policymakers can influence. The next section considers the methodological challenges that such a research agenda faces. These challenges are in some ways similar but in other ways profoundly distinct from the methodological challenges faced by the rich cousin of educational research: medical research.

Field Trials of Instructional Regimes

The argument here is that evaluating alternative instructional regimes ought to be at the core of the enterprise of educational policy research. How, then, shall we carry out such evaluations?

The clinical trial in medicine immediately emerges as a model for testing instructional regimes. Indeed, a number of educational randomized trials are now in the field (Spybrook, 2007), reflecting a dramatic shift in research policy at the U.S. Department of Education's Institute for Educational Sciences. That shift in policy itself reflects a kind of "sea change" in thinking about educational research (Mosteller & Boruch, 2002) favoring a much stronger emphasis on the use of randomized experiments in education.

Despite important similarities between experimentation in medicine and education, the social structure of instruction—the fact that it occurs within classrooms nested within schools—negates the validity of canonical assumptions underlying the clinical trial in medicine. This distinction has fundamental implications for the design of research—experimental and nonexperimental—in education.

To clarify the differences between clinical trials in medicine and education, let us now consider the key statistical assumptions underlying a paradigm case: the randomized drug trial. It will become apparent that some of the key assumptions invoked in such a trial will generally not apply in education, and that new assumptions are needed as the logical basis of experiments in education.

To motivate the argument, consider a drug trial in which each patient is assigned to receive a "new drug" ($Z=1$) or the "standard treatment" ($Z=0$) with the aim of discerning the impact of the new drug relative to the standard treatment on some outcome of interest (e.g., blood pressure), denoted Y.

According to current thinking about causation in statistical science (Holland, 1986), we can then define for each patient two potential outcomes of treatment. If the patient is assigned to the new drug, we will observe the outcome $Y(1)$, literally the value Y will take on if $Z=1$. In contrast, if the patient is assigned to the standard treatment, we will instead observe the outcome $Y(0)$. This set-up then defines for each patient the causal effect as:

$$\Delta = Y(1) - Y(0) \qquad (1)$$

The idea of causal effects as person-specific differences between potential outcomes is at the heart of what statisticians now call the "Rubin causal model" because of Donald Rubin's highly influential work (Rubin, 1978) and that of his colleagues (Holland, 1986; Rosenbaum & Rubin, 1983). However, the roots of this approach may be found in earlier statistical work (Neyman, 1935) and in economics (Haavelmo, 1943; Heckman, 1979).

The crucial point is that Δ is a patient-specific causal effect that, by definition, can never be observed: If the patient is assigned to $Z=1$, we will observe $Y(1)$ but not $Y(0)$. For that patient, $Y(0)$ is known as the "counter-factual outcome." In contrast, if the patient is assigned to $Z=0$, we will observe $Y(0)$, but not the counter-factual outcome $Y(1)$.

Holland (1986), who provides an exceptionally lucid presentation of this approach, has termed "the fundamental problem of causal inference" the fact that the key quantity of interest (the causal effect Δ) cannot be observed. However, under certain key assumptions, we can estimate the "population-average" causal effect, which I will label as:

$$\delta = E[Y(1) - Y(0)] = E[Y(1)] - E[Y(0) =] \tag{2}$$

The Stable Unit-Treatment Value Assumption (SUTVA)

We have thus far assumed that each patient possesses two and only two potential outcomes, one for each possible treatment assignment. Rubin (1986) called this the "stable unit treatment value assumption" (SUTVA). The basic idea is that one's potential outcome under a given treatment remains stable regardless of (a) the treatment assignment of other persons and (b) the mechanism used to assign the treatment. Part (a) requires that a patient's potential outcome under each treatment does not depend on the treatment assignment of other patients. Part (b) requires that the doctor who prescribes the drug and the pharmacist who fills the prescription do not affect the patient's outcome.

Suppose now that we uncritically import the paradigm of the randomized drug trial into research on instruction—that is, we assume stable units, and randomly assign students to receive a novel instructional regime, say $Z = 1$, or to the standard approach to instruction, that is, the control condition $Z =$. Although some might regard such a study as meeting a gold standard for causal inference, we would immediately confront the implausibility of SUTVA. It will not generally be reasonable to assume (a), that each student's outcome is unaffected by the treatment assignment of other students. Indeed, teachers must take into account the entire composition of the classroom when making instructional decisions; and peers can affect a student's motivation. Moreover, importing (b) into the educational context would require that the teacher to whom that student is assigned be irrelevant once the regime is in place.

We can conclude that SUTVA, as invoked in the clinical trial in medicine, is not generally plausible in the context of research on instruction. This conclusion has profound implications for how we define causal effects and for how we design research, both experimental and non-experimental. Hong (2004) and Hong and Raudenbush (2006) consider these issues in detail, and the following discussion draws heavily on their work.

Implications for Experimental Design

The first implication of the failure of SUTVA as invoked in medical trials is that we will often prefer to assign entire classes or even whole schools to alternative instructional regimes rather than assigning individual students. This shift in unit of randomization leads, however, to new assumptions.

Suppose that classrooms within schools are assigned to regimes. We must then assume that there is "no interference" between the intact classrooms—that the treatment assignment of one teacher does not affect the outcomes of another teacher's class. If we are unwilling to make this assumption—if we assume that teachers and students share important information across classrooms within a school—then the unit of random assignment must be the entire school or district. We also must assume "intact classrooms" (if classrooms are the unit of randomization) or "intact schools" if schools are the unit. This means that the treatment assignment will not cause teachers or kids to move from one classroom to another or from one school to another. If a new regime is particularly effective, such an assumption may break down over time. For example, if "at-risk" kids benefit from a novel regime, their parents may keep them in their school rather than move them to a new school. The same parents might well move their children to a different school if these at-risk students do not have the benefit of the new regime and instead encounter trouble in learning.

Cost and Statistical Power

The forgeoing discussion suggests that "group-randomized" experiments (where groups are classrooms or schools) will typically be the "gold standard" for assessing the impact of instructional regimes on cognitive outcomes. This finding is consistent with current practice in that the vast majority of experiments supported by the U.S. Department of Education have in fact used randomization by group (Spybrook, 2007).

A question that immediately arises is how many classrooms or schools are required to achieve adequate statistical power in such studies and how much it will cost to run each study. The answer is disconcerting—unless great care is taken in the planning of these studies (Raudenbush, Martinez, & Spybrook, 2007).

Consider a fairly typical scenario in which schools will be assigned to a new instructional regime or a standard regime. We want to know how many schools must be studied in order to achieve an adequate statistical power of .80, meaning that our study will have a probability of .80 of detecting the true positive effect of the intervention. Assume that the expected effect size is .30 in standard deviation units, somewhat large but within a realistic range, and that about 20 percent of the variation in the outcome lies between schools within treatments, a fairly typical finding (Bloom, Richburg-Hayes, & Black, 2007; Hedges & Hedberg, 2007). Suppose further that we shall sample three classrooms per school with 20 children in each classroom. Under these assumptions, we would need 86 schools—43 in each treatment condition—a daunting result!

Suppose, however, that there is a pretreatment covariate, the prior mean achievement in a school, and that the correlation between this covariate and the outcome increases is .80, a plausible correlation when the outcome is also a measure of achievement (Bloom et al., 2007). By using this covariate in the analysis, the required number of schools is sharply reduced to 32, that is, 16 per treatment—still a fairly large study, but not implausibly so (see Borman et al., 2005).

An alternative way to use prior information is to match pairs of schools on one or more such pretreatment covariates. This approach will produce power that typically approximates that in the case of analysis of covariance (Raudenbush, Martinez, & Spybrook, 2007). An advantage is that matching can insure that the two groups are closely balanced on salient characteristics such as ethnic or social class composition.

A key conclusion from this line of work is that randomized studies of alternative instructional regimes are likely to be fairly large. Even then, careful exploitation of prior information is needed to insure that those studies will have adequate power. A fleet of such studies, as recommended here, will require a fairly substantial investment. Given that many nonexperimental studies will also be needed to support this research agenda (see Raudenbush, 2005; and also the next section of this chapter), we can anticipate that a focus on comparing instructional regimes as a strategy for improving educational policy will require sustained financial support at a reasonably high level.

Intended and Enacted Regimes and the Contributions of Nonexperimental Research

The foregoing section suggests the need for a fleet of randomized studies evaluating the impacts and costs of alternative instructional regimes. However, a successful agenda for research on the impact of instructional regimes actually requires a well-orchestrated interplay between experimental and nonexperimental research.

We will inevitably see differences, often substantial, between an *intended* regime and the regime actually *enacted.* If so, two new research questions immediately emerge. First, how does the planned intervention (that is, the "intended regime") affect the regime children actually experience (the "enacted regime")? Second, how does the enacted regime affect student learning? Randomization helps us with the first question, but not the second. Let us consider why each of these questions is important before taking up the methodological challenges entailed in answering them.

Why Study the Impact of the Intended Regime on the Enacted Regime?

In carefully controlled laboratory experiments, it is often possible to insure a very high level of fidelity of the implementation of the experimental conditions. In contrast, in large-scale trials of instructional regimes (where one must depend on administrators and teachers to enact new instruction under complex and sometimes chaotic conditions), one cannot safely make such an assumption. As a result, in field trials of instructional regimes, it will generally be essential to study the extent to which assignment to a new regime actually affects practice in the ways intended.

The claim here is that knowing how an intervention affects instruction is essential to interpreting impacts on student learning. This is mostly clearly the case when experiments show no effects, but is also true in the presence of a positive effect.

Interpreting Null Results

Consider a study in which the assignment of schools or classrooms to an intended regime is found to have no significant impact on student learning. Assume that the study design was unbiased and provided adequate statistical power to detect a non-negligible effect. Two explanations immediately arise.

First, it may be that adoption of the intended regime changed classroom instruction in the ways intended, but that those classroom changes made no difference in student learning. Program evaluators use the term "theory failure" to describe this scenario because the theory that links intended changes in instruction to intended student outcomes will have proven incorrect.

Second, the instructional regime may never have been effectively implemented in classrooms. Perhaps the innovators lacked skill in working with teachers or perhaps the teachers lacked the skill, knowledge, or motivation to put the innovative ideas to work in their teaching. In any case, program theory about the relationship between instruction and student outcomes was never tested, leading to what evaluators often call "implementation failure."

Without valid assessments of the enacted regime, it would be impossible to distinguish between these two explanations, severely limiting the study's contribution to knowledge. One would never know whether the theory underlying the program had in fact been tested—a big problem given the likely cost of such an experiment.

Interpreting Nonnull Results

Suppose instead that assignment to the intended regime did produce gains in student learning. One might then assume that the intended regime "worked" by improving instruction according to its theory. But without valid measurements of the instruction children actually experienced, that is, without measurement of the enacted regime, this conclusion would be unwarranted. Perhaps the innovation "worked" in other ways, an assertion that could not be probed without studying the impact of the innovation on instruction. Once again, a failure to measure key aspects of classroom life yields major ambiguities in the findings, leaving open key questions about how to use the findings to improve teaching and learning.

Why Study the Impact of the Enacted Regime on Student Learning?

Suppose now that a new regime, if enacted faithfully, effectively boosts student learning. If all teachers faithfully followed an experimental instructional regime, the intended and enacted regimes would be identical and, of course, their impacts on student learning would be identical as well. Suppose, however, that some teachers assigned to a new regime fail to enact it. Then the experimental estimate of the impact of the intended regime would underestimate the impact of the instruction children actually experience—that is, the enacted regime. The effect of the intended regime would be attenuated further if some of the control teachers had, contrary to expectation, adopted and used the new experimental methods in their classroom. This poses serious, but not fatal, challenges to the validity of evaluation research.

Such a scenario is not hypothetical. The "Success for All" (SFA) regime specifies an approach to reading comprehension that Rowan, Camburn, and Correnti (2004) refer to as "skill based." These authors also studied an alternative regime, "America's Choice" (AC), that uses a "literature-based" approach to reading comprehension using considerable writing, emphasizing what children read. A third set of teachers in comparison schools had adopted neither SFA nor AC. Rowan et al. collected detailed information on classroom practices of all three kinds of schools using teacher logs of daily activities. The results showed that about 80 percent of the SFA teachers adhered quite faithfully to the skill-based approach, while another 20 percent did not. Similarly, about 70 percent of the AC teachers relied heavily on the literature-based approach. Interestingly, quite a large minority of comparison teachers, who ostensibly had no relation to either SFA or AC, appeared to adhere quite faithfully to an approach that closely resembled one of the approaches. It is clear in this setting that the effects of the intended regimes would underestimate the effect of the enacted regimes.

One might argue that only the impact of the intended regime is important to the policymaker. Policymakers and administrators can encourage adoption of an intended regime, but they have no direct control over teacher behavior and therefore cannot insure that the intended regime will be enacted. If the intended regime is taken to scale, one would expect slippage in implementation, and the resulting attenuation should be reflected in any calculation of the expected benefit of adopting the intended regime.

Yet understanding the impact of the enacted regime is also important for policy. Once the enacted regime's full benefit becomes visible, incentives shift. Teachers may become more inclined to give a new approach a try if evidence shows it to be especially effective, and policy may directly or indirectly reward them for doing so. Moreover, knowing the impact of the enacted regime tests the instructional theory underlying the regime.

Methodological Challenges

Once we accept the rationale for studying the causes and consequences of the enacted regime, important methodological entailments follow. First, we have to develop technologies for measuring instruction as it is enacted in classroom settings. Inevitably, we will measure classroom instruction with error. Quantifying the degree of measurement error then becomes essential in research design and statistical analysis. Second, studies of the impact of enacted regimes on student learning will not benefit from random assignment. Sound methods of causal inference in nonrandomized studies therefore become essential. Let us consider how these methodological issues play out as we pursue the two new questions of interest.

Studying the Effect of the Intended Regime on the Enacted Regime

We first consider the case in which the aim is to study the impact of assigning a classroom to a new regime on the enacted instructional approach. A key problem is that the enacted instruction, whether measured through direct classroom observation, teacher logs, interviews, or questionnaires, will generally be measured with some degree of error. These errors of measurement do not cause bias, but they do reduce precision. To obtain adequate power requires sampling a larger number of classrooms than would be required if the enacted instruction were measured with perfect reliability.

Raudenbush, Martinez, Bloom, Zhu, and Lin (2007) have shown that the impact of reliability on statistical power can be substantial. In the case where classroom observations assess the enacted instruction, reliability can be increased by means of better training of observers, use of more observers per classrooms, or observing each classroom on more occasions. A trade-off arises between investing resources in increasing the reliability of measurement and investing resources in recruiting and sustaining the involvement of more classrooms and schools. Raudenbush, Martinez, Bloom, et al. (2007) consider these trade-offs. This work demonstrates the importance of careful studies to discern the reliability and validity of measures of classroom instruction within the framework of a research agenda focusing on instructional regimes.

Studying the Effect of the Enacted Regime on Student Learning

Next we consider the case in which the aim is to assess the impact of the enacted regime on student learning. Once again, the enacted regime is measured with error.

The key result in this case is that measurement error creates bias. Specifically, if the effect of the enacted regime on student outcomes is positive, this effect based on a fallible measure will be negatively biased. One of the benefits of conducting a study of the reliability of the classroom measurement is that information from such a study can be used to correct the bias that arises from measurement error, as explained in detail by Raudenbush and Sadoff (2008).

The Challenge of Studying Multiyear Instructional Sequences

Discovering the impact of multiyear sequences of instruction might well be the fundamental aim of policy-relevant research on instruction. Whether children can read or reason mathematically is the cumulative result of sequences of instructional experiences over several years. To understand the impact of sequences of instructional regimes, however, poses a special challenge to valid causal inference, particularly in light of the fact that children's experiences in instruction occur in multiple social settings. This requires what might be called a "metaregime," that is, a sequence of regimes in which the aim of each new regime is to build on the consequences of the past regime.

Background

Causal comparative studies of instruction tend to be bounded by a single academic year. Although understanding instructional effects during a single year is necessary, it is essential to understand how sequences of instruction extending over two or more years cumulatively affect learning. The effect of a multiyear sequence of instructional experiences cannot logically be equated to the sum of the effects of instruction occurring each year. Early reading instruction provides a useful example.

We now have considerable confidence in our methods to teach a broad range of children how to decode familiar text during grade 1. Suppose we had equal confidence that certain strategies for building comprehension during grade 2 are also effective. To maximize comprehension by the end of grade 2, however, does not necessarily imply simply maximizing the additive effects of these grade 1 and grade 2 approaches. The problem is that many children come to school with comparatively weak vocabulary, general knowledge, and fluency in standard English (National Institute of Child Health and Human Development, 2000). These limitations plausibly constrain learning in grades 3–5 when comprehension ("reading-to-learn") replaces the goal decoding of familiar text ("learning-to-read"). It may well be that, during grade 1, an immersion in language, ideas, oral comprehension, and text is necessary along with a strong emphasis on explicit instruction in decoding in order to maximize the child's gains in comprehension observable in later grades. Such an effect cannot be discerned from even the best studies of teaching and learning during a single academic year.

The Problem of Time-Varying Confounding

If we could randomly assign students to sequences of instructional experiences across year and if students' treatment assignment remained intact, we could obtain unbiased inferences about the impact of a sequence of instructional experiences using standard and quite simple methods of analysis. The essential problem is that students will tend to migrate across classrooms and even across schools from one year to the next. In many urban areas, such school mobility is remarkably high. And, in a study where instructional sequences differ within schools, parents and teachers may find reasons to opt into treatment conditions not assigned. In these cases, the randomization will tend to break down. If so, this scenario generates the problem of *time-varying confounding*.

Once again, we can learn important lessons from public health research about causal inference, in this case using weighting (Robins, Hernán, & Brumback, 2000). Hong and Raudenbush (2008) showed how to adapt Robins's methods to the social setting in which instruction occurs. We know that children will migrate across teachers within a school and even between schools as they pass through the elementary grades. A child will have one set of classmates within grade 1 and then will have a somewhat different set of classmates in year 2. Hong and Raudenbush (2008) employed a special four-level hierarchical model, in which children are cross-classified by teachers who themselves are nested within schools. They proved that Robins's weighting approach can apply in this setting. However, great care is needed in defining and applying the weights in this multilevel setting. They applied the method to study the impact of "intensive math instruction" experienced over two years (grades 4 and 5) based on the Longitudinal Evaluation of School Policy and Change (LESCP). They defined intensive math instruction as characterized by a high level of conceptual demand combined with a large amount of time devoted to math.

Their results are graphed in Figure 21.2. It turned out that experience in year 2 was decisive. Although there was some evidence of a positive effect of year 1 intensive math, this estimated effect did not significantly differ from zero. However, the effect of year 2 instruction was significant and positive. The data were not fully adequate for estimating the joint effect of receiving two years of intensive math instruction because

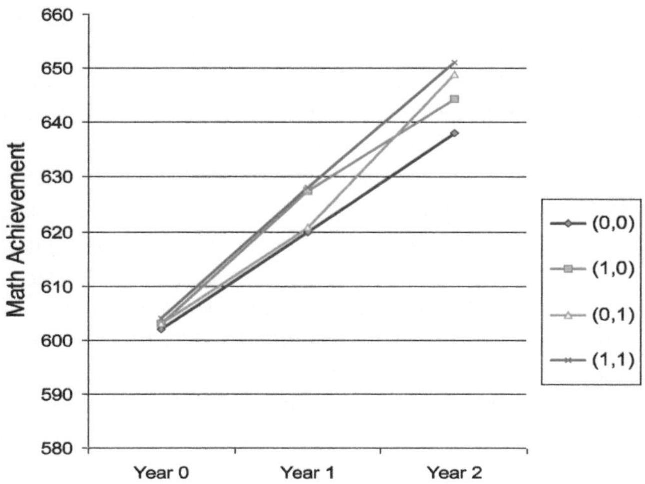

Figure 21.2 Fitted Model for a Sequence of Two Treatments. Adapted from Hong, G., & Raudenbush, S.W. (2008). Causal inference for time-varying instructional treatments. *Journal of Educational and Behavioral Statistics*, 33(3), 333–362. Published by Sage. © 2008 AERA. http://jebs.aera.net

the sample size of students who switched from intensive to nonintensive math or from nonintensive to intensive math was not large enough to support a sufficiently precise statistical inference, a result that has implications for research design on sequences. Every sequence must be adequately represented in the sample to uncover "amplifying effects."

Conclusions

The approach recommended here asks how instructional regimes influence student learning. Questions for policy immediately follow: What resources, incentives, and governance structures cost-effectively support the best regimes? Sharply constraining these policy questions is the reality that the logic of the instructional regime has profound implications for notions of teacher expertise, differentiated status of teachers, conceptions of teacher advancement, and notions of school organization and leadership.

In pursing a program of research on instructional regimes, methodological questions loom large: How does one design research to define and answer questions about the impacts of instructional regimes on student learning, and how does one study the moderating effect of resources, incentives, and governance?

I have argued that research on instruction has similarities but also fundamental differences from research on clinical practice in medicine. These differences arise from the central role of the teacher as the enactor of any instructional regime, and from the intrinsically social nature of classroom instruction. These differences have fundamental implications for research design and analysis.

Research on instructional regimes also requires a deliberate interplay between experimental and nonexperimental research. Experimental research can reveal the impact of intended instructional regimes on student learning. However, sound nonexperimental methods are needed to understand the impact on children of the instruction they actually experience. Moreover, to study such enacted regimes requires good measures of classroom instruction. When the aim is to study the effect of the intended regime on the enacted regime, low reliability of measurement causes no bias but will weaken statistical power to detect effects. However, when we turn to studies of how enacted instructional regimes affect learning, errors of measurement of the enacted instruction will lead to bias. This bias can be corrected, however, if we have good knowledge of the properties of our measures of instruction.

Finally, research on sequences of instruction is central to this agenda. Most of the skills and knowledge we hope children will develop emerge from the cumulative effects of sequences of instruction as strongly suggested by the chapters in this volume. Again there are important parallels with clinical research in medicine but there is also a fundamental difference. Unlike the physician, who takes a patient's history and develops a plan of treatment tailored to that history, the teacher must consider the instructional histories of the entire class and then develop instructional approaches tailored to the whole class or to subgroups of children. The child's response will, in principle, depend not only on that child's own instructional history but also on the instructional history of treatment that child's classmates. The social nature of instruction makes the teacher's task complicated and also complicates the researcher's attempt to draw causal inferences about the impact of instructional sequences.

If these arguments make sense, it is essential that educational researchers develop distinctive models and methods for studying causal effects on student development as it unfolds in the social settings of classrooms and schools. This effort will require an infusion of well-trained, creative social scientists in education who have strong methodological skills and who know how to integrate their methods with the best available thinking about how children learn and teachers can organize classrooms to promote learning.

Acknowledgment

The research reported here was supported by a grant from the Spencer Foundation entitled "Improving Research on Instruction: Models, Designs, and Analytic Methods."

References

Bloom, H. S., Richburg-Hayes, L., & Black, A. (2007). Using covariates to improve precision for studies that randomize schools to evaluate educational interventions. *Educational Evaluation and Policy Analysis*, 29(1), 30–59.

Borman, G. D., Slavin, R. E., Cheung, A., Chamberlain, A., Madden, N., & Chambers, B. (2005). Success for all: First-year results from the National Randomized Field Trial. *Educational Evaluation and Policy Analysis*, 27, 1–22.

Clotfelter C. T., Ladd, H. F., & Vigdor, J. L. (2007). How and why do teacher credentials matter for student achievement (NBER Working Papers 12828). National Bureau of Economic Research.

Cohen, D. K., Raudenbush, S. W., & Ball, D. B. (2003). Resources, instruction, and research. *Educational Evaluation and Policy Analysis*, 25, 119–142.

Connor, C. M., Morrison, F. J., Fishman, B. J., Schatschneider, C., & Underwood, P. (2007). Algorithm-guided individualized reading instruction. *Science*, 315, 464–465.

Finn, J. D., & Achilles, C. M. (1990). Answers and questions about class size: A statewide experiment. *American Educational Research Journal*, 27(3), 557–577.

Haavelmo, T. (1943). The statistical implications of a system of simultaneous equations. *Econometrica*, 11, 1–12.

Hanushek, E. A. (1989). The impact of differential expenditures on school performance. *Educational Researcher*, 18(4), 45–62.

Hanushek, E. A. (1996). School resources and student performance. In G. Burtless (Ed.), *Does money matter? The effect of school resources on student achievement and adult success.* Washington, DC: Brookings Institution Press. pp. 43–73.

Heckman, J. (1979). Sample selection bias as a specification error. *Econometrica*, 47, 153–161.

Hedges, L. V., & Hedberg, E. C. (2007). Intraclass correlation values for planning group-randomized trials in education. *Educational Evaluation and Policy Analysis*, 29(1), 60–87.

Hedges, L. V., Laine, R. D., & Greenwald, R. (1994). An exchange: Part I: Does money matter? A meta-analysis of studies of the effects of differential school inputs on student outcomes. *Educational Researcher*, 23(3), 5–14.

Holland, P. (1986). Statistics and causal inference. *Journal of the American Statistical Association*, 81, 945–960.

Hong, G. (2004). Causal inference for multilevel observational data with application to kindergarten retention. Ann Arbor: University of Michigan, unpublished doctoral dissertation.

Hong, G., & Raudenbush, S. W. (2006). Evaluating kindergarten retention policy: A case study of causal inference for multilevel observational data. *Journal of the American Statistical Association*, 101, 901–910.

Hong, G., & Raudenbush, S. W. (2008). Causal inference for time-varying instructional treatments. *Journal of Educational and Behavioral Statistics*, 33(3), 333–362.

Kerbow, D. (2006). *Strategic teaching and evaluation of progress*. University of Chicago: Center for Urban School Improvement.

Krueger, A. B., & Whitmore, D. M. (2001). The effect of attending a small class in the early grades on college-test taking and middle school test results: Evidence from Project STAR. *Economic Journal*, 111(468), 1–28.

Mosteller, F., & Boruch, R. (Eds.) (2002). *Evidence matters: Randomized trials in education research*. Washington, DC: Brookings Institution Press.

Murphy, S. (2003). Optimal dynamic treatment regimes. *Journal of the Royal Statistical Society*, Series B, 65(2), 331–366.

National Institute of Child Health and Human Development (2000). Report of the National Reading Panel. *Teaching children to read: An evidence-based assessment of the scientific research literature on reading and its implications for reading instruction* (NIH Publication No. 00-4769). Washington, DC: U.S. Government Printing Office.

Neal, D. (2008). Designing incentive systems for schools. In M. Spring (Ed.), *Performance incentives: Their growing impact on American K–12 education*. Washington, DC: Brookings Institution Press.

Neyman, J. (1935). Statistical problems in agricultural experimentation. *Journal of the Royal Statistical Society*, Supplement, 2(2), 107–180. Online. Available at: http://www.jstor.org/action/showPublication?journalCode=suppjroyastatsoc.

Nye, B., Hedges, L. V., & Konstantopoulos, S. (2000). The effects of small classes on academic achievement: The results of the Tennessee class size experiment. *American Educational Research Journal*, 37(1),123–151.

Raudenbush, S. W. (2005). Learning from attempts to improve schooling: The contribution of methodological diversity. *Educational Researcher*, 34(5), 25–31.

Raudenbush, S. W., Martinez, A., Bloom, H., Zhu, P., & Lin, F. (2007). The reliability of group-level measures and the power of group-randomized studies. Unpublished manuscript, University of Chicago.

Raudenbush, S. W., Martinez, A., & Spybrook, J. (2007). Strategies for improving precision in group-randomized experiments. *Educational Evaluation and Policy Analysis*, 29, 5–29.

Raudenbush, S. W., & Sadoff, S. (2008). Statistical inference when classroom quality is measured with error. *Journal of Research on Educational Effectiveness*, 1, 138–154.

Robins, J., Hernán, M., & Brumback, B. (2000). Marginal structural models and causal inference in epidemiology. *Epidemiology*, 11, 550–560.

Rosenbaum, P., & Rubin, D. B. (1983). The central role of the propensity score in observational studies for causal effects. *Biometrika*, 69, 41–55.

Rowan, B. (1990). Commitment and control: Alternative strategies for the organizational design of schools. In C. Cazden (Ed.), *Review of research in education* (Vol. 16). Washington, DC: American Educational Research Association. pp. 353–389.

Rowan, B. E., Camburn, E., & Correnti, R. (2004). Using teacher logs to measure the enacted curriculum: A study of literacy teaching in third-grade classrooms. *Elementary School Journal*, 105(1), 75–101.

Rubin, D. B. (1978). Bayesian inference for causal effects: The role of randomization. *Annals of Statistics*, 6, 34–58.

Rubin, D. B. (1986). Statistics and causal inference: Comment: Which ifs have causal answers? *Journal of the American Statistical Association*, 81(396), 961–962.

Spybrook, J. (2007). Examining the experimental designs and statistical power of group randomized trials funded by the Institute of Education Sciences. Ann Arbor: University of Michigan, unpublished doctoral dissertation.

Volume Contributors

Dor Abrahamson, Ph.D., University of California, Berkeley, Graduate School of Education, Berkeley, California

Martha W. Alibali, Ph.D., University of Wisconsin, Madison, Department of Psychology, Madison, Wisconsin

Florencia K. Anggoro, Ph.D., College of the Holy Cross, Department of Psychology, Worcester, Massachusetts

Jason L. G. Braasch, Ph.D., University of Illinois, Chicago, Department of Psychology, Chicago, Illinois

Kimberly Brenneman, Ph.D., Rutgers University, Department of Psychology, Piscataway, New Jersey

Timothy Burke, Ph.D., University of California, Los Angeles, Department of Psychology, Los Angeles, California

Jennifer L. Coffman, Ph.D., University of North Carolina, Chapel Hill, Center for Developmental Science, Chapel Hill, North Carolina

Carol M. Connor, Ph.D., Florida State University, Florida Center for Reading Research, Tallahassee, Florida

Thomas D. Cook, Ph.D., Northwestern University, Institute for Policy Research, Evanston, Illinois

Scotty Craig, Ph.D., University of Memphis, Department of Psychology, Memphis, Tennessee

Don Franceschetti, Ph.D., University of Memphis, Department of Physics, Memphis, Tennessee

Yuka Fujimori, Carnegie-Mellon University, Department of Psychology, Pittsburgh, Pennsylvania

Rochel Gelman, Ph.D., Rutgers University, Center for Cognitive Science, Piscataway, New Jersey

Barry Gholson, Ph.D., University of Memphis, Department of Psychology, Memphis, Tennessee

Susan R. Goldman, Ph.D., University of Illinois, Chicago, Learning Sciences Research Institute, Chicago, Illinois

Kimberley W. Gomez, Ph.D., University of Pittsburgh, School of Education, Pittsburgh, Pennsylvania

Arthur C. Graesser, Ph.D., University of Memphis, Department of Psychology, Memphis, Tennessee

Elizabeth A. Gunderson, University of Chicago, Department of Psychology, Chicago, Illinois

Catherine A. Haden, Ph.D., Loyola University Chicago, Department of Psychology, Chicago, Illinois

Marc W. Hernandez, Ph.D., National Opinion Research Center (NORC), Academic Research Centers, Chicago, Illinois

Janellen Huttenlocher, Ph.D., University of Chicago, Department of Psychology, Chicago, Illinois

Philip J. Kellman, Ph.D., University of California, Los Angeles, Department of Psychology, Los Angeles, California

Kimberly A. Lawless, Ph.D., University of Illinois, Chicago, Educational Psychology, Chicago, Illinois

Susan C. Levine, Ph.D., University of Chicago, Department of Psychology, Chicago, Illinois

Joan Lucariello, Ph.D., City University of New York, University Dean for Education, Office of Academic Affairs, New York, New York

Flori H. Manning, University of Illinois, Chicago, Learning Sciences Research Institute, Chicago, Illinois

Christine M. Massey, Ph.D., University of Pennsylvania, Institute for Research in Cognitive Science, Philadelphia, Pennsylvania

Frederick J. Morrison, Ph.D., University of Michigan, Department of Psychology, Ann Arbor, Michigan

Mitchell J. Nathan, Ph.D., University of Wisconsin, Madison, School of Education, Madison, Wisconsin

Peter A. Ornstein, Ph.D., University of North Carolina, Chapel Hill, Department of Psychology, Chapel Hill, North Carolina

Yasuhiro Ozuru, Ph.D., University of Alaska Anchorage, Department of Psychology, Anchorage, Alaska

James W. Pellegrino, Ph.D., University of Illinois, Chicago, Learning Sciences Research Institute, Chicago, Illinois

Kristin R. Ratliff, Ph.D., University of Chicago, Department of Psychology, Chicago, Illinois

Stephen W. Raudenbush, Ed.D., University of Chicago, Department of Sociology, Chicago, Illinois

Lindsey E. Richland, Ph.D., University of California, Irvine, Department of Education, Irvine, California

Nancy R. Romance, Ed.D., Florida Atlantic University, Charles E. Schmidt College of Science, Boca Raton, Florida

Zipora Roth, University of Pennsylvania, Institute for Research in Cognitive Science, Philadelphia, Pennsylvania

Christopher A. Sanchez, Ph.D., Arizona State University, Department of Applied Psychology, Mesa, Arizona

Michael J. Slanovits, University of Illinois, Chicago, Learning Sciences Research Institute, Chicago, Illinois

Catherine E. Snow, Ph.D., Harvard University, School of Education, Cambridge, Massachusetts

Nancy L. Stein, Ph.D., University of Chicago, Department of Psychology & National Opinion Research Center, Chicago, Illinois

Michele Tine, Ph.D., Boston College, Lynch School of Education, Chestnut Hill, Massachusetts

Michael R. Vitale, Ph.D., East Carolina University, AA Curriculum and Instruction, Greenville, North Carolina

Claire White, Ed.D., Harvard University, School of Education, Cambridge, Massachusetts

Jennifer Wiley, Ph.D., University of Illinois, Chicago, Department of Psychology, Chicago, Illinois

Friedrich Wilkening, Ph.D., University of Zurich, Department of Psychology, Zurich, Switzerland

Manyee Wong, Ph.D., Northwestern University, Institute for Policy Research, Evanston, Illinois

Vivian C. Wong, Northwestern University, Institute for Policy Research, Evanston, Illinois

Index